VOLUME ONE HUNDRED AND TWELVE

VITAMINS AND HORMONES

Aquaporin Regulation

VOLUME ONE HUNDRED AND TWELVE

VITAMINS AND HORMONES

Aquaporin Regulation

Series Editor

GERALD LITWACK, PhD

Toluca Lake, North Hollywood, California

ELSEVIER

ACADEMIC PRESS

An imprint of Elsevier

Cover photo credit:
Free energy profiles for water transport through aquaporin AQP2 as a function of water position along pore axis and angle between water dipole and pore axis. Orientations of water corresponding to 0°, 90° and 180° are shown alongside y-axis labels. The free energy profile on the top corresponds to the wild-type and that at the bottom corresponds to the N184A mutant. The figure on the left hand side depicts the pore axis. *Reproduced with permission from Padhi, S., & Priyakumar, U. D. (2017). Microsecond simulation of human aquaporin 2 reveals structural determinants of water permeability and selectivity.* Biochimica et Biophysica Acta (BBA)-Biomembranes, 1859(1), 10–16.
Vitamins and Hormones (2020) Volume 112.

Academic Press is an imprint of Elsevier
50 Hampshire Street, 5th Floor, Cambridge, MA 02139, United States
525 B Street, Suite 1650, San Diego, CA 92101, United States
The Boulevard, Langford Lane, Kidlington, Oxford OX5 1GB, United Kingdom
125 London Wall, London, EC2Y 5AS, United Kingdom

First edition 2020

ISBN: 978-0-12-820809-0
ISSN: 0083-6729

For information on all Academic Press publications
visit our website at https://www.elsevier.com/books-and-journals

Publisher: Zoe Kruze
Acquisitions Editor: Ashlie M. Jackman
Editorial Project Manager: Joanna Collett
Production Project Manager: Denny Mansingh
Cover Designer: Miles Hitchen

Typeset by SPi Global, India

Contents

Contributors

Mariana Álvarez-Pérez
Laboratorio de Rumiología y Metabolismo Nutricional (RuMeN), Facultad de Estudios Superiores Cuautitlán, UNAM, Cuautitlán, Mexico

Marco G. Alves
Department of Microscopy, Laboratory of Cell Biology, Unit for Multidisciplinary Research in Biomedicine (UMIB), Institute of Biomedical Sciences Abel Salazar (ICBAS), University of Porto, Porto, Portugal

Raquel L. Bernardino
Department of Microscopy, Laboratory of Cell Biology, Unit for Multidisciplinary Research in Biomedicine (UMIB), Institute of Biomedical Sciences Abel Salazar (ICBAS), University of Porto, Porto, Portugal

Jason P. Breves
Department of Biology, Skidmore College, Saratoga Springs, NY, United States

Laura Campderrós
Department of Biochemistry and Molecular Biomedicine, Institute of Biomedicine of the University of Barcelona; CIBER Fisiopatología de la Obesidad y Nutrición, Barcelona, Spain

David F. Carrageta
Department of Microscopy, Laboratory of Cell Biology, Unit for Multidisciplinary Research in Biomedicine (UMIB), Institute of Biomedical Sciences Abel Salazar (ICBAS), University of Porto, Porto, Portugal

Rubén Cereijo
Department of Biochemistry and Molecular Biomedicine, Institute of Biomedicine of the University of Barcelona; CIBER Fisiopatología de la Obesidad y Nutrición, Barcelona, Spain

Hyo-Jung Choi
Department of Biochemistry and Cell Biology, School of Medicine, Kyungpook National University, Taegu, South Korea

Alicia E. Damiano
Laboratorio de Biología de la Reproducción, Instituto de Fisiología y Biofísica Bernardo Houssay (IFIBIO)-CONICET-Facultad de Medicina; Cátedra de Biología Celular y Molecular, Departamento de Ciencias Biológicas, Facultad de Farmacia y Bioquímica, Universidad de Buenos Aires, Buenos Aires, Argentina

Annarita Di Mise
Department of Biosciences, Biotechnologies and Biopharmaceutics, University of Bari, Bari, Italy

Valeria Dietrich
Cátedra de Biología Celular y Molecular, Departamento de Ciencias Biológicas, Facultad de Farmacia y Bioquímica, Universidad de Buenos Aires, Buenos Aires, Argentina

José L. Dueñas
Obstetrics and Gynecology Department, Virgen Macarena University Hospital, Medical School, University of Seville, Seville, Spain

Laura González-Dávalos
Laboratorio de Rumiología y Metabolismo Nutricional (RuMeN), Facultad de Estudios Superiores Cuautitlán, UNAM, Cuautitlán, Mexico

Pilar Guadix
Obstetrics and Gynecology Department, Virgen Macarena University Hospital, Medical School, University of Seville, Seville, Spain

Kenichi Ishibashi
Division of Pathophysiology, Meiji Pharmaceutical University, Tokyo, Japan

L. Jaskiewicz
Department of Human Physiology, School of Medicine, University of Warmia and Mazury in Olsztyn, Olsztyn, Poland

Bhanu P. Jena
Department of Physiology, Wayne State University School of Medicine, Detroit, MI, United States

Hyun Jun Jung
Division of Nephrology, Department of Medicine, Johns Hopkins University School of Medicine, Baltimore, MD, United States; Department of Biochemistry and Cell Biology, School of Medicine, Kyungpook National University, Taegu, South Korea

Tae-Hwan Kwon
Department of Biochemistry and Cell Biology, School of Medicine, Kyungpook National University, Taegu, South Korea

Ofelia Mora
Laboratorio de Rumiología y Metabolismo Nutricional (RuMeN), Facultad de Estudios Superiores Cuautitlán, UNAM, Cuautitlán, Mexico

Yoshiyuki Morishita
Division of Nephrology, Saitama Medical Center, Jichi Medical University, Saitama, Japan

Henricus A.M. Mutsaers
Department of Clinical Medicine, Aarhus University, Aarhus, Denmark

Rikke Nørregaard
Department of Clinical Medicine, Aarhus University, Aarhus, Denmark

Pedro F. Oliveira
Department of Microscopy, Laboratory of Cell Biology, Unit for Multidisciplinary Research in Biomedicine (UMIB), Institute of Biomedical Sciences Abel Salazar (ICBAS); Department of Genetics, Faculty of Medicine; i3S-Instituto de Investigação e Inovação em Saúde, University of Porto, Porto, Portugal

Siladitya Padhi
TCS Innovation Labs—Hyderabad (Life Sciences Division), Tata Consultancy Services Limited, Hyderabad, India

Ras Pandey
School of Mathematics and Natural Sciences, University of Southern Mississippi, Hattiesburg, MS, United States

Eui-Jung Park
Department of Biochemistry and Cell Biology, School of Medicine, Kyungpook National University, Taegu, South Korea

Antonio Pérez-Pérez
Department of Clinical Biochemistry, Virgen Macarena University Hospital, Medical School, University of Seville, Seville, Spain

Enrique Piña
Departamento de Bioquímica, Facultad de Medicina, UNAM, Ciudad de México, Mexico

U. Deva Priyakumar
Center for Computational Natural Sciences and Bioinformatics, International Institute of Information Technology, Hyderabad, India

Tania Quesada-López
Department of Biochemistry and Molecular Biomedicine, Institute of Biomedicine of the University of Barcelona; CIBER Fisiopatología de la Obesidad y Nutrición, Barcelona, Spain

Warin Rangubpit
Center of Excellence in Computational Chemistry, Department of Chemistry, Chulalongkorn University, Bangkok, Thailand

Marianna Ranieri
Department of Biosciences, Biotechnologies and Biopharmaceutics, University of Bari, Bari, Italy

Víctor Sánchez-Margalet
Department of Clinical Biochemistry, Virgen Macarena University Hospital, Medical School, University of Seville, Seville, Spain

Armando Shimada
Laboratorio de Rumiología y Metabolismo Nutricional (RuMeN), Facultad de Estudios Superiores Cuautitlán, UNAM, Cuautitlán, Mexico

A. Skowronska
Department of Human Physiology, School of Medicine, University of Warmia and Mazury in Olsztyn, Olsztyn, Poland

M.T. Skowronski
Veterinary Center, University of Nicolaus Copernicus, Torun, Poland

Pornthep Sompornpisut
Center of Excellence in Computational Chemistry, Department of Chemistry, Chulalongkorn University, Bangkok, Thailand

Thomas M. Stulnig
Clinical Division of Endocrinology and Metabolism, Department of Medicine III, Medical University of Vienna, Vienna, Austria

Grazia Tamma
Department of Biosciences, Biotechnologies and Biopharmaceutics, University of Bari, Bari, Italy

Yasuko Tanaka
Division of Pathophysiology, Meiji Pharmaceutical University, Tokyo, Japan

D. Tanski
Department of Anatomy and Animal Physiology, Faculty of Biology and Biotechnology, University of Warmia and Mazury in Olsztyn, Olsztyn, Poland

Matteo Tardelli
Division of Gastroenterology and Hepatology, Joan & Sanford I. Weill Cornell Department of Medicine, Weill Cornell Medical College, New York, NY, United States; Clinical Division of Endocrinology and Metabolism, Department of Medicine III, Medical University of Vienna, Vienna, Austria

Stine Julie Tingskov
Department of Clinical Medicine, Aarhus University, Aarhus, Denmark

Susanna Törnroth-Horsefield
Department of Biochemistry and Structural Biology, Lund University, Lund, Sweden

Giovanna Valenti
Department of Biosciences, Biotechnologies and Biopharmaceutics, University of Bari, Bari, Italy

Alfredo Varela-Echavarria
Laboratorio de Diferenciación Neural y Axogénesis, Instituto de Neurobiología, UNAM, Querétaro, Mexico

Cecilia L. Varone
Department of Biological Chemistry, Facultad de Ciencias Exactas y Naturales, Universidad de Buenos Aires, Buenos Aires, Argentina

Teresa Vilariño-García
Department of Clinical Biochemistry, Virgen Macarena University Hospital, Medical School, University of Seville, Seville, Spain

Francesc Villarroya
Department of Biochemistry and Molecular Biomedicine, Institute of Biomedicine of the University of Barcelona; CIBER Fisiopatología de la Obesidad y Nutrición, Barcelona, Spain

Preface

Aquaporins (AQPs) are molecular channels comprised of subunits that are usually assembled into the apical cell membrane after hormone-induced phosphorylation of the subunits in the epithelial cytoplasm. Aquaporins occur in most cell types. There are 13 different aquaporins; they are numbered AQP-0 through AQP-12. AQP-1 and AQP-2 occur in the apical membranes of certain kidney cells, for example, and AQP-3 and AQP-4 are fixed in the basolateral membrane of such cells to transport water molecules to the extracellular space and eventually into the bloodstream. AQP-1 occurs in the red blood cell. AQP-4 is expressed in the blood–brain barrier and is involved in the formation of the cerebrospinal fluid. Whereas AQP-9 is involved in brain edema, AQP-4 eliminates water after brain injury. These are only a few examples of the multiple involvements of aquaporins in cell physiology.

This volume begins with basic studies on aquaporins followed by their specific types of regulation, including the situation in fishes, existing in a total aqueous environment and finally in specific areas of human physiology. Therefore, the opening chapter deals with evolution as described in "Perspectives on the evolution of aquaporin superfamily" by K. Ishibashi, Y. Tanaka, and Y. Morishita. This is followed by "Structure and dynamics of aquaporin-1" by W. Jetsadawisut, P. Sompornpisut, and R.B. Pandey. In Chapter 3, S. Padhi and U.D. Priyakumar report on "Selectivity and transport in aquaporins from molecular simulation studies." M. Tardelli and T.M. Stulnig discuss "Aquaporin regulation in metabolic organs." Biochemical regulation is exemplified by "Phosphorylation of human AQP2 and its role in trafficking" by S. Törnroth-Horsefield and then by "Regulation of aquaporin-2 by RNA interference" by H.J. Jung, E.-J. Park, H.-J. Choi, and T.-H. Kwon. Continuing in this vein, B.P. Jena offers "Aquaporin regulation: Lessons from secretory vesicles." "CFTR regulation of aquaporin-mediated water transport" is the contribution of D.F. Carrageta, R.L. Bernardino, M.G. Alves, and P.F. Oliveira. In the area of specific regulators, "Glucocorticoid gene regulation of aquaporin-7" is reported by L. González-Dávalos, M. Álvarez-Pérez, T. Quesada-López, R. Cereijo, L. Campderrós, E. Piña, A. Shimada, F. Villarroya, A. Varela-Echavarria, and O. Mora. Then, A. Skowronska, D. Tanski, L. Jaskiewicz, and M.T. Skowronski elaborate on "Modulation by steroid hormones and other factors on the expression of aquaporin-1 and aquaporin-5." S.J. Tingskov, H.A.M. Mutsaers, and R. Nørregaard discuss "Estrogen regulates aquaporin-2 expression in the kidney." In another species living in water, J.P. Breves reports on "Hormonal regulation of aquaporins in fishes." Moving back to the human is the contribution, "Calcium-sensing receptor exerts a negative regulatory action toward vasopressin-induced aquaporin-2 expression and trafficking in renal collecting duct," from M. Ranieri, A. Di Mise, G. Tamma, and G. Valenti. The final two chapters relate to human pregnancy in the form of "Aquaporins and placenta" by A. Pérez-Pérez, T. Vilariño-García, V. Dietrich, P. Guadix, J.L. Dueñas, C.L. Varone, A.E. Damiano, and V. Sánchez-Margalet. A.E. Damiano finally reports on "Aquaporins during pregnancy."

Joanna Collett of Elsevier, Oxford, UK, faithfully manages the initial phases of the publication process for which I am most grateful.

The illustration on the cover is reproduced from Fig. 2 of Chapter 3: "Selectivity and transport in aquaporins from molecular simulation studies" by S. Padhi and U.D. Priyakumar showing the structure of an aquaporin.

Gerald Litwack
Toluca Lake, California, United States
June 17, 2019

CHAPTER ONE

Perspectives on the evolution of aquaporin superfamily

Kenichi Ishibashi[a,*], Yasuko Tanaka[a], Yoshiyuki Morishita[b]
[a]Division of Pathophysiology, Meiji Pharmaceutical University, Tokyo, Japan
[b]Division of Nephrology, Saitama Medical Center, Jichi Medical University, Saitama, Japan
*Corresponding author: e-mail address: kishiba@my-pharm.ac.jp

Contents

Abstract

Aquaporins (AQPs) belong to a transmembrane protein superfamily composed of an internal repeat of a three membrane-spanning domain and each has a highly conserved NPA box. Based on the more variable carboxyl-terminal NPA box, AQPs can be divided into three subfamilies: (1) glycerol-channel aquaglyceroporin (gAQP) (2) water-selective AQP (wAQP), and (3) deviated superaquaporin (sAQP) in the order of passible evolution. This classification has functional and localization relevance: most wAQPs transports water selectively whereas gAQPs and sAQPs also transport small molecules with sAQPs mostly localized inside the cell. As this classification is not based on the function, some wAQPs functioning as glycerol channels will not be included in gAQPs. AQP ancestors may have first originated in eubacteria as gAQPs to transport small molecules such as glycerol. Later some of them may have acquired a water-selective filter to become wAQPs. Although AQPs are absent in many bacteria, especially in archaea, both gAQPs and wAQPs may have been carried over to eukaryotes or horizontally transferred. Finally, multicellular organisms have obtained new sAQPs, which are curiously absent in fungi and plants. Interestingly, both plants and higher insects independently have lost gAQPs,

Vitamins and Hormones, Volume 112
ISSN 0083-6729
https://doi.org/10.1016/bs.vh.2019.08.001

whose functions, however, have been taken over by functionally modified wAQPs partly obtained by horizontal gene transfers from bacteria. This evolutionary viewpoints on AQPs will facilitate further functional analysis of AQP-like sequences and expand our viewpoints on AQP superfamily.

1. Introduction: Aquaporin (AQP) superfamily

Aquaporin (AQP) belongs to an ancient channel family, which transports water and small molecules including glycerol, urea, nutrients, metabolic precursors, waste products, toxins and even gases (reviewed in Ishibashi, Hara, & Kondo, 2009; Laloux et al., 2018; Verma, Gupta, & Sankararamakrishnan, 2015). Because of their fundamental roles in essential water homeostasis, the distribution of AQPs were initially regarded ubiquitous from prokaryotes to eukaryotes. Indeed, a pair of AQPs: GlpF and later AQPZ, have been identified in *E. coli*, *Shigella, Pseudomonas*, *Salmonella,* and so on. However, many bacteria, and even some protozoa and fungi have no AQPs as revealed by accumulated genome projects. In fact, the plasma membrane is highly permeable to small sized non-ionic water molecules even in the absence of AQPs except at low temperature. Moreover, other membrane proteins may facilitate the diffusion of water across the plasma membrane. Thus, AQPs may not be essential for most cells and organisms, which is further shown by relatively benign phenotypes of AQP-null mice and humans (Calvanese, D'Auria, Vangone, Falcigno, & Oliva, 2018). Moreover, they are even harmful as shown by AQP4 to worsen the brain edema after brain infarction (Filippidis, Carozza, & Rekate, 2017; Manley et al., 2000).

On the other hand, some AQPs are beneficial and even critical for the survival as shown by the disease, nephrogenic diabetes insipidus, in which the failure to reabsorb urine rapidly through AQP2 in the kidney leads to severe polyuria (Milano, Carmosino, Gerbino, Svelto, & Procino, 2017). As AQPs are highly expressed in almost all fluid-related structures, the evolution of AQPs seems to be tightly linked to fluid homeostasis and osmoregulation. In fact, AQPs can be a transmembrane osmosensor by facilitating cell volume changes (Hill & Shachar-Hill, 2015). Moreover, the transport of small molecules by AQPs also adds another layer of complexities as transporters of osmolytes and nutrients relevant to even fat metabolism (da Silva & Soveral, 2017).

Genome projects have accumulated many AQP-like sequences containing a conserved signature sequence for AQP superfamily, a pair of

NPA (Asn-Pro-Ala) boxes, even with the overall homology in less than 30%. Although most of their functions are unknown, the analysis of their primary sequences is instrumental to speculate their function and adaptive roles in altered environments. Moreover, evolutional viewpoints of AQPs will shed lights on the distribution and regulation of AQPs (Di Giorgio et al., 2014). This analysis, however, may contain caveats of lineage-independent gene duplications and horizontal gene transfers.

We have been interested in this aspects of AQPs and published a couple of review articles before (Ishibashi, Kondo, Hara, & Morishita, 2011; Ishibashi, Morishita, & Tanaka, 2017). Well organized public data bases such as NCBI are useful for this analysis, which was employed in this review using protein-protein blast (basic local alignment search tool) (blastp) against non-redundant protein sequences of specific organisms using model sequences of AQP subfamilies as queries.

2. Classification of AQPS

Although many sophisticated phylogenetic analyses of AQPs using whole amino acid sequences have been reported (Abascal, Irisarri, & Zardoya, 2014; Di Giorgio et al., 2014; Finn & Cerdà, 2015; Soto, Alleva, Amodeo, Muschietti, & Ayub, 2012), our simple method of AQP classification based on the key sequences is easier to assign AQP-like proteins into three AQP subfamilies: (1) bacterial glycerol channel-like aquaglyceroporin (gAQP) (2) "water-selective" AQP (wAQP), and (3) "intracellular" super-aquaporin (sAQP) with only 20% of overall sequence homology (Ishibashi et al., 2017). The order follows the suspected appearances in the evolution and the abbreviations are simple and easy to associate with their typical functions and structures.

Our classification keys are illustrated in Fig. 1 and the diagram for classification is depicted in Fig. 2, which simulates the hypothesized evolution of AQP superfamily (Fig. 3). As AQPs have been produced by tandem repeats of an identical or similar three-spanning hemipore-channel membrane protein (Fig. 3), they have highly conserved two NPA boxes, which are partly embedded in the plasma membrane to form the constriction of the pore. Although NPA boxes are the key for the identification of AQPs, the first NPA boxes of sAQPs are not well conserved and sometimes difficult to identify, especially in lower animals such as nematodes (CeAQP9, 10, 11 in Table 1). Therefore, we have employed the second NPA box as a key sequence to classify AQPs.

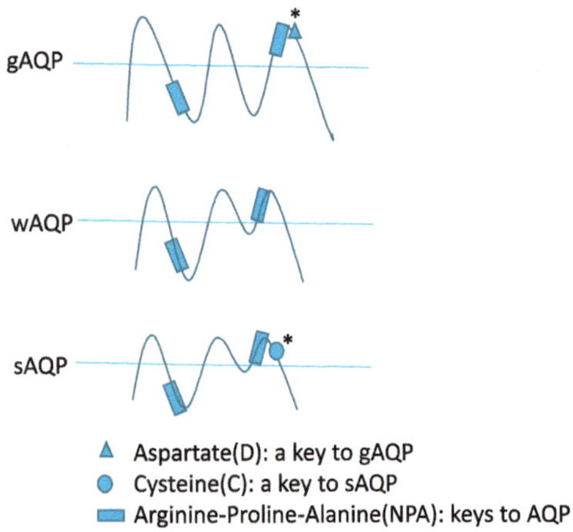

Fig. 1 The transmembrane structure of each AQP subfamily with keys for the classification. The signature sequence for gAQP is the aspartic acid residue (D) in the second NPA box. The signature sequence for sAQP is the cysteine residue (C) at the downstream of the second NPA box. The absence of these D and C residues indicates wAQP.

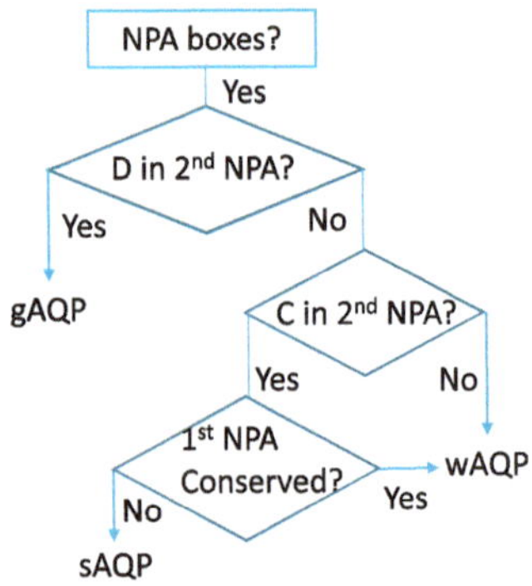

Fig. 2 The diagram for the classification of AQPs into three subfamilies. The characteristics of AQP subfamily shown in Fig. 1 is the basis for this diagram. See text for more details.

After the identification of the second NPA box, the aspartic acid residue (D) in the second NPA box should be looked for (Fig. 2) (bolds in Tables 2 and 3). This D is the signature key for gAQPs and located just the downstream of the arginine forming the Ar/R (aromatic residues/arginine) narrowest filter for the selective water permeation (Verma et al., 2015). As this D residue enlarges this pore constriction and makes more hydrophobic, gAQPs permeate small molecules larger than water. Interestingly, a tight correlation is present between this D residue and a longer fourth loop (loop D) located in the cytoplasm, which may be another key for gAQPs.

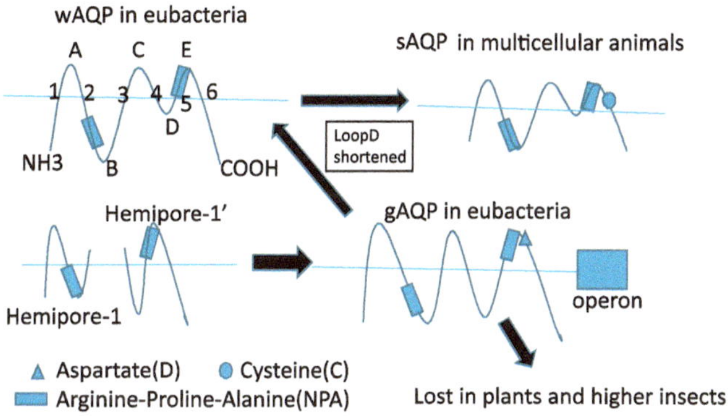

Fig. 3 A putative evolutional pathway of AQP subfamilies. AQPs may have been formed by duplication of a hemipore gene or a fusion of two similar genes (hemipore-1 and hemipore-1′). gAQP may have been transformed to wAQP by the shortening of loop D and the disrupting mutation of the signature D residue of gAQP in eubacteria. sAQP may have been produced by the transformation of wAQP in multicellular animals. gAQPs have been lost in plants and higher insects.

Table 1 Sequence of sAQPs at the first and the second NPA boxes.

	First NPA box	Second NPA box
Invertebrates		
CeAQP9	-IEFQRDAVAHPCPLVTNCYR-	-GINYTGMYANPIVAWACTFN**C**L-
CeAQP10	-NIFNRGAMTNCAPIFEQFVF-	-LYVVGVPGLNPIVATARLYG**C**R-
CeAQP11	-ALCNRTAFCSPLAPIEQYLF-	-VTFVGDQALDPLVASTLFFG**C**R-
H.C	-SSTGRGSYGTPLSPIELLYY-	-ILFIGDAGLDPVVASSLFFG**C**N-
H.C.2	-MVKYYGPMAFFFTVTSLLTV-	-LIYVGVPGLNPVVASSRLFG**C**E-
Toxo	-SFVNRGAFVSPLKPVELFYF-	-LAFIGVVGLNPTVSSSRMQG**C**D-
Toxo2	-NCYFLRDAIAHPCPLITSAT-	-GINFTGMYANPIVAWACTFN**C**E-
Loa	-GIMNREAYVSPLTPIELYYK-	-LLVVGVPGLNPTVASSRLHG**C**D-
Loa2	-AFFLRDAIAHPCPLVTTASR-	-GINYTGMYANPIVAWACTFN**C**T-
Urch11	-LTFDGDSTANTCMIWQSMLK-	-GLEWTGMMFNPALAAGITLN**C**G-
Urch12	-NEELSNAGDAPLGQAVQVQP-	-GLEYTGAPMNPILGFASGWG**C**K-
Int7	-FQTFALRNSVPAFQNTSYCK-	-ALATDALSLVLSPVGLRCWK**C**H-

Continued

Table 1 Sequence of sAQPs at the first and the second NPA boxes.—cont'd

	First NPA box	Second NPA box
Ticks and insects		
L.salm7	–GTRWTDETACPYTHFEECLL-	-AFDYSGGYYNPVLATGLKYG**C**K-
L.salm8	–SSRWSTETACPYTHFEECLI-	-SFNYSGGYLNPVLATSLKYG**C**R-
MtAQP11	–TFTFQDGTCDPSECYEKFCK-	-GLFVSGGYFNPTLSFAMEYG**C**Q-
Dros	-GRVWGDASACPYTHMEDVVE-	-AFNFSGGYFNPVLATALKWG**C**R-
L.hespe	-SRNWGDATACPYTHFEDVAE-	-AFNYSGGYFNPALATSLKLG**C**D-
P.h.c.	-SNDWGNANGAPNVQLDDVVL-	-TFDRTGGYYNPVLATSHKFG**C**E-
C.lect	–SKNWGDATACPYTHFEDVAE-	-AFNYSGGYFNPALATSLKYG**C**H-
Vertebrates		
Callo10	-EATFDGAEANPLVTMQELLR-	-VGPYTAALFNPALAFSFTFH**C**S-
ZF11	-GFSFRGAICNPTGALELLSR-	-GGRLTGAVFNPALAFSIQFP**C**P-
ZF12	-TAVMQDVSGNPAVTLLRLLQ-	-ANNYTSGYVNPALAYAVTLT**C**P-
Salm1	-VITFHGAFANPNGAIENVYR-	-GGHISGAVFNPIMAFSVQFP**C**T-
Salm2	-GLTFHGALGNPCSTLEYTYR-	-GGKMTGAVFNPALALSTQFS**C**S-
Salm3	-GVICAGATGNPSLTLMNFLL-	-ASNYTMAFLNPSLAYGLTFY**C**P-
Xen11	-HGLTARAVCNPSGSLERYVR-	-GGSLTGAVFNPALAYSLLFH**C**E-
Xen12	-GFTFNKASGNSAVSLQDFLL-	-AGSYTGAFFNPTLAAALTFQ**C**S-
Tur11	-GLTLTGSTCNPCGTLQQILD-	-SEALSEGCCNPIQTTETQAF**C**I-
Tur12	-GVSFDGASANPTVTLQEFLV-	-AGPFTGAFFNPTLASMVTFH**C**S-
Tur13	-GLTLTGSTCNPCGTLQQILD-	-GGNLTGAIFNPALAFSLHAS**C**F-
Chic11	-GLTLPGSTCNPCGTLQPLWG-	-GGNLTGAIFNPALAFSLHPH**C**F-
Chic12	-AACANGAASNPTVSLQEFLL-	-AAPATGAFFNPALATASTFL**C**A-
AQP11	-GLTLVGTSSNPCGVMMQMML-	-GGSLTGAVFNPALALSLHFM**C**F-
AQP12	-GVTLDGASANPTVSLQEFLM-	-AGPFTSAFFNPALAASVTFA**C**S-

CeAQP: *Caenorhabditis elegans*: 9NP_001021552.2, 10NP_496105.1, 11NP_499821.2; H.C.: *Haemonchus contortus*: 1CDJ83340.1, 2CDJ81492.1; Toxo: *Toxocara canis*: 1KHN86094.1, 2KHN78654.1; Loa: *Loa loa*: 1XP_020303707.1 2XP_003142473.1; Urch11: *Strongylocentrotus purpuratus*: sea urchin: 11XP_780933.1, 12XP_003731685.2; Int7: *Ciona intestinalis*: XP_026690588.1; L.salm: *Lepeophtheirus salmonis*: louse; MtAQP: *Milnesium tardigradum*; Dros: *Drosophila melanogaster*: fruit fly; L.hespe: *Lygus hesperus*: western tarnished plant bug; P.h.c: *Pediculus humanus corporis*: louse; C.lect: *Cimex lectularius*: bed bug; Callo: *Callorhinchus milii*: sharks; Salm: *Salmo salar*: salmon; Xen: *Xenopus laevis*; Tur: *Chelonia mydas*: sea turtle: 11EMP28605.1, 12EMP38415.1, 13XP_007067464.1, Chic: chichen: *Gallus gallus*, AQP11,12: human.
The key Cys (C) residues for sAQPs are indicated in bold.

Table 2 Sequence of gAQPs at the first and the second NPA boxes.

	First NPA box	Second NPA box
Bacteria		
GlpF	-TAGVSGAHLNPAVTIALWLF-	-MGPLTGFAMNPAR**D**FGPKVFAW-
Salmon	-TAGISGGHLNPAVTIALWLF-	-TGPLTGFAMNPAR**D**FGPKLFTW-
Entero	-LFVFGGVCINPAMALAQAIL-	-LGGTTGFAMNQAR**D**LGPRIAYQ-
Staphy1	-VGQFSGAHLNPAVTLALALD-	-LGGATGYAINPAR**D**LGPRIAHA-
Staphy2	-SGKIAPAHLNPAVSLAFAMK-	-LGGTTGYAINPAR**D**LGPRLIHA-
Liste	-AGAYSGAHLNPAVTIALAVG-	-FGPTTGYAINPAR**D**LGPRLAYF-
Pseudo	-SAGVSGAHLNPAVSIALWLF-	-MGPLTGFAMNPAR**D**FGPKLMTY-
LACTO	-LFIFGNVCINPAMVLAQCLL-	-LGGPTGFAMNLAR**D**MGPRIAHA-
LACTO	-SSFMSPAHLNPAVSLGMAVA-	-LGGTTGYAINPAR**D**LGPRIAHA-
LACTO	-GQFGSDGHLNPAVTVGFALF-	-LGGTTGFAINPAR**D**WAPRLAYT-
LACTO	-FGSISGNHLNPAITIGQAVI-	-LGGPTGPALNPAR**D**LGPRLVHA-
AQPV1	-MFGFVSAHLNPAMCLALFIL-	-MGGVTSIAANPAR**D**FSPRLAHF-
Fungi		
U.MY2	-AGGISGGMINPAVTIALACF-	-LGWQTAYAVNPAR**D**LGPRIMLY-
U.MY3	-SGGISGGHCNPSVTIALALF-	-LGYNTGYAINGAR**D**TGPRIALW-
FPS1	-GSAISGAHLNPSITLANLVY-	-MAYQTGTAMNLAR**D**LGPRLALY-
AQY3P	-AGGISGGHINPAVTISMAIF-	-LGYQTSFTINPAR**D**LGPRIFAS-
POMBE1	-AGGISGGHVNPAVTISLAIF-	-LGYQTSFTLNPAR**D**LGPRMFAW-
TRICH4	-VGGKSGGHLNPAVTFANCLY-	-FGWETGYAINLAR**D**FGPRLVSF-
TRICH5	-CGGVSGAHLNPAVTLANCIF-	-FGSNTGYAINPAR**D**LGPRIMSH-
TRICH6	–VGGKSGGHLNPAVTFGNCLY-	-FGWETGYAINLAR**D**FGPRLVSF-
TRICH7	–SGGVSGAHLNPTITIMLWFF-	-FANQTGAALNPSR**D**FGPRLALL-
TRICH8	-CSGVSGAHLNPAVTLANCIF-	-FGSNTGYAINPAR**D**LGPRIMSH-
MOSS		
GIP	-FVGHISGFFNPAVALAAAVV-	-GGGMTGPALNPAR**D**LGPALVSG-
Protozoa		
P.viv	-AAKLSGAHLNLAVTVGFATI-	-FGGNTGFALNPSR**D**LGARLLSL-
P.falc	-SAKLSGAHLNLAVSIGLSSI-	-FGGNTGFALNPSR**D**LGSRFLSL-
T.BR1	-SLGISGGHLNSAVTVGNAVF-	-FGLASPLAMNPSL**D**FGPRVFSS-
T.BR2	-SLGISGGHLNSAVTVGNAVF-	-IGYSTGYAINPAR**D**FGPRVFSS-
T.BR3	-SLGISSGHLNPAVTVGNAVF-	-IGYSTGYAINPAR**D**FGPRVFSS-

GlpF: *E. coli*; Salmo, *Salmonella enterica*; Entero: *enterococcus*; staphy: *Staphylococcus aureus* (1: WP_124421540.1; 2: SUL88767.1); LACTO: *Lactobacillus* (KRL33321.1, KYK52904.1, WP_021730460.1, WP_070083648.1); Liste: *Listeria*; Pseudo: *Pseudomonas aeruginosa*; AQPV1: *chlorella virus*, U.MY: *Ustilago maydis* (2:XP_011387976.1, 3: XP_011387547.1); FPS1 and AQY3: *Saccharomyces cerevisiae*; POMBE: *S. pombe*; TRICH, *Trichoderma harzianum*.

GIP: *Physcomitrella patens*.

P.viv: *Plasmodium vivace*; P.falc: *Plasmodium falciparum*; T.BR: *Trypanosoma brucei*.

The key Asp (D) residues for gAQPs are indicated in bold.

After gAQPs have been identified, the remaining AQPs will be wAQPs or sAQPs. The signature sequence for sAQP is the cysteine residue (C) at the nine residues downstream of the C-terminal of the second NPA, NPAxxxxxxxxC (x is any amino acid residue) (bolds in Table 1). This C is exposed on the surface of the protein at the periplasmic side of the membrane (Calvanese, Pellegrini-Calace, & Oliva, 2013) and may be necessary to form 3D structure by S-S bondage with an uncharacterized upstream cysteine residue. As in the case with D in gAQPs, the functional importance of this C in sAQPs has been shown by its point mutation in mouse AQP11, a sAQP, which has caused a similar phenotype to AQP11-null mice (Tchekneva et al., 2008). As sAQPs display highly deviated NPA boxes even among sAQPs, this C is helpful to identify sAQPs. However, some eubacterial AQPs have this C such as *Chlorobaculum* (bold in Table 4) but does not belong to sAQP as the overall sequence is much more similar to wAQPs. Accordingly, we have modified the previous diagram to avoid this miss classification by further inspecting the upstream sequence from the first NPA which is highly conserved in wAQPs and gAQPs, but not in sAQPs.

AQPs without these D or C residues in the second NPA box will be classified as wAQPs since the key sequence for wAQPs has not been identified (Tables 4 and 5). However, wAQPs have been further divided into several subfamilies based on the overall sequences especially in plants. The following subfamilies have been proposed in plants: PIPs (Plasma membrane Intrinsic Protein), TIPs (Tonoplast Intrinsic Proteins), NIPs (NOD-26 like Intrinsic Protein), SIPs (Short Intrinsic basic Proteins), XIPs (uncategorized X Intrinsic Proteins), HIP (Hybrid Intrinsic Protein) and LIP (Large Intrinsic Protein) (Danielson & Johanson, 2008; Khabudaev, Petrova, Grachev, & Likhoshway, 2014). These subfamilies from wAQPs beyond plants and their signature sequences require further analysis. Beyond plant subfamilies from wAQPs, a new subfamily has been proposed for the combined plant TIPs and vertebrate AQP8s although the signature key sequence is not clear (Kirscht, Sonntag, Kjellbom, & Johanson, 2018).

So far, the above simple method to classify AQPs into three subfamilies has been as efficient as those using whole amino acid sequences, possibly due to the robust basic structure of AQP subfamilies conserved throughout the evolution. As our classification is not based on the function, some glycerol-permeable wAQPs should not be classified as gAQPs. For example, based on the function alone, an AQP of malaria carrier *Anopheles* has been named AgAQP3 (aquaglyceroporin) (Liu, Tsujimoto, Huang, Rasgon, & Agre, 2016) similar to entomoglyceroporin (Eglyp) (Finn, Chauvigne, Stavang, Belles, & Cerda, 2015). As both are wAQPs, such naming will be confusing and sometimes misleading.

Table 3 Sequence of gAQPs at the first and the second NPA boxes.

	First NPA box	Second NPA box
Invertebrates		
CeAQP1	-TARMSGGHLNPAVSLLLWSL-	-FGMNIGYPINPAR**D**LGPRLFSY-
CeAQP2	-SAKLSGGHINPAVSFAFLSV-	-FGYNCGYPVNPAR**D**FAPRLFTS-
CeAQP3	-GSKISGAHLNPAVSFFQLTQ-	-LALNAGYAINPAR**D**FAPRLFNL-
CeAQP7	-CSKTSGGHFNPAVSIAFLTL-	-YGMNLGYPINPAR**D**LGPRLFSF-
CeAQP8	-AASVSGGHLNPAISVAQSIL-	-FGANGGFAINPAR**D**FGPRVFCL-
Urch6	-FSHVSGAHFNPSVTVPLVIY-	-FSGADRRGTNPCM**D**EIDVVVQW-
Urch7	-AAGVSGAHLNPAVTLAFACL-	-FGYNCGYAINPAR**D**LGPRLFTA-
Urch8	-TGGVSGGHINPAVTLSMCSL-	-YGYNAGAAVNPAR**D**FSPRLFTA-
Urch9	-SFGVSGGHINLAVSIGVAVL-	-FGHNAGAPLNPSI**D**LSGRLCLG-
Urch10	-GFGVSGGHVNLALSIGIAAI-	-YGYNAGAPLNPSI**D**LSGRLLLA-
Int6	-TGKVSGAHINPAVSVAQCAF-	-FGLNCGYAINPAR**D**FGPRLFTF-
Ticks and insects		
MtAQP1	–GAGVSGGIINPALTFAVAL-	-FSYSAGAAMNPAR**D**LSPRLWSL-
MtAQP2	-AGGVTGAFLNPAIAVAFSV-	-LAYNAGAALNPSR**D**LAPRLFTY-
MtAQP7	-AGGVSGGHLNPAVTLAFAV-	-FGFNCGYPINPAR**D**FGPRLFTA-
MtAQP10	-AGGISGGHLNPAVSLAFAA-	- YGFNCGYPINPAR**D**MGPRLFTL-
MTAQP3	-GGISGAHLNPAVTTTMLVM-	-FSYNAGAALNPAR**D**LAPRLFLW-
MTAQP4	-GGISGGLLNPAITFTIAFL-	-FSYSAGAAMNPAR**D**ISPRLWTL-
MTAQP8	-GGVSGAFLNPAVALAFAVL-	-LSYNAGAAMNPSR**D**LAPRFFSY-
MTAQP9	-GAMSGAIINPALTLAVALL-	-FSYSTGAAMNPAR**D**FSPRLWSL-
L.SALM1	–GKISGGHMNPAVTIAFAIF-	-FSFNAGAALNPAR**D**LGPRLFTA-
L.SALM2	–GGVSGGHLNPAVTLAMASI-	-FGVNAGTAINPAR**D**FSPRLFTI-
L.SALM3	–GGVSGGHLNPAVTLAMASI-	-FGVNAGTAINPAR**D**FAPRLFTF-
Vertebrates		
Callo6	-AGQVSGAHLNPAVTFALCLL-	-MGFNCGYAVNPAR**D**FGPRLFTS-
Callo7	-TGKVSGAHLNPAVTFALCLL-	-MGFNSKYCLNPAR**D**IGPRLFSA-
Callo8	-TIGVSGAHLNPAISLSMCVL-	-MSFNCGGSINPAR**D**LGPRLFTA-
Callo9	-AGGVSGAHLNPAISICFCVL-	-MGSNCGYAINPAR**D**FGPRLFSY-

Continued

Table 3 Sequence of gAQPs at the first and the second NPA boxes.—cont'd

	First NPA box	Second NPA box
AQP3	-AGQVSGAHLNPAVTFAMCFL-	-MGFNSGYAVNPAR**D**FGPRLFTA-
AQP7	-AGRISGAHMNAAVTFANCAL-	-LGMNTGYAINPSR**D**LPPRIFTF-
AQP9	-AGGVSGGHINPAVSLAMCLF-	-LGLNSGCAMNPAR**D**LSPRLFTA-
AQP10	-GGNVSGAHLNPAFSLAMCIV-	-MGANCGIPLNPAR**D**LGPRLFTY-

CeAQP: *Caenorhabditis elegans*:1NP_495510.1, 2NP_495973.1, 3NP_502044.1, 7NP_508515.2, 8NP_001024758.1.
Urch: *Strongylocentrotus purpuratus*: sea urchin: 6XP_001198238.1, 7XP_011673265.1, 8XP_789770.3, 9XP_011679508.1, 10XP_792142.4.
Int6: *Ciona intestinalis*: 6XP_002131906.1.
MtAQP: *Milnesium tardigradum*.
L.SALM: *Lepeophtheirus salmonis*: louse.
Callo: *Callorhinchus milii*: shark.
AQP0–12: human.
The key Asp (D) residues for gAQPs are indicated in bold.

The structures of wAQPs and gAQPs have been found to be very similar, whereas the structure of sAQPs has not yet been examined (Frick et al., 2014; Kirscht et al., 2016, 2018). Therefore, the structural basis for this classification awaits further structural studies (Calvanese et al., 2013). Furthermore, the bases for the function and the subcellular localization of sAQPs also need to be clarified.

3. Distribution of AQP subfamilies

3.1 Bacterial AQPs: Origin of AQPs

As bacteria such as *E. coli* have a set of two AQPs: GlpF and AQPZ, corresponding to gAQP and wAQP, respectively (Soupene, King, Lee, & Kustu, 2002), both gAQPs and wAQPs most likely have originated in prokaryotes. Since some GlpF gene is located in the functional gene complex, glpFK operon encoding glycerol kinase, gAQPs may has a solid functional role in glycerol metabolism suggesting that gAQPs may have originated earlier than wAQPs which are not included in any operons (Abascal et al., 2014). In fact, some bacteria have multiple gAQPs such as *Lactobacillus plantarum* and *Mycobacterium tuberculosis* with four gAQPs (Table 6).

Although gAQPs have been reported to transport water in the *Xenopus* oocyte expression system, bacterial GlpF seems to transport little water. Furthermore, single channel water permeabilities by gAQPs are not as efficient as wAQPs even in the *Xenopus* oocyte expression system (Yang & Verkman, 1997). In fact, a recent reexamination of human gAQPs by a new Nuclear Magnetic Resonance approach has also found little water transport when

Table 4 Sequence of wAQPs at the first and the second NPA boxes.

	First NPA box	Second NPA box
Bacteria		
AQPZ	-VGHISGGHFNPAVTIGLWAG-	-SIPVTNTSVNPARSTAVAIFQG-
Chl.P1	-MGTVSGAHLNPAVTIAFAMR-	-AAPISGASMNPVRSLAPALV**C**G-
CHL.P2	-VGNVSGAHLNPAVTLGFVFA-	-GGPLTGASMNPARSLAPALLSG-
Meth	-FGRISGCHINPAVTIALFAT-	-IGNLTGASLNPARTFGPYLGDW-
Fungi		
AQY1	-FAGVSGGALNPAVSLSLCLA-	-LTAYTGTGVNPARSLGAAVAAR-
AQY2	-FAGVSGGALNPAVSLSLCLA-	-LTGYTGTGVNPARSLGAAVAAR-
U.MY1	-FYRVSGGLFNPAVTLALFLT-	-GTNYTGGSLNPARTLGPNTVLA-
Cript	-FFRVSGGLFNPAVSLGMVLA-	-GVPYSGGALNPVRSLGPAVVTH-
TRICH1	-TAPASGGHLNPMITFSAILT-	-IPGYAGAQMNPSRCLAFGIARR-
TRICH2	-FYRVTGGMFNPAVTLALVLV-	-GTNFTGTGINPARSFGPAVVTD-
TRICH3	-FYRVTGGMFNPAVTLGLVLV-	-GTNFTGTGINPVRSLGPAVVTG-
Protozoa		
Tryp1	-FGYISGGHFNPAVTMAVFLV-	-VGRISGGAFNPAAATGLQLALC-
Tryp2	-FGYISGAHFNPAITFATFIN-	-VGGFTGGAFNPAVATGTQLVGC-
Tryp3	-FGYISGGHFNPAVTFSAMLI-	-VGSVSGGAFNPAVATGLQLTKC-
Tryp4	-FGYMSGGHLNPMVSFAVFLS-	-CGSISGGAFNPSVATALIVVRC-
T.gondi	-FKDISGAHFNPAVSCTIYMT-	-AGAFAGATMNPAVTFGIFISNI-
D.disA	-VSGVSGCNLNPAVTLANLLS-	-GFNFSGGALNPVRVLGPSIISG-
D.disB	-ISGISGCQLNPAVTVGCVTT-	-LNLFTGGSLNPARSFGPAVFSD-
D.disC	-FADVSGAHFNPAVTFATCVT-	-GGSVSGGAFNPARVFGTALVGN-
D.disD	-CAPVSGGHLNPSITLATFFA-	-IAPNYIFGFNIARCLSPAIVLS-
D.disE	-CAPVSGGHLNPSITIATFFS-	-ISPNYIFGFNMARCLCPAIVTG-
Plants		
XIP1.1	-APATSGGHVNPCITWTEMLT-	-FSGYGGAGINPGRCIGPAVVLG-
TIP1.1	-GANISGGHVNPAVTFGAFIG-	-GGAFSGASMNPAVAFGPAVVSW-
PIP2.6	-TAGISGGHINPAVTFGLFLA-	-TIPITGTGINPARSFGAAVIYN-
HIP1.1	-TGAISGGHINPAVTLAFVVA-	-GVPYTGASMNPARSFGPALVSG-

Continued

Table 4 Sequence of wAQPs at the first and the second NPA boxes.—cont'd

	First NPA box	Second NPA box
NIP1.2	-LGHISGAHFNPAVTIAFASC-	-AGPVSGASMNPGRSLGPAMVYS-
SIP1.1	-TVIFGSASFNPTGSAAFYVA-	-GSKYTGPAMNPAIAFGWAYMYS-
LIP	-DIVSGGSQVNPSVSVAMFVH-	-GTPYTGPAMNPMIAFGWAVQSD-

AQPZ: *E. coli*; Chl.p: *Chlorobaculum parvum* (1WP_012502795.1, 2WP_012501697.1); Meth: *Methanothermobacter marburgensis*: Archaea.
AQY1,2: *Saccharomyces cerevisiae*, U.MY1: *Ustilago maydis*,1:XP_011389539, Cript: *Cryptococcus neoformans*, TRCH: *Trichoderma harzianum*; Tryp: *Trypanosoma cruzi*, T.gondi: *Toxoplasma gondii*, D.dis: *Dictyostelium discoideum*.
PIPs (Plasma membrane Intrinsic Protein), TIPs (Tonoplast Intrinsic Proteins), NIPs (NOD-26 like Intrinsic Protein), SIPs (Short Intrinsic basic Proteins), XIPs (uncategorized X Intrinsic Proteins), HIP (Hybrid Intrinsic Protein), LIP (Large Intrinsic Protein).

Table 5 Sequence of wAQPs at the first and the second NPA boxes.

	First NPA box	Second NPA box
Invertebrates		
CeAQP4	-FGHISGGHFNPAVSWAIAGA-	-TGSITGASMNPARSLGPSIIGS-
CeAQP5	-FGKISGGHFNPVVSWAMVLC-	-TASITGTAMNPVRALSPNIVGE-
CeAQP6	-FGGVSGAHINPAVTFGIALV-	-AGAISGASMNPARSFGPNIMGQ-
CeAQP12	-VANLTPAHLNPAVTLLQWFR-	-NFRQIGNLPNMKVSGKFGKPTI-
Urch1	-FSHVSGAHFNPSVTVPLVIY-	-GIPYTGASMNPARSFGPALVAN-
Urch2	-TAHISGGHLNPAVSVAFLLL-	-AIQYSGASMNPARSLGSAVVGG-
Urch3	-FLNISGGLFNPGLTFAMALS-	-AYHSSGGSFNPARSFGPAVISG-
Urch4	-FDAISGSHLNPVVTMGFIIT-	-GYAFTQASMNPARSLAAAVITN-
Urch5	-FVNISGGHFNPAVSLATFVS-	-SFYTAGAGFNPAVSFGSAIVSQ-
Int1	-FEKISGGHMNPAVTFAMVIA-	-AGRLTGSSMNPVRSLPPALLSK-
Int2	-IGHISGAHVNFAVTTAVFIG-	-GFNISGACMNPTLSFGPAVVSG-
Int3	-FFKICVVHFNPAISVGFLVS-	-GKWIGAGCLNPARWFGPAVMAG-
Int4	-FWNICVINFNPAISVGLAVA-	-GKWIGAGCLNPSRAFGPAVVLG-
Int5	-FFNISVFHFNPAVSVGFLVS-	-GRWVGAGCLNPARWFGPALVTG-
Ticks and insects		
MtAQP5	-ICGVSGGHINPAVSLGFLVT-	-AIPLTGTSMNPARSLGPAVLIG-
MtAQP6	-CFKSISAHFNPVITIAALLL-	-ARQATGGFLNPMRAFSLALFEL-
L.salm4	-FGHISGCHVNPAVTAGLFCA-	-AIPFTGSSMNPARTFGPAVILN-
L.salm5	-FASISDCHANPSMTIARTFT-	-ALVAWKGSLNPARALGSVFVSK-
L.salm6	-FLDVTVAHFNPAITLSAMVF-	-GASSSGGAYNPALVFAPVLLTG-
Bib	-FLHISGAHINPAVTLALCVV-	-CCFVSMPYLNPARSLGPSFVLN-

Table 5 Sequence of wAQPs at the first and the second NPA boxes.—cont'd

	First NPA box	Second NPA box
DRIP	-LGHLSGCHINPAVTLGFLIV-	-AIKLSGASMNPARSFGPAVVQG-
Eglp1	-FGFVSGAHSNPCISISCYLM-	-GIQLTGASMNPAKTLVPAIFYG-
Eglp2	-FGCVCGAHLNPAVTLATYVY-	-AGQLTGASMNPVRSFAPAIWNG-
Eglp3	-FGSVSGAHLNPAITLAAWLY-	-AGLFTGASMNPTRSLGPAVWND-
Eglp4	-FGCVSGAHLNPAVTVAAYIY-	-AGPFTGGSMNPARSFAPALWNK-
Vertebrates		
Callo1	-FGHISGGHVNPAVTAAMVCT-	-AIDYTSSSMNPARSFGPAVIMG-
Callo2	-TRGVSGGHLNPALTLALLAR-	-TIGFTGCSMNPARSLGPAVVTS-
Callo3	-VGHVSGAHLNPAVTFAYLLG-	-GIYYTGCGMNPARSFAPAVLTR-
Callo4	-FGEISGAHVNPAVTVAFLCT-	-AIGFTGCGMNPARSFGPAVVTG-
Callo5		-FGEISGAHVNPAVTVAFLCTRK-
AQP0	-VGHISGAHVNPAVTFAFLVG-	-GMYYTGAGMNPARSFAPAILTR-
AQP1	-VGHISGAHLNPAVTLGLLLS-	-AIDYTGCGINPARSFGSAVITH-
AQP2	-LGHISGAHINPAVTVACLVG-	-GIHYTGCSMNPARSLAPAVVTG-
AQP4	-FGHISGGHINPAVTVAMVCT-	-AINYTGASMNPARSFGPAVIMG-
AQP5	-LGPVSGGHINPAITLALLVG-	-GIYFTGCSMNPARSFGPAVVMN-
AQP6	-TWKASGAHANPAVTLAFLVG-	-GIHFTGCSMNPARSFGPAIIIG-
AQP8	-LGNISGGHFNPAVSLAAMLI-	-GGPVSGGCMNPARAFGPAVVAN-

CeAQP: *Caenorhabditis elegans*: 4NP_505512.3, 5NP_505691.2, 6NP_505727.1, 12NP_001022480.1.
Urch: *Strongylocentrotus purpuratus*: sea urchin: 1XP_011673761.1, 2XP_799266.2, 3XP_794577.1, 4XP_001190612.1, 5XP_001198238.1.
Int: *Ciona intestinalis*: 1XP_002124819.1, 2XP_018670041.1, 3XP_002128950.1, 4XP_002131543.1, 5XP_004226091.2.
MtAQP: *Milnesium tardigradum*; L.salm, *Lepeophtheirus salmonis*: louse; Bib, Drip, Eglp: *Drosophila melanogaster*: fruit fly; Callo: *Callorhinchus milii*: shark; AQP0–12: human.

expressed in native membranes (Palmgren et al., 2017). Therefore, gAQPs may primarily function as transporters of osmolytes, nutrients or toxins of small molecular sizes but not water.

The speculated significance of gAQPs for bacterial cellular osmoregulation is to facilitate the extrusion of glycerol or osmolytes to prevent water movement into the cell along inward osmotic gradients when bacteria are suddenly transferred to hypotonic environments. At the same time, the facilitated water entry through wAQPs will enhance cell swelling and leading to more rapid cell rupture. Accordingly, low water permeability through gAQPs may be important to prevent the cell rupture when exposed to a hypotonic environment.

In contrast, when exposed to hyperosmotic conditions, water exit through wAQPs may be detrimental. In fact, *Vibrio vulnificus*, a halophilic pathogenic bacterium of marine environments showed a 2.3-fold decrease of the expression of AQPZ (Rao, Shashidhar, & Bandekar, 2013). This decrease in water loss coincided with an enhanced tolerance to hyperosmotic stress demonstrated by an increased growth rate.

The speculated significance of wAQPs in bacteria is to prevent the icing damage of the cell (Tanghe, Van Dijck, & Thevelein, 2006). In the event of a sudden cold exposure though rare in nature, wAQP may be necessary to prevent intracellular icing by increasing intracellular osmolality to enhance the freeze tolerance with rapid extrusion of water even at low temperature where the plasma membrane alone transports less water. On the other hand, the extrusion of glycerol as an anti-freezer through gAQP will decrease cold tolerance. Therefore, gAQP and wAQP may function in a contradictory way to osmotic and freezing environmental challenges.

Bacterial wAQPs also transport H_2O_2 to detoxify the endogenous H_2O_2 in *Streptococci* family (Tong et al., 2019). A cyanobacterium, *Synechocystis* performs oxygenic photosynthesis and has two wAQP but no gAQP, one of which is important for glucose metabolism (Akai et al., 2011) as well as cell volume regulation (Akai et al., 2012). Therefore, some wAQPs may support bacteria to have intra- and inter-species competitive advantages.

Reflecting controversial importance of AQPs in bacteria, many bacteria do not have AQPs such as *Helicobacter pylori*, which suggests that AQP may have been lost due to their detrimental effects (Abascal et al., 2014). Indeed, water fluxes through plasma membrane by passive diffusion is sufficient for smaller organisms due to the higher surface to volume ratio and much so for bacteria living in higher temperature with higher water permeability. Moreover, the osmolyte in most bacteria is potassium or amino acids and not glycerol unlike fungi (Ahmadpour, Geijer, Tamás, Lindkvist-Petersson, & Hohmann, 2014), which may further have facilitated the loss of gAQP. As gAQPs transport metalloids (Zangi & Filella, 2012), the deletion of gAQPs may prevent the entry of toxic substances.

As a major fraction of the genes in the prokaryotic genomes have been acquired by horizontal gene transfer (Koonin, Makarova, & Aravind, 2001), it is difficult to identify the evolutional pathway for AQPs in bacteria which was also often associated with the lifestyle. For example, nitrogen-fixing bacteria have wAQPs alone which transport small molecules. These wAQPs may have been transferred to plants to become NIPs as ammonium channels (Frare, Ayub, Alleva, & Soto, 2018) associated with symbiotic relationship with legumes (Soto et al., 2012; Zardoya, Ding, Kitagawa, & Chrispeels, 2002).

Table 6 The distribution of aquaporins in three subfamilies.

Organisms	Total	gAQP	wAQP	sAQP
Bacteria				
E. coli	2	1	1	0
P. aeruginosa	2	1	1	0
S. oligofermentans	3	2	1	0
S. aureus	2	2	0	0
L. plantarum	4	4	0	0
M. tuberculosis	4	4	0	0
H. pylori	0	0	0	0
M. marburgensis	1	0	1	0
Protozoa				
T. gondii	1	0	1	0
P. falciparum	1	1	0	0
D. discoideum	5	0	5	0
L. major	5	1	4	0
T. cruzi	4	0	4	0
T. brucei	3	3	0	0
Fungi				
C. neoformans	1	0	1	0
S. cerevisiae	4	2	2	0
S. pombe	1	1	0	0
A. nidulans	5	4	1	0
T. harzianum	8	5	3	0
Plants				
Moss	23	1	22 (8, 4, 5, 2, 2,1)	0
Rice	33	0	33 (11, 10, 10, 2)	0
Arabidopsis	35	0	35 (13, 10, 9, 3)	0
Poplar	55	0	55 (15, 17, 11, 6, 6)	0
Bamboo	63	0	63 (22, 20, 17, 4)	0
Cotton	71	0	71(28, 23, 12, 7, 1)	0
Soybean	72	0	72 (22, 23, 17, 8, 2)	0

Continued

Table 6 The distribution of aquaporins in three subfamilies.—cont'd

Organisms	Total	gAQP	wAQP	sAQP
Invertebrates				
C. elegans	12	5	4	3
B. improvsius	8	2	5	2
S. purpuratus	12	5	5	2
C. intestinalis	7	1	5	1
Ticks and insects				
D. farinae	6	3	2	1
L. salmonis	8	3	3	2
M. tardigradum	11	8	2	1
D. melenogaster	8	0	7	1
A. gambiae	7	0	6	1
G. morsitans	10	0	9	1
C. lectularius	11	0	10	1
Vertebrates				
Shark	10	4	5	1
Zebrafish	20	7	11	2
Salmon	42	13	23	6
Carp	37	14	19	4
Coelacanth	15	4	9	2
Xenopus	15	5	8	2
Green anole	16	5	9	2
Snake	15	4	9	2
Zebra finch	13	4	7	2
Mouse	12	3	7	2
Pig	13	4	7	2
Human	14	4	7	3

S. oligofermentans: *Streptococcus oligofermentans*; *L. plantarum*: *Lactobacillus plantarum*; *H. pylori*: *Helicobacter pylori*; *M. marburgensis*: *Methanothermobacter marburgensis*; *A. niulans*: *Aspergillus nidulans*; *T. harzianum*: *Trichoderma harzianum*; *S. purpuratus*: *Strongylocentrotus purpuratus*: sea urchin; *C. intestinalis*: *Ciona intestinalis*: ascidian; *D. farinae*: *Dermatophagoides farina*: house dust mite; *L. salm*: *Lepeophtheirus salmonis*: louse; *M. tardigradum*: *Milnesium tardigradum*: tardigrade; *A. gambiae*: *Anopheles gambiae*: mosquito; *G. morsitans*: *Glossina morsitans*: Tsetse flies; *C. lectularius*: *Cimex lecturarius*: bed bug; *B. improvius*: *Balanus improvisus*: barnacle; *moss*: *Physcomitrella patens*; *D. melenogaster*: *Drosophila melanogaster*. fruit fly;
The number of plant wAQP subfamilies are in the order of PIP, TIP, NIP, SIP, XIP, HIP.

The limited presence of AQPs in eubacteria may have led to much less AQPs in more advanced archaea especially in thermophilic archaea since higher ambient temperature facilitates transmembrane water permeability even without AQPs. Exceptionally, *Methanothermobacter marburgensis*, has a single wAQP, AqpM accepting larger molecules such as H_2S for possible competitive advantages. It could have been obtained by a horizontal gene transfer from eubacteria (Kozono et al., 2003). The fact that a *chlorella virus* has a gAQP, AQPV1, suggests a horizontal AQP gene transfer by viruses (Gazzarrini et al., 2006).

3.2 Unicellular eukaryotic AQPs: Variable wAQPs/gAQPs ratios

As eukaryotes have evolved from symbiosis of eubacteria within archaea, they may have inherited several AQPs from eubacteria. Eukaryotes may need AQPs for their larger cellular sizes which require more efficient water and solutes transport systems at the plasma membrane and possibly at the membrane of intracellular organelles. Even though few AQPs have been derived from bacteria, the expansion of AQP genes by gene duplication and horizontal gene transfer may have occurred among eukaryotes.

Accordingly, huge variations of AQPs in numbers and subfamily ratios are present in protozoa. For example, *Leishmania major* has one gAQP and four wAQPs, whereas *Trypanosoma brucei* has only three gAQPs and *Trypanosoma cruzi* has only four wAQPs. *Dictyostelium discoideum* (*Amoebozoa*) have five wAQPs alone (Table 6). Some protozoa have only one AQPs, wAQP for *Toxoplasma* and gAQP for *Plasmodium*. A cyst-forming symbiotic *Cryptosporidium parvum* has no AQP. Therefore, environmental constraints such as nutrients and toxins may have produced huge variations by expanding or losing gAQPs and wAQPs in protozoa through evolution (Uzcátegui et al., 2018; Von Bülow & Beitz, 2015).

3.3 Fungal AQPs: Osmoregulatory AQPs

Although fungi started as unicellular organisms, they include multicellular organisms as well, which may have diversified the distribution of AQPs. For example, a budding yeast, *Saccharomyces cerevisiae*, has four AQPs: two gAQPs and two wAQPs, which may reflect the previous multicellular life style as well as a whole genome duplication. One gAQP, FPS1, accumulates glycerol as an osmolyte by closing the channel at hypertonic challenge, whereas it releases glycerol in mating to protect the cell against hypotonicity produced by the osmotically ruptured plasma membrane (Tamás et al., 1999).

On the other hand, the expression of wAQPs seems to be related to spore formation and wAQPs may enhance water content of the cell wall to become more hydrophobic whose roles remain to be clarified (Hohmann, 2002).

Interestingly, many yeasts have only one or no AQPs. For example, much distant fission yeast, *S. pombe* has only one gAQP related to AQY3, another gAQP in *S. cerevisiae* but no wAQPs. In fact, AQY3 members are widely distributed in filamentous fungi. The predominant expression of gAQP than wAQP in fungi may indicate that glycerol and nutrient imports or toxin exports may be the major function of AQPs rather than water transport although the active glycerol transporters, GUP1 and GUP2, play major roles in glycerol influx in yeasts (Holst et al., 2000). Alternatively, gAQPs may be involved in the resistance to toxic compounds by exporting them through FSP1 transporting metalloids (Wysocki et al., 2001).

Notably soil grown *Trichoderma harzianum* has eight AQPs, five gAQPs and three wAQPs (Table 6). Even in such a fungus with multiple AQPs, no sAQP has been identified. As fungi have no sAQPs, absence of sAQPs in moss and plants may be a lineage dependent phenomenon, whereas the loss of gAQPs in plants may not be the case as gAQPs are the predominant AQP subfamily in fungi.

3.4 Plant AQPs: Diversified wAQPs after losing gAQPs

Plants express much greater numbers of AQPs than any other organisms, possibly due to the greater duplication rates of plant genomes (Fox, Maistriaux, & Chaumont, 2017) and to the functional advantages during static plant lifestyle: 71 AQPs in upland cotton (*Gossypium hirsutun*) (Park, Scheffler, Bauer, & Campbell, 2010) and 121 AQPs in oil seed rape *Brassica napus* (Yuan et al., 2017). However, the diversity is curiously limited to wAQPs with complete absence of gAQPs and sAQPs. For example, *Arabidopsis thaliana* has 35 AQPs, all wAQPs, which are classified into four groups: 13 PIPs, 10 TIPs, 9 NIPs, and 3 SIPs (Table 6). Interestingly, poplar trees have another group of wAQPs, XIPs (Danielson & Johanson, 2008; Lopez et al., 2012). Both XIPs and SIPs have less conserved NPA boxes than other wAQPs but not so much as sAQPs. Moreover, algae have more complex wAQP members; HIP similar to PIPs and TIPs, and LIP similar to SIP (Danielson & Johanson, 2008; Khabudaev et al., 2014).

An exceptional gAQP, GLpF-like intrinsic proteins (GIPs) identified in a moss *Physcomitrella patens* is speculated to have been derived from bacteria by a horizontal gene transfer to this primitive land plant (Gustavsson, Lebrun, Norden, Chaumont, & Johanson, 2005). Therefore, plants already have lost gAQPs at the separation from fungi, and further loss of GIPs could have been caused by the dominant expression of NIPs although green algae has both GIP and NIPs (Anderberg, Danielson, & Johanson, 2011). The more diversified function of NIPs to transport small molecules such as glycerol, silicon and boron (Noronha et al., 2016) may also have facilitated the loss of gAQPs in plants.

As the subfamilies of mammal and plant AQPs appear to be correlated, vertical transfers of four ancestral AQP subfamilies independently to mammals and plants have been proposed; AQP1-like vs. PIP, AQP8-like vs. TIP, AQP3-like vs. NIP, and finally AQP11-like vs. SIP (Soto et al., 2012). Although this hypothesis has some functional relevance, the evolutionary pathway of the AQP family may be different between plants and mammals as NIPs and SIPs belong to wAQPs in our classification. Furthermore, there are no supportive studies on the evolutionary integrity of the multiple plant wAQPs subfamilies.

3.5 Invertebrate AQPs: Acquisition of sAQPs

Invertebrates include lower animals such as nematodes, ticks and insects, which represent about 97% of all Animalia (Campbell, Ball, Hoppler, & Bowman, 2008). It is noteworthy that sAQPs first appeared in multicellular animals but not in unicellular protozoa nor in multicellular fungi and plants. For example, a nematode, *Caenorhabditis elegans* has three sAQPs (Table 6). The controversial CeAQP12 (NP_001022480.1) is classified here as wAQP rather than sAQP because the signature Cys for sAQP is absent (Table 5). However, the upstream sequence from the first NPA is deviated and the second NPA is unusual NMK rather than canonical NPA, which has led to its previous classification as sAQP (Finn & Cerdà, 2015). As CeAQP12 challenges our classification, further collection of its EST sequences and the protein identification will be required to verify the presence of CeAQP12 as it can be a pseudogene.

The number of 12 AQPs in nematode is relatively large for the smaller body size, which may suggest the importance of AQPs in this small animal. However, their roles are still unclear as multiple AQP knockouts has produced no abnormal phenotypes although knockouts of sAQPs have not yet been conducted (Huang, Lamitina, Agre, & Strange, 2007). Interestingly, a gAQP, CeAQP1 (Table 3) has been shown to lengthen the life span in a low-sugar diet as a feedback regulator of the insulin/IGF-1-signaling (Lee, Murphy, & Kenyon, 2009). The importance of multiple AQPs in invertebrates is still unclear and requires further studies.

Much higher invertebrates such as sea urchin also has 12 AQPs despite their larger sizes (Table 6). The absence of the expansion of AQPs from nematodes may indicate the evolution of AQPs has reached the summit in smaller animals. Another renovation may be necessary to expand the repertoire of AQPs such as vertebrates which have undergone two rounds of whole genome duplications.

3.6 Tick and insect AQPs: Eventual loss of gAQPs

All three AQP subfamilies are present in early arthropods with eight legs (Ecdysozoa) such as louse, *Lepeophtheirus salmonis*, with eight AQPs: three gAQPs, three wAQPs, and two sAQPs (Table 6). Our search has identified a new wAQP as L.salm6 which was not previously reported (Stavang et al., 2015). House dust mite, *Dermatophagoides farina*, has six AQPs (three gAQPs, two wAQPs and one sAQP) (Peng et al., 2018). Tardigrade also has multiple AQPs, eight gAQPs, two wAQPs and a single sAQP, which seem to be important for the survival in drying and dehydration by anhydrobiosis (Grohme et al., 2013), in which glycerol transport will be important suggested by predominant gAQPs.

Interestingly, more advanced insects with six legs have lost gAQP. For example, the bed bug *Cimex lectularius* has 10 wAQPs and no gAQP (Tsujimoto, Sakamoto, & Rasgon, 2017). A fruit fly, *Drosophila melanogaster* has no gAQP in eight AQPs: seven wAQPs and one sAQP. One of wAQPs has larger NH3– and COOH– ends named Bib (Big Brain), which has no transport function but important for neuron differentiation. Similarly, tsetse fly, *Glossina morsitans*, a vector for African trypanosomiasis, has 10 AQPs: nine wAQPs and one sAQP. The malaria vector mosquito *Anopheles gambiae* has seven AQPs (one sAQP and six wAQPs including two Bibs) and the dengue virus mosquito vector, *Aedes aegypti*, has six AQPs (one sAQP and five wAQPs including one Bib). All winged insects have similar number and ratios of AQPs. Similar to plants, the loss of gAQP in higher insects seems to be compensated by the functional conversion of wAQPs to glycerol and urea channels named entomoglyceroporin (Eglp) (Finn et al., 2015). Insect wAQPs can be classified into four subfamilies: a glycerol-permeable

Rpip, a water-selective Drip and Prip, and finally Bib (Xia et al., 2017). Other than water transport in intestines and malpighian corpuscle to cope with high-volume liquid diets, wAQPs in higher insects may also play an important role in cryoprotection and anhydrobiosis as gAQPs do in early arthropods (Maruyama, Kambara, Naka, & Azuma, 2015).

3.7 Vertebrate AQPs: More wAQPs?

All three subfamilies of AQPs have been carried over to vertebrates from multicellular animals. For example, pigs have 13 AQPs: four gAQPs (AQP3, 7, 9, 10); 7 wAQPs (AQP0, 1, 2, 4, 5, 6, 8); and 2 sAQPs (AQP11, 12) (Table 6). As humans have two isoforms of AQP12, the total number is 14 (Calvanese et al., 2013). As mouse authentic AQP10 has turned to the pseudogene, the total number is 12 (Tanaka, Morishita, & Ishibashi, 2015).

The ancestor for vertebrates is chordate such as ascidian (*Ciona intestinalis*) which has seven AQPs (five wAQPs, one gAQPs, and one sAQP). However, the sequence of C. int7 (XP_026690588.1) as sAQP is problematic as it has unusual NPAs, VPA and VLS. Further analysis will be necessary as is the case with CeAQP12. The decreased number of gAQPs is noteworthy as other lower animals have almost equal number of gAQPs and wAQPs. The development of other solute transporters in higher invertebrates may have decreased the need for gAQPs as solute carriers, whereas the need for wAQPs as water-selective carriers may have been increased to meet the demand of efficient water transport in their larger body sizes.

Two distinct whole genome duplication (WGD) events early in vertebrate evolution have been proposed to produce four-way paralogous regions in the genome, which have expanded gene families from a single member in invertebrates to four members in vertebrates (Dehal & Boore, 2005). However, most gene pairs formed by a WGD have a short life to leave a single-copy locus. Accordingly, even after two rounds of WGDs, sharks, chondrichthyes, have only 10 AQPs. However, four gAQPs seems to be compatible with two WGDs.

Zebrafish, a teleost has undergone another round of WGD to increase the total AQP number to 20. Some fish such as salmon and carp have undergone another round of WGD to have the total AQPs to 42 and 37, respectively, with compatible subfamily distributions.

Through the evolution of vertebrates, complex changes of AQPs have been observed, which may reflect adaptive evolution of species to environmental changes. For example, as four wAQPs, AQP0, 2, 5, and 6, are located at the same chromosome, 12q13.3-12 in humans, it has been speculated that AQP2 and 5 have been produced by the local gene duplication of AQP0 for the terrestrial adaptation of amphibians and further AQP6 has been added in reptiles (Suzuki, Hasegawa, Ogushi, & Tanaka, 2007). Similarly, as AQP3 and AQP7 are located on the same chromosome, 9p13.3 in humans, both gAQPs may have been produced by local gene duplication. Interestingly, AQP12 is duplicated specifically in humans at the chromosome 2q37.3 to produce two isoforms, AQP12A and AQP12B, whose sequences are 99% identical, with only three substitutions in non-pore-lining positions (Calvanese et al., 2013).

On the other hand, some AQPs have been lost as pseudogenes. AQP10 has been lost in turtles (Finn, Chauvigne, Hlidberg, Cutler, & Cerda, 2014) or turned to the pseudogene in rodents and cows (Tanaka et al., 2015). AQP10 may be dispensable as three more gAQPs (AQP3, 7, 9) are present in mammals. Since the role of glycerol as an energy source is limited in higher vertebrates, gAQPs in mammals may not be as much important in lower animals. In fact, little abnormal phenotypes have been reported in gAQP-nulls in mice and humans whereas wAQP-nulls have produced diseases such as cataracts, nephrogenic diabetes insipidus and sicca syndrome.

The limited number of sAQPs even after WGD may reflect that the significance of this family is also limited as suggested by almost no abnormalities of AQP12-null mice (Ohta et al., 2009). Although AQP11-null mice develop dysfunctional dilated proximal tubules leading to only one-month life (Tanaka, Watari, Saito, Morishita, & Ishibashi, 2016), a conditional AQP11-knock-out after birth did not produce much changes in the kidney (Rützler et al., 2017).

4. Summary and perspectives

An ancient AQP may be a member of gAQP subfamily with larger pores permitting the uptake of nutrients and release of waste products but shutting off inorganic cations, proton and water by forming a special filter. The mutation of the critical aspartate (D) residue and other structural changes have turned gAQPs to wAQPs by narrowing the pore to become selective to water. Since bacteria have few wAQPs, wAQPs may have been transferred among eukaryotes to produce the diversity of eukaryotic wAQPs.

At the single-cell level, intracellular AQPs may have become necessary in multicellular larger organisms as efficient water and solute transporters to perform more complex cellular roles such as secretion, endocytosis, and accumulation of substances. As sAQPs are mostly located inside the cell, sAQPs may have evolved from wAQPs to keep intracellular water and solute homeostasis leading to differentiation after cell division (Nozaki, Ishii, & Ishibashi, 2008). Although fungi and plants are multicellular organisms, they are surrounded by the cell wall which hinders the communication between the cells. To overcome this problem, plants developed plasmodesmata which are connections across the cell wall to allow a cytoplasmic continuity between cells. The plasmodesmata are composed of desmotubules, a portion of endoplasmic reticulum (ER), which conveys proteins and mRNAs. This single-cell like structure in plants may explain the absence of sAQPs which are commonly expressed in the ER. For unknown reasons, the presence of cell wall may also have suppressed the evolution of sAQPs from wAQPs. Alternatively, plants have multiple intracellular wAQPs such as TIPs, NIPs and SIPs, which may have decreased the need for sAQPs.

Although all three AQP subfamilies are present in animals, subsequent evolution has changed their balance and even gAQP loss in higher insects and plants, as gAQPs were not efficient water transporters and compensated by wAQPs converted to functional glycerol channels.

In vertebrates, the evolution of AQPs may have been linked to the evolution of tissues and regulating hormones. For example, osmoregulatory hormone (vasopressin-like), its receptor (V2-type-like), cAMP signaling and wAQP (AQP2-like) developed co-evolutionary from the urinary bladder of frogs to distal kidney segments of mammals as a crucial part of the urine-concentrating mechanism (Ogushi et al., 2010).

The clarification of cooperative evolutions of AQPs and regulating hormones as well as associated proteins and relevant transporters and channels will be an important step toward understanding the role of AQPs counteracting against the environmental challenges from the perspective of evolution. Such challenges may also be relevant to the disease states of patients with local or systemic edema and dehydration. As gAQPs transport glycerol, fat metabolism by the adipocyte and liver through AQPs may also have evolutional relevance. As AQPs are widely expressed in many organisms with conserved structures, the adapting strategies taken by lower organisms using AQPs may provide new insights into the development of novel therapies and prevention of the diseases.

Acknowledgment

This work was supported by JSPS KAKENHI Grant Number 15K09302.

Conflict of interest

The authors declare that there is no conflict of interest regarding the publication of this article.

References

Abascal, F., Irisarri, I., & Zardoya, R. (2014). Diversity and evolution of membrane intrinsic proteins. *Biochimica et Biophysica Acta*, *1840*, 1468–1481.

Ahmadpour, D., Geijer, C., Tamás, M. J., Lindkvist-Petersson, K., & Hohmann, S. (2014). Yeast reveals unexpected roles and regulatory features of aquaporins and aquaglyceroporins. *Biochimica et Biophysica Acta*, *1840*, 1482–1491.

Akai, M., Onai, K., Kusano, M., Sato, M., Redestig, H., Toyooka, K., et al. (2011). Plasma membrane aquaporin AqpZ protein is essential for glucose metabolism during photomixotrophic growth of *Synechocystis* sp. PCC 6803. *Journal of Biological Chemistry*, *286*, 25224–25235.

Akai, M., Onai, K., Morishita, M., Mino, H., Shijuku, T., Maruyama, H., et al. (2012). Aquaporin AqpZ is involved in cell volume regulation and sensitivity to osmotic stress in *Synechocystis* sp. strain PCC 6803. *Journal of Bacteriology*, *194*, 6828–6836.

Anderberg, H. I., Danielson, J. A., & Johanson, U. (2011). Algal mips, high diversity and conserved motifs. *BMC Evolutionary Biology*, *11*, 110.

Calvanese, L., D'Auria, G., Vangone, A., Falcigno, L., & Oliva, R. (2018). Structural basis for mutations of human aquaporins associated to genetic diseases. *International Journal of Molecular Sciences*, 19. pii: E1577.

Calvanese, L., Pellegrini-Calace, M., & Oliva, R. (2013). In silico study of human aquaporin AQP11 and AQP12 channels. *Protein Science*, *22*, 455–466.

Campbell, E. M., Ball, A., Hoppler, S., & Bowman, A. S. (2008). Invertebrate aquaporins: A review. *Journal of Comparative Physiology*, *178*, 935–955.

da Silva, I. V., & Soveral, G. (2017). Aquaporins in obesity. *Advances in Experimental Medicine and Biology*, *969*, 227–238.

Danielson, J. Å., & Johanson, U. (2008). Unexpected complexity of the aquaporin gene family in the moss Physcomitrella patens. *BMC Plant Biology*, *8*, 45.

Dehal, P., & Boore, J. L. (2005). Two rounds of whole genome duplication in the ancestral vertebrate. *PLoS Biology*, *3*, e314.

Di Giorgio, J. P., Soto, G., Alleva, K., Jozefkowicz, C., Amodeo, G., Muschietti, J. P., et al. (2014). Prediction of aquaporin function by integrating evolutionary and functional analyses. *Journal of Membrane Biology*, *247*, 107–125.

Filippidis, A. S., Carozza, R. B., & Rekate, H. L. (2017). Aquaporins in brain edema and neuropathological conditions. *International Journal of Molecular Science*, *18*, 55.

Finn, R. N., & Cerdà, J. (2015). Evolution and functional diversity of aquaporins. *Biological Bulletin*, *229*, 6–23.

Finn, R. N., Chauvigne, F., Hlidberg, J. B., Cutler, C. P., & Cerda, J. (2014). The lineage-specific evolution of aquaporin gene clusters facilitated tetrapod terrestrial adaptation. *PLoS One*, *9*, e113686.

Finn, R. N., Chauvigne, F., Stavang, J. A., Belles, X., & Cerda, J. (2015). Insect glycerol transporters evolved by functional co-option and gene replacement. *Nature Communications*, *6*, 7814.

Fox, A. R., Maistriaux, L. C., & Chaumont, F. (2017). Toward understanding of the high number of plant aquaporin isoforms and multiple regulation mechanisms. *Plant Science, 264*, 179–187.
Frare, R., Ayub, N., Alleva, K., & Soto, G. (2018). The ammonium channel NOD26 is the evolutionary innovation that drives the emergence, consolidation, and dissemination of nitrogen-fixing symbiosis in angiosperms. *Journal of Molecular Evolution, 86*, 554–565.
Frick, A., Eriksson, U. K., de Mattia, F., Oberg, F., Hedfalk, K., Neutze, R., et al. (2014). X-ray structure of human aquaporin 2 and its implications for nephrogenic diabetes insipidus and trafficking. *Proceedings of the National Academy of Sciences of the United States of America, 111*, 6305–6310.
Gazzarrini, S., Kang, M., Epimashko, S., Van Etten, J. L., Dainty, J., Thiel, G., et al. (2006). Chlorella virus MT325 encodes water and potassium channels that interact synergistically. *Proceedings of the National Academy of Sciences of the United States of America, 103*, 5355–5360.
Grohme, M. A., Mali, B., Wełnicz, W., Michel, S., Schill, R. O., & Frohme, M. (2013). The aquaporin channel repertoire of the Tardigrade *Milnesium tardigradum*. *Bioinformatics and Biology Insights*, 7, 153–165.
Gustavsson, S., Lebrun, A. S., Norden, K., Chaumont, F., & Johanson, U. (2005). A novel plant major intrinsic protein in *Physcomitrella patens* most similar to bacterial glycerol channels. *Plant Physiology, 139*, 287–295.
Hill, A. E., & Shachar-Hill, Y. (2015). Are Aquaporins the missing transmembrane osmosensors? *Journal of Membrane Biology, 248*, 753–765.
Hohmann, S. (2002). Osmotic adaptation in yeast-control of the yeast osmolyte system. *International Review of Cytology, 215*, 149–187.
Holst, B., Lunde, C., Lages, F., Oliveira, R., Lucas, C., & Kielland-Brandt, M. C. (2000). GUP1 and its close homologue GUP2, encoding multimembrane-spanning proteins involved in active glycerol uptake in *Saccharomyces cerevisiae*. *Molecular Microbiology, 37*, 108–124.
Huang, C. G., Lamitina, T., Agre, P., & Strange, K. (2007). Functional analysis of the aquaporin gene family in *Caenorhabditis elegans*. *American Journal of Physiology-Cell Physiology, 292*, C1867–C1873.
Ishibashi, K., Hara, S., & Kondo, S. (2009). Aquaporin water channels in mammals. *Clinical and Experimental Nephrology, 13*, 107–117.
Ishibashi, K., Kondo, S., Hara, S., & Morishita, Y. (2011). The evolutionary aspects of aquaporin family. *American Journal of Physiology-Regulatory, Integrative and Comparative Physiology, 300*, R566–R576.
Ishibashi, K., Morishita, Y., & Tanaka, Y. (2017). The evolutionary aspects of aquaporin family. *Advances in Experimental Medicine and Biology, 969*, 35–50.
Khabudaev, K. V., Petrova, D. P., Grachev, M. A., & Likhoshway, Y. V. (2014). A new subfamily LIP of the major intrinsic proteins. *BMC Genomics, 15*, 173.
Kirscht, A., Kaptan, S. S., Bienert, G. P., Chaumont, F., Nissen, P., de Groot, B. L., et al. (2016). Crystal structure of an ammonia-permeable aquaporin. *PLoS Biology, 14*, e1002411.
Kirscht, A., Sonntag, Y., Kjellbom, P., & Johanson, U. (2018). A structural preview of aquaporin 8 via homology modeling of seven vertebrate isoforms. *BMC Structural Biology, 18*, 2.
Koonin, E. V., Makarova, K. S., & Aravind, L. (2001). Horizontal gene transfer in prokaryotes: Quantification and classification. *Annual Review of Microbiology, 55*, 709–742.
Kozono, D., Ding, X., Iwasaki, I., Meng, X., Kamagata, Y., Agre, P., et al. (2003). Functional expression and characterization of an archaeal aquaporin. AqpM from Methanothermobacter marburgensis. *Journal of Biological Chemistry, 278*, 10649–10656.
Laloux, T., Junqueira, B., Maistriaux, L. C., Ahmed, J., Jurkiewicz, A., & Chaumont, F. (2018). Plant and mammal Aquaporins: Same but different. *International Journal of Molecular Sciences, 8*, 19.

Lee, S. J., Murphy, C. T., & Kenyon, C. (2009). Glucose shortens the life span of *C. elegans* by downregulating DAF-16/FOXO activity and aquaporin gene expression. *Cell Metabolism, 10*, 379–391.
Liu, K., Tsujimoto, H., Huang, Y., Rasgon, J. L., & Agre, P. (2016). Aquaglyceroporin function in the malaria mosquito *Anopheles gambiae*. *Biology of the Cell, 108*, 294–305.
Lopez, D., Bronner, G., Brunel, N., Auguin, D., Bourgerie, S., Brignolas, F., et al. (2012). Insights into Populus XIP aquaporins: Evolutionary expansion, protein functionality, and environmental regulation. *Journal of Experimental Botany, 63*, 2217–2230.
Manley, G. T., Fujimura, M., Ma, T., Noshita, N., Filiz, F., Bollen, A. W., et al. (2000). Aquaporin-4 deletion in mice reduces brain edema after acute water intoxication and ischemic stroke. *Nature Medicine, 6*, 159–163.
Maruyama, M., Kambara, K., Naka, H., & Azuma, M. (2015). Insect water-specific aquaporins in developing ovarian follicles of the silk moth Bombyx mori: Role in hydration during egg maturation. *Biological Bulletin, 229*, 58–69.
Milano, S., Carmosino, M., Gerbino, A., Svelto, M., & Procino, G. (2017). Hereditary nephrogenic diabetes insipidus: Pathophysiology and possible treatment. An update. *International Journal of Molecular Sciences, 10*, 18.
Noronha, H., Araújo, D., Conde, C., Martins, A. P., Soveral, G., Chaumont, F., et al. (2016). The grapevine uncharacterized intrinsic protein 1 (VvXIP1) is regulated by drought stress and transports glycerol, hydrogen peroxide, heavy metals but not water. *PLoS One, 11*, e0160976.
Nozaki, K., Ishii, D., & Ishibashi, K. (2008). Intracellular aquaporins: Clues for intracellular water transport? *Pflügers Archiv European Journal of Physiology, 456*, 701–707.
Ogushi, Y., Akabane, G., Hasegawa, T., Mochida, H., Matsuda, M., Suzuki, M., et al. (2010). Water adaptation strategy in anuran amphibians: Molecular diversity of aquaporin. *Endocrinology, 151*, 165–173.
Ohta, E., Itoh, T., Nemoto, T., Kumagai, J., Ko, S. B., Ishibashi, K., et al. (2009). Pancreas-specific aquaporin 12 null mice showed increased susceptibility to caerulein-induced acute pancreatitis. *American Journal of Physiology-Cell Physiology, 297*, C1368–C1378.
Palmgren, M., Hernebring, M., Eriksson, S., Elbing, K., Geijer, C., Lasič, S., et al. (2017). Quantification of the intracellular life time of water molecules to measure transport rates of human aquaglyceroporins. *The Journal of Membrane Biology, 250*, 629–639.
Park, W., Scheffler, B. E., Bauer, P. J., & Campbell, B. T. (2010). Identification of the family of aquaporin genes and their expression in upland cotton (*Gossypium hirsutum* L.). *BMC Plant Biology, 13*(10), 142.
Peng, J., Zhou, Y., Jia, H., Li, L., Qian, J., Han, F., et al. (2018). Transcriptomics-based identification of aquaporin diversity in the house dust mite *Dermatophagoides farinae* (Acariformes: Pyroglyphidae). *Journal of Insect Science, 1*, 18.
Rao, N. V., Shashidhar, R., & Bandekar, J. R. (2013). Comparative analysis of induction of osmotic-stress-dependent genes in Vibrio vulnificus exposed to hyper- and hypo-osmotic stress. *Canadian Journal of Microbiology, 59*, 333–338.
Rützler, M., Rojek, A., Damgaard, M. V., Andreasen, A., Fenton, R. A., & Nielsen, S. (2017). Temporal deletion of Aqp11 in mice is linked to the severity of cyst-like disease. *American Journal of Physiology-Renal Physiology, 312*, F343–F351.
Soto, G., Alleva, K., Amodeo, G., Muschietti, J., & Ayub, N. D. (2012). New insight into the evolution of aquaporins from flowering plants and vertebrates: Orthologous identification and functional transfer is possible. *Gene, 503*, 165–176.
Soupene, E., King, N., Lee, H., & Kustu, S. (2002). Aquaporin Z of *Escherichia coli*: Reassessment of its regulation and physiological role. *Journal of Bacteriology, 184*, 4304–4307.

Stavang, J., Chauvigne, C., Kongshaug, H., Cerda, J., Nilsen, F., & Finn, R. N. (2015). Phylogenomic and functional analyses of salmon lice aquaporins uncover the molecular diversity of the superfamily in Arthropoda. *BMC Genomics*, *16*, 618.

Suzuki, M., Hasegawa, T., Ogushi, Y., & Tanaka, S. (2007). Amphibian aquaporins and adaptation to terrestrial environments: A review. *Comparative Biochemistry and Physiology Part A: Molecular & Integrative Physiology*, *148*, 72–81.

Tamás, M. J., Luyten, K., Sutherland, F. C., Hernandez, A., Albertyn, J., Valadi, H., et al. (1999). Fps1p controls the accumulation and release of the compatible solute glycerol in yeast osmoregulation. *Molecular Microbiology*, *31*, 1087–1104.

Tanaka, Y., Morishita, Y., & Ishibashi, K. (2015). Aquaporin10 is a pseudogene in cattle and their relatives. *Biochemistry and Biophysics Reports*, *1*, 16–21.

Tanaka, Y., Watari, M., Saito, T., Morishita, Y., & Ishibashi, K. (2016). Enhanced autophagy in polycystic kidneys of AQP11 null mice. *International Journal of Molecular Sciences*, *17*, 1993.

Tanghe, A., Van Dijck, P., & Thevelein, J. M. (2006). Why do microorganisms have aquaporins? *Trends in Microbiology*, *14*, 78–85.

Tchekneva, E. E., Khuchua, Z., Davis, L. S., Kadkina, V., Dunn, S. R., Bachman, S., et al. (2008). A newly identified ENU-induced single amino acid mutation in aquaporin-11 resulting in perinatal kidney failure in mice. *Journal of the American Society of Nephrology*, *19*, 1955–1964.

Tong, H., Wang, X., Dong, Y., Hu, Q., Zhao, Z., Zhu, Y., et al. (2019). A Streptococcus aquaporin acts as peroxiporin for efflux of cellular hydrogen peroxide and alleviation of oxidative stress. *Journal of Biological Chemistry*, *294*, 4583–4595.

Tsujimoto, H., Sakamoto, J. M., & Rasgon, J. L. (2017). Functional characterization of aquaporin-like genes in the human bed bug *Cimex lectularius*. *Scientific Reports*, *7*, 3214.

Uzcátegui, N. L., Figarella, K., Segnini, A., Marsiccobetre, S., Lang, F., Beitz, E., et al. (2018). Trypanosoma brucei aquaglyceroporins mediate the transport of metabolic end-products: Methylglyoxal, D-lactate, L-lactate and acetate. *Biochimica et Biophysica Acta (BBA)—Biomembranes*, *1860*, 2252–2261.

Verma, R. K., Gupta, A. B., & Sankararamakrishnan, R. (2015). Major intrinsic protein superfamily: Channels with unique structural features and diverse selectivity filters. *Methods in Enzymology*, *557*, 485–520.

Von Bülow, J., & Beitz, E. (2015). Number and regulation of protozoan aquaporins reflect environmental complexity. *Biological Bulletin*, *229*, 38–46.

Wysocki, R., Chery, C. C., Wawrzycka, D., Van Hulle, M., Cornelis, R., Thevelein, J. M., et al. (2001). The glycerol channel Fps1p mediates the uptake of arsenite and antimonite in *Saccharomyces cerevisiae*. *Molecular Microbiology*, *40*, 1391–1401.

Xia, W., Zhao, P., Yi, Z., & Cui, Y. (2017). Phylogenetic and specific sequence analysis of four paralogs in insect Aquaporins. *Molecular Medicine Reports*, *16*, 4903–4908.

Yang, B., & Verkman, A. S. (1997). Water and glycerol permeabilities of aquaporins 1-5 and MIP determined quantitatively by expression of epitope-tagged constructs in Xenopus oocytes. *Journal of Biological Chemistry*, *272*, 16140–16146.

Yuan, D., Li, W., Hua, Y., King, G. J., Xu, F., & Shi, L. (2017). Genome-wide identification and characterization of the aquaporin gene family and transcriptional responses to boron deficiency in *Brassica napus*. *Frontiers in Plant Science*, *8*, 1336.

Zangi, R., & Filella, M. (2012). Transport routes of metalloids into and out of the cell: A review of the current knowledge. *Chemico-Biological Interactions*, *197*, 47–57.

Zardoya, R., Ding, X., Kitagawa, Y., & Chrispeels, M. J. (2002). Origin of plant glycerol transporters by horizontal gene transfer and functional recruitment. *Proceedings of the National Academy of Sciences of the United States of America*, *99*, 14893–14896.

CHAPTER TWO

Structure and dynamics of aquaporin-1

Warin Rangubpit[a], Pornthep Sompornpisut[a], Ras Pandey[b,*]
[a]Center of Excellence in Computational Chemistry, Department of Chemistry, Chulalongkorn University, Bangkok, Thailand
[b]School of Mathematics and Natural Sciences, University of Southern Mississippi, Hattiesburg, MS, United States
*Corresponding author: e-mail address: Ras.Pandey@usm.edu

Contents

Abstract

Structural response of a AQP1 is examined by a coarse-grained model with a phenomenological interaction potential with a knowledge-based residue–residue interaction (derived from an ensemble of protein structures in PDB). The thermal response of the protein chain exhibits an unexpected characteristics in its native phase where the radius of gyration of the protein decreases on raising the temperature. The radius of gyration of AQP1 increases on increasing the temperature before saturating to a random-coil morphology in denatured phase at high temperatures. Three regions of persistent globularization are identified, toward the end segments 1M-25V and 250V-269K and a narrow region in the middle 155A-163D along the backbone. Varying the temperature leads to a systematic redistribution of self-organizing residues with globular and fibrous morphologies with an effective dimension $D \sim 2$ (random coil) at high temperature and $D \sim 3$ (globular conformation) in native phase. A preliminary analysis is also presented on the effect of a crowded membrane environment on the protein structure by incorporating effective solute constituents. Conformation of the protein is found to be pinned by selective binding of solute to specific targets; the matrix directed structure differs considerably from that of a protein in a generic solvent. The structure of AQP1 can be controlled by temperature and constitutive elements of the underlying matrix.

Vitamins and Hormones, Volume 112
ISSN 0083-6729
https://doi.org/10.1016/bs.vh.2019.09.001

1. Introduction

Aquaporins (AQPs) are known for transporting water (Aquaporin, 2018) in and out of the membrane selectively while preventing ions and other solutes from passing. AQPs are found in bacteria, plants, bacteria, tissues, etc. AQPs are critical in many physiological processes (Papadopoulos & Saadoun, 2015; Pelagalli, Squillacioti, Mirabella, & Meli, 2016; To & Torres, 2015; Tomita et al., 2017; Wang et al., 2015) and therefore, associated with health disorders such as diabetes, inflammation and demyelination of optic nerves and spinal cord. Different types of AQPs perform specific functions, e.g., AQP1, 2, 4, 5 are known for their selective transport of water while AQP3, 7, 9, 10 for the permeation of water and urea among other solutes, etc. Further, the range of permeation of water depends on the size of these AQPs. Investigation of AQPs has exploded ever since the 2003 Nobel prize in chemistry (Knepper & Nielsen, 2004); there are too many papers to cite though there are excellent reviews focus on its role in 'health and diseases' particularly related to guts (Pelagalli et al., 2016), signals in the onset of cancer (Tomita et al., 2017), etc. We focus on AQP1 which controls the transport and flow of water in and out of the cell membrane and are associated with kidney, and various membranes (epithelial cells), red blood cells, sweat glands, lungs and gastro-intestinal track. AQP1s are believed to form homo-tetramer with six transmembrane alpha-helical domains each of which channel water independently. In a crowded cell, AQP1s are embedded in lipid membrane to perform such a complex task of selective transport (Boassa, Stamer, & Yool, 2006). Understanding the dynamics of conformational response of AQP1 in a membrane environment is crucial in identifying the pathways for the water transport in and out of the cell. However, it is a challenge to probe how the structure of protein responds to dynamic environment involving fluctuating membrane thickness, pore sizes, and interactions with the solvent and solute constituent at multiple scales (molecular to cellular). Numerous experiments resort to reconstituted environments due to difficulties in probing the structural dynamics in a complex real cellular environment.

Because of the specific selectivity, AQPs have become prime candidates in designing custom membranes from molecular-scale for water filtration along with such smart materials as graphene, block-copolymers, and other nano-channels (Kumar, Grzelakowski, Zilles, Clark, & Meier, 2007; Werber, Osuji, & Elimelech, 2016). Weber and Elimelech have recently

(Werber & Elimelech, 2018) pointed out some of the challenges in designing a biomimetic desalination membrane typically used in reverse-osmosis. One of the main problem is the 'inadequate rejection' of undesirable ions and neutral solute by many potential membranes. AQPs, as an excellent selective water channel, constitute a natural choice in developing a viable membrane, e.g., for reverse-osmosis. Designing such a membrane and assessing its performance and effectiveness depends on how to incorporate AQPs into the membrane appropriately. A good understanding of structural dynamics of AQPs is therefore highly desirable.

As pointed above, enormous efforts have been made to measure the water permeability through the reconstituted membrane with aquaporin channels. Connecting the laboratory observations with the structural dynamics of the protein via elusive hypothesis remains highly speculative. Because of the complexity of the underlying crowded but cooperative environment involving molecular components of various shapes, sizes, interactions, etc. with multi-scale relaxations it is rather difficult to identify the role of critical components such as AQPs. The fundamental issue in 'protein folding' still remains an open question despite enormous efforts for decades.

A number of computer simulation studies (Ash, Zlomislic, Oloo, & Tieleman, 2004; Bond, Holyoake, Ivetac, Khalid, & Sansom, 2007; Gumbart, Wang, Aksimentiev, Tajkhorshid, & Schulten, 2005; Hashido, Ikeguchi, & Kidera, 2005; Hashido, Kidera, & Ikeguchi, 2007; Jensen & Mouritsen, 2006; Kong & Ma, 2001; Lindahl & Sansom, 2008; Wang & Tajkhorshid, 2007; Yu, Yool, Schulten, & Tajkhorshid, 2006) have been carried out via simplified models with reconstituted membrane environments to assess permeability through AQPs channels. A thorough exploration of the enormous conformational phase space of a protein such as AQP1 with a traditional MD simulation involving atomistic force field may not be feasible particularly in probing the large-scale response properties. Some degree of coarse-graining (Bond et al., 2007; de Jong et al., 2013; Ercolessi & Adams, 1994; Haliloglu & Bahar, 1998; Lindahl & Sansom, 2008; Liwo, Czaplewski, Oldziej, & Scheraga, 2008; Lyubartsev & Laaksonen, 1995; Pandey & Farmer, 2010; Reith, Putz, & Muller-Plathe, 2003; Sorensen, Xavier, Skeby, Marrink, & Schiott, 2011; van Giessen & Straub, 2005; Zhou, Chen, & Jiang, 2003; Zhou, Thorpe, Izvekov, & Voth, 2007) is almost unavoidable to probe desirable questions that require large-scale structural dynamics. We focus on the structural variability of a free aquaporin AQP1 (Pandey & Farmer, 2017) first before exploring the effects of interacting elements of the underlying matrix systematically.

In this article, we examine the thermal response of AQP1 in its native and denature phases (free from the complexity of membrane environment) via a coarse-grained model. Impacts of an effective environment with explicit aquatic solute will then be considered. In particular, the effect of the solute–residue interaction on the conformation of the protein will be examined. Even though the model is simplified (see Section 2), we hope that results (Section 3) would be helpful in postulating the hypothesis generally invoked in interpreting the laboratory observations. Lessons learnt from a simplified computer simulation model are presented in Section 4 (conclusion and outlook).

2. Model and method

It is common to use popular software (e.g., CHARMM, AMBER, LAMMPS, etc., and a combination thereof with additional tools and tricks) in protein modeling which generally involves some degree of atomistic details in molecular dynamics (MD) simulations. All atom MD simulations are generally performed in continuum space with huge spatial degrees of freedom involving various force fields. Such simulations become computationally expensive particularly in exploring large-scale structures of complex protein systems. Coarse-grained approaches (Bond et al., 2007; de Jong et al., 2013; Ercolessi & Adams, 1994; Haliloglu & Bahar, 1998; Lindahl & Sansom, 2008; Liwo et al., 2008; Lyubartsev & Laaksonen, 1995; Pandey & Farmer, 2010; Reith et al., 2003; Sorensen et al., 2011; van Giessen & Straub, 2005; Zhou et al., 2003, 2007) are frequently used to examine protein structures in such complex system; most of these simplified models involve coarsening of force-fields and combining several atoms into one unit (a kind of united atom representation) while retaining the nature of MD simulations in continuum space. In order to develop an efficient model that can capture the specificity of the amino acid while accelerating their dynamic movement, it may be natural to represent each amino acid as a united unit in a coarse-grained lattice space (Mirau, Farmer, & Pandey, 2015; Pandey, Farmer, & Gerstman, 2015). Such a simplified 'bond-fluctuation' approach has been used extensively in studying complex problems in polymers (Binder, 1995). We use a 'bond-fluctuation' coarse-grained approach with ample degrees of freedom for each amino acid to move and peptide bonds to vary while capturing specificity of each residue to examine the structural dynamics of aquaporin. One of advantages in our coarse-grained approach is that it does not require a reference conformation of the protein generally needed in popular 'protein folding' modeling since all microstates (involving protein configurations) of an ensemble should be accessible from any other microstate, a basic premise of statistical physics.

The protein (AQP1) is represented by a set of 269 nodes (amino acids) tethered together by peptide bonds where each bond can fluctuate in its length (Pandey & Farmer, 2017). A cubic cell (with eight lattice sites) of the lattice represents a node, a united atoms approximation where the structural details of the amino acids are neglected but the specificity is captured by its unique interaction (residue-residue, residue-solute, etc.). Note that a unit cell has 26 adjacent cells to move on a cubic lattice. Initially, the protein chain is placed in a random configuration with the covalent (peptide) bond length between consecutive nodes varying randomly between 2 and $\sqrt{(10)}$ in unit of the lattice constant. The degrees of freedom for nodes to move and bonds to fluctuate are comparatively limited (with respect to off-lattice) but are still plenty, unlike the minimalist protein chain models on a discrete lattice. Such a bond-fluctuation approach provides ample phase space coverage to recover the results of simulations in continuum space remarkably well in simple polymer systems with advantage of its utility in tackling problems with additional complexity such as melt. Moreover the degrees of freedom can be enhanced by fine-graining (Pandey & Farmer, 2013) while maintaining the efficiency of discrete matrix. In addition to protein chain, one can incorporate solute explicitly, i.e., by representing a solute constituent by a cubic cell (similar to that of a protein node) in a coarse-grained representation (Pandey, Jcobs, & Farmer, 2017) considered here; a solute with different shapes and sizes can however be designed if needed. The volume fraction of the effective solute can be controlled as well as its unique interactions. The effective solute constituents are randomly placed on the lattice (Pandey et al., 2017), one cell at a time, with excluded volume constraints initially.

Each constitutive element, i.e., residue and solute molecule interacts with a generalized Lennard-Jones potential (U_{ij}) within a cut-off range (r_c),

$$U_{ij} = \left[\varepsilon_{ij}\left(\frac{\sigma}{r_{ij}}\right)^{12} + \varepsilon_{ij}\left(\frac{\sigma}{r_{ij}}\right)^{6}\right], \quad r_{ij} < r_c \tag{1}$$

where r_{ij} is the distance between the interacting elements at site i and j; $r_c = \sqrt{8}$ and $\sigma = 1$ in units of lattice constant. Residue-residue and residue–solute interactions are unique for each residue via specific values for ε_{ij}. A knowledge–based interaction matrix, derived from a large ensemble of protein structures (PDB) (Betancourt & Thirumalai, 1999; Godzik, Kolinski, & Skolnick, 1996; Huang & Xiaoqin, 2011; Miyazawa & Jernigan, 1985, 1996; Tanaka & Scheraga, 1976), is used for the residue–residue pair interaction (ε_{ij}) which is unique for each interaction pair with appropriate positive (repulsive) and negative (attractive) values (Fritsche, Pandey, Farmer, & Heermann, 2012; Pandey & Farmer, 2012). It is worth pointing out that effect of solvent environment that controls the structures of a diverse range of proteins in PDB is implicitly included in such knowledge-based interactions.

In order to customize the matrix, we incorporate explicit solute molecules (Pandey et al., 2017) to represent an effective aquatic environment which requires parametrization of each pair interaction (residue-solute, solute-solute). For a specific single component effective matrix, 21 parameters are needed to describe solute–solute interaction and solute–residue interactions to capture its specificity. For purpose of demonstration and to simplify the procedure, we use the hydropathy index (Haliloglu & Bahar, 1998) of each residue to parametrize its interaction with the aquatic solute molecule via $\varepsilon_{ij} = f_w\, \varepsilon_r\, A_{h/p/e}$. The hydropathy index characterizes the specificity (i.e., attractive or repulsive) of interaction between the solute molecule and a residue (ε_r). The residue-solute interaction is repulsive ($\varepsilon_r = 0.1$) for all hydrophobic (H) residues, attractive ($\varepsilon_r = -\ 0.2$) for all polar (P) residues, and is even more attractive ($\varepsilon_r = -\ 0.3$) to all electrostatic (E) residues. Even though the hydropathy index is broadly divided into three groups (H, P, E), the weight $A_{h/p/e}$ of a residue varies within each group according to its relative hydropathy index (Pandey et al., 2017). The interactions among different types of residues and solvent constituents are thus unique. The empirical parameter f_w introduced above modulates solvent quality of the underlying matrix. The interaction between solute molecules is only excluded volume.

The number of solute molecules in a simulation box in presence of the protein chain is fixed throughout the simulation. Each constituent (residue and solute molecule) executes its stochastic motion with the Metropolis algorithm (Pandey et al., 2017) within the constraints of excluded volume and the flexibility of the bond length in the protein. Temperature, time step, and spatial length scales are reported in natural units for the simulation since our focus is on changes in physical quantities in response to changing the temperature and the solute interaction strength. Simulations are carried out for sufficiently long time steps to reach steady-state and/or thermodynamic equilibrium, typically 5–10 million time steps here. Such simulations are performed for each solute interaction strength f_w for a range of low to high temperatures to capture the structural dynamics in both native and denatured phases of the protein. A number of local and global physical quantities such as energy, mobility, and contact profiles, radius of gyration and its structure factor are examined, some of these are presented in the following section.

3. Results and discussion

The conformation of the protein is critical in its prolific yet specific response as pointed in the introduction. Let us examine the conformational response with the analysis of some of our computer simulation data: first on a free AQP1, i.e., in absence of explicit matrix interference as a function of temperature followed by the effects of its interaction with its constitutive environment. Competing and cooperative effect of thermal agitation is a natural characteristics encountered in structural evolution of physical and chemical phenomena. At low temperatures (with low thermal energy), interactions among the constitutive elements dominate over the thermal noise while opposite is the case at high temperatures. The residue–residue interactions in a free protein provides the specificity of its native structure at low temperature and become irrelevant at high temperatures where the protein conforms to a random-coil structure in its denatured phase. Analysis of the structures over a wide range of temperature may provide a broad insight into the stability of the protein conformation due to interplay between the thermal noise and interactions (residue-residue, residue-matrix) along with the steric constraints. How does structure of a protein respond to thermal noise may depend on the unique make-up of the protein and the underlying matrix.

Fig. 1 shows representative snapshots of a free AQP1 at a range of increasing temperatures in its native phase where residue–residue interactions play a dominant role in orchestrating the structural evolution. We see a systematic increase in segmental aggregation or globularization and decrease in the overall size of the protein on raising the temperature, an unusual negative thermal response. Note that each conformation (Fig. 1) represents only one of a large set of structures in conformational ensemble at each temperature. Therefore, the general trend of thermal response should not be assessed by few snapshots. Even though these snapshots provide some insight into the thermal response, its reliability must be assessed by analyzing an ensemble averaged quantity (see below). Further, it is not feasible to compare these snapshots with the protein structure in PDB (typically an X-ray

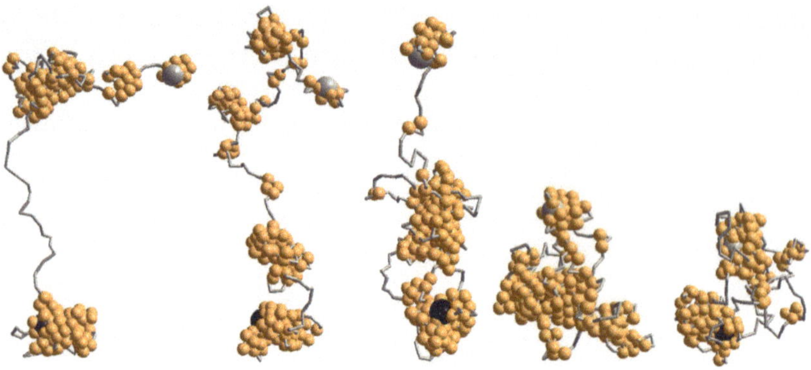

Fig. 1 Snapshots of AQP1 structure at the end of 10 million time steps ($T=0.020$ (left), 0.060 (right)). Larger spheres represent the end residues (1M, black) and (269K, gray). The golden sphere represents the presence of neighboring residues (contact map). Sequence of AQP1: 1MASEFKKKLF 11WRAVVAEFLA 21TTLFVFISIG 31SALGFKYPVG 41NNQTAVQDNV 51KVSLAFGLSI 61ATLAQSVGHI 71SGAHLNPAVT 81LGLLLSCQIS 91IFRALMYIIA 101QCVGAIVATA 111ILSGITSSLT 121GNSLGRNDLA 131DGVNSGQGLG 141IEIIGTLQLV 151LCVLATTDRR 161RRDLGGSAPL 171AIGLSVALGH 181LLAIDYTGCG 191INPARSFGSA 201VITHNFSNHW 211IFWVGPFIGG 221ALAVLIYDFI 231LAPRSSDLTD 241RVKVWTSGQV 251EEYDLDADDI 261NSRVEMKPK.

snapshot in laboratory) due to difficulties in identifying one structure from millions of conformations of a large ensemble. In order to identify a reliable overall structural response, one should analyze the ensemble average of local and global physical quantities.

The number of surrounding residues with a range of interaction of each residue shows the degree of local globularity (Fig. 1). Average number (Nr) of surrounding residue of each along the protein backbone, the contact profile, is one of the local physical quantities that provides some insight into the segmental self-assembly which may be useful in assessing its rigidity, flexibility and stabilities. The contact profile of the protein is presented in Fig. 2 at representative low (native phase) and high (denatured phase) temperatures. High values of Nr in native phase (i.e., $T=0.015$) for many residues throughout along the backbone (i.e., 12R, 18F, 24F, 98I, 225L, etc.) shows that AQP1 adopts to compact (globular) conformations overall. The segmental globularization becomes localized (e.g., around 1M-25V, 81L-100A, 147L-187T, 224V-269K) on raising the temperature ($T=0.020$). Further increasing the temperature ($T=0.030$) however reduces the globularity considerably as the protein denatures and become random coil at high temperature ($T=0.040$) with a relatively low value

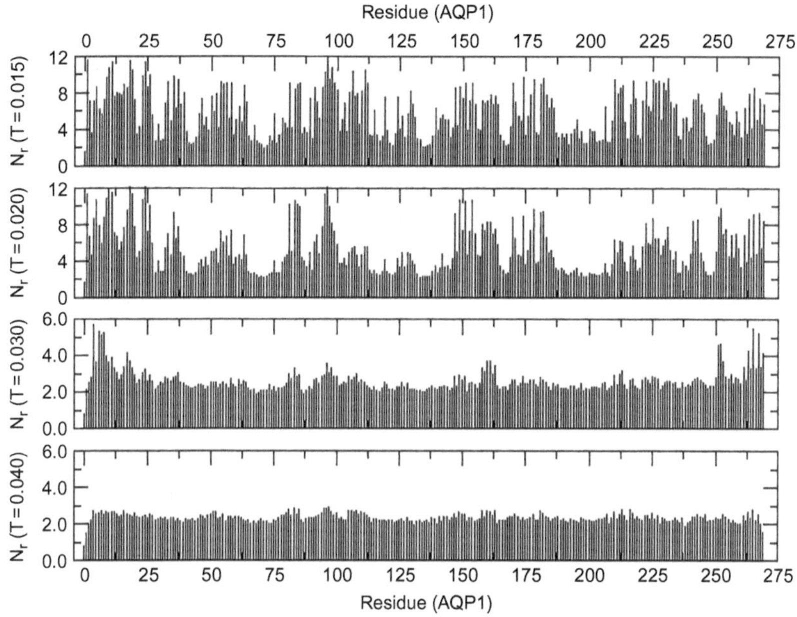

Fig. 2 Average number of residues within the range of interaction around each residues at $T=0.015$, 0.020, 0.030, and 0.040.

of contact profile index Nr ~ 2–3. Thus, the self-organizing structures of different segments of AQP1 vary considerably from low temperature regime (the residue–residue interaction controlled native phase) to high temperature denatured phase where thermal agitations control the fate. Higher values of the contact profile index generally implies higher rigidity and lower mobility as observed (Pandey & Farmer, 2017) in corresponding mobility profile of AQP1.

How do the stochastic movements, aggregation and dispersion of residues lead to their overall collective conformation? Careful analysis of the average radius of gyration (R_g), a global physical quantity, is in order to assess the conformational spread of the protein. Fig. 3 shows the variation of the radius of gyration of the protein with the temperature in native (low T) and denature (high T) phases. In denatured phase, the radius of gyration increases on increasing the temperature before reaching a somewhat saturation ($T=0.024$–0.028) which is not unexpected as the thermal energy takes over the residue–residue interaction. In contrast, the radius of gyration decreases in native phase ($T=0.012$–0.024), a thermal-induced globularization. Increase in thermal agitation generally reduces the 'order' (a common order-disorder phenomena), instead it enhances the 'order' by organizing

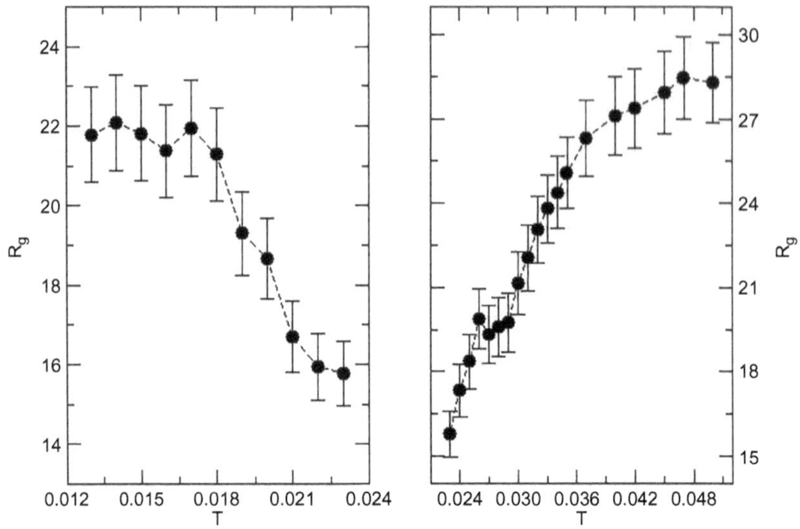

Fig. 3 Variation of the radius of gyration (R_g) of AQP1 with the temperature. R_g is averaged over the number of independent samples (Pandey & Farmer, 2017).

into compact globular morphology. Such a negative thermal response has been recently observed in other membrane proteins (Kitjaruwankul et al., 2018) although one has to reserve the claim on the universal characterization of such thermal-induced transition for all membrane proteins without analyzing their response properties.

It should be pointed out that the knowledge-based residue–residue interactions used here are derived from a huge ensemble of a range of proteins structures perhaps in different solvents reported in PDB. Therefore, the analysis presented above should be considered as the structure of a free AQP1 in an effective solvent medium. A protein in membrane encounters a range of additional matrix components which are difficult to incorporate in a computer simulation to examine its large-scale structural dynamics. Resorting to a simplified representation of the crowded matrix environment is a viable approach as in experiments with reconstituted environments. The constitutive matrix elements (solute) of such a cellular matrix are represented here by interactive particles with size comparable to that of a residue and perform their stochastic motion. Note that in addition to excluded volume (hard-core) interaction, solute molecules interact with each residue with specific interactions (attractive or repulsive with appropriate strength, see Section 2). The dynamics of protein is controlled by the temperature, the residue–residue interactions (as above) and the residue–solute interactions.

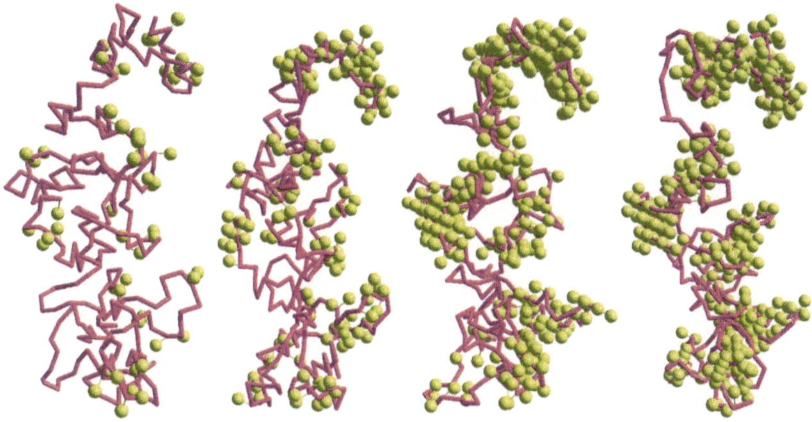

Fig. 4 Snapshot of AQP1 (backbone) and bonded solute particles (sphere) at time step $t=10^3$, 3×10^3, 2×10^4, 10^6.

The global dynamics of AQP1 is governed by the stochastic movement of each residue, some of which may be more constrained if surrounded by solute molecules than others.

A set of time series snapshots of AQP1 along with interacting solute molecules are presented in Fig. 4. Unless captured by the targeted residue, solute particles are relatively more mobile than the covalently bonded residue. Solute particles seem to reach their targets with increasing numbers in time and pin down the protein conformation rather quickly. How does the targeted binding (and consequently the local and global response properties of the protein) depend on temperature, concentration, and interaction strength is a huge undertaking which is out of scope at present. Based on our limited investigation however, we have made some interesting observations some of which are reported here particularly in the native phase.

Average number Nw of solute around each residue as a result of targeted binding could be a measure of matrix-contact map of the protein. The matrix-contact profile of AQP1 in its native phase ($T=0.020$, with overwhelming residue–residue interaction in comparison the thermal noise) is presented in Fig. 5 at a range of residue–solute interaction strength f_w. A well-defined pattern of the contact map with increasing interaction shows targeted binding along the backbone of the protein. For example, the segment with residues 158D-163D appears to be the most targeted region of the protein. Apart from the binding targets such as end segments (around 1M-19L and 225L-269K) pockets of local segments (e.g., 35F-52V) appear throughout the protein backbone particularly with high residue–solute

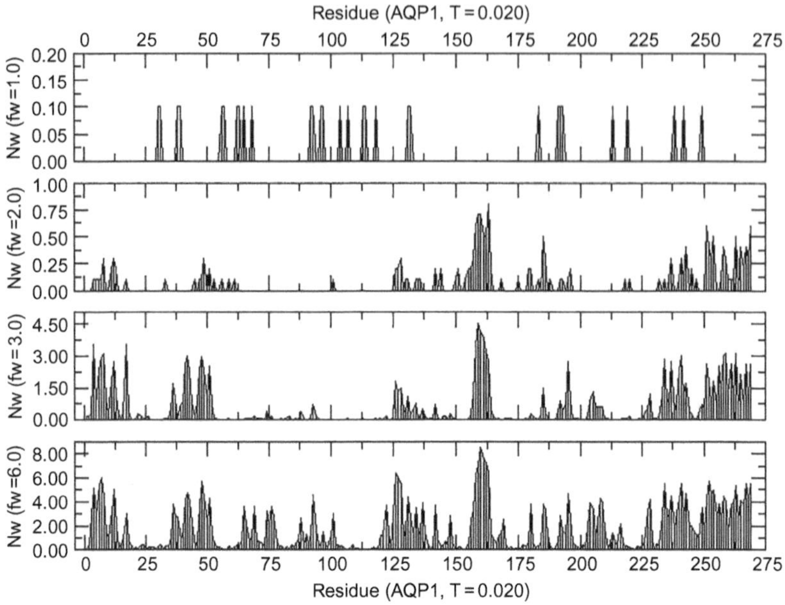

Fig. 5 Average number of solute molecules around each residue with the solute–residue interaction strength $f_w = 1.0, 2.0, 3.0, 6.0$.

aquatic interaction. Fig. 6 shows the variation of the radius of gyration (a global physical quantity) of the protein with the residue–solute interaction. Despite large error bars, it is difficult to identify a well-defined trend except reconciling to a nearly constant R_g with a considerable fluctuation of course. Targeted binding of the dynamic matrix constituents (Fig. 5) pins down the morphology of the protein (Fig. 6) in an almost frozen state in an effective matrix environment considered here. A huge variability in modeling the reconstituted matrix environment remains to be explored.

In order to quantify the overall structural spread of the protein, it may be useful to analyze the structure factor (S(q)),

$$S(q) = \left\langle \frac{1}{N} \left| \sum_{j=1}^{N} e^{-i\vec{q} \times r_j} \right|^2 \right\rangle_{||\vec{q}||} \tag{2}$$

where r_j is the position of each residue of the protein chain and $|q| = 2\pi/\lambda$ is the wave vector of wavelength λ. Using the scaling of the structure factor with the wavelength λ,

$$S(\lambda) \propto \lambda^{1/\gamma} \tag{3}$$

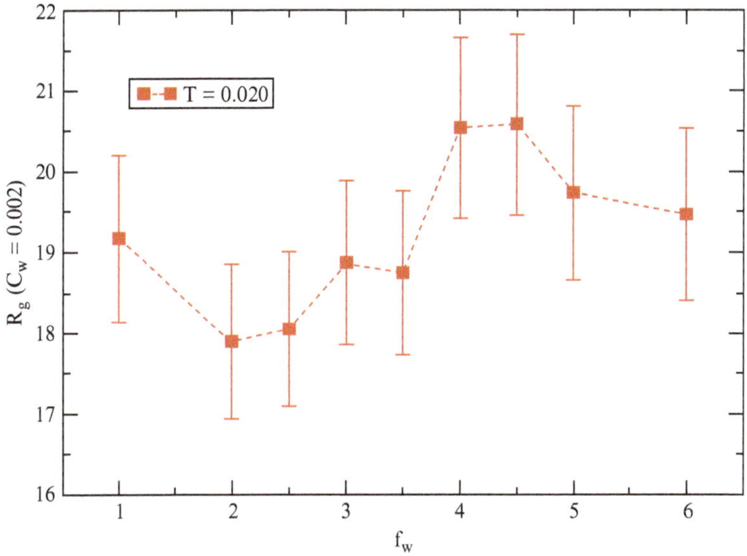

Fig. 6 Variation of the radius of gyration with the solute–residue interaction strength f_w at the temperature $T=0.020$ with the solute concentration $C_w=0.020$.

one may study the spread of residues over the length scale λ by evaluating the exponent γ which describes the mass (residue) distribution. Since the radius of gyration is a measure of the spatial spread of the protein, it can be comparable to the wavelength. This provides the scaling of the radius of gyration with the number N of residue, $R_g \propto N^{\gamma}$. In other words, the effective dimension D of the distribution of residues over its radius of gyration (R_g) is given by $D=1/\gamma$.

Variation of the structure factor with the wave vector is presented in Fig. 7 for a range of residue–solute interaction strength f_w (0.0–6.0) in its native phase. The range of the wave vector $q\approx 0.2$–0.6 corresponds to $R_g \approx 18$–20 suitable for $f_w=0.0$, 1.0, and 6.0 (inset in Fig. 7). Estimates of the effective dimension of the protein shows that AQP1 adopts to a compact globular structure ($D\approx 3$) in absence of reconstituted matrix ($f_w=0.0$) and conforms to a random-coil structure ($D\approx 2$) in a matrix with a relatively strong interaction ($f_w=6.0$). Thus the matrix environment in which the protein is embedded can orchestrate the structure of the protein.

4. Conclusions and outlook

Structural dynamics of AQP1 is analyzed by a coarse-grained model with a phenomenological interaction potential based on a knowledge-based

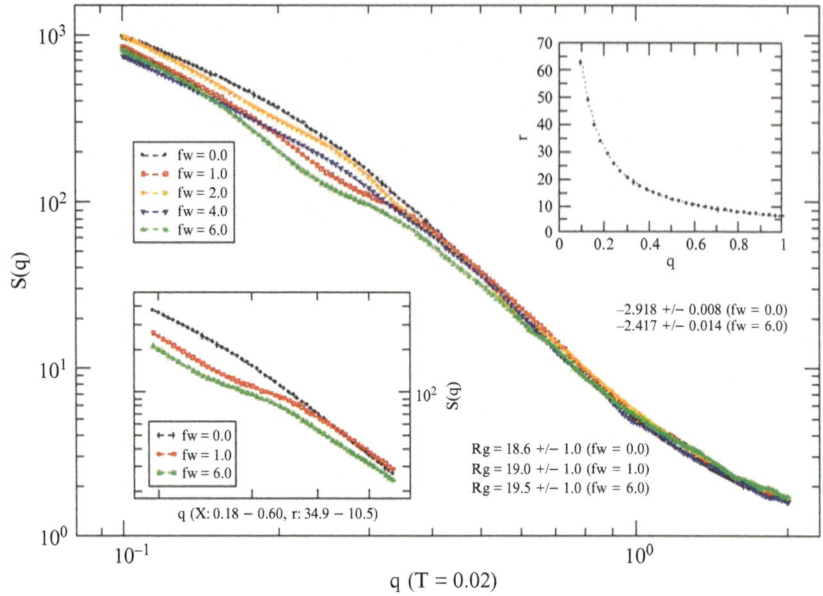

Fig. 7 Variation of the structure factor *S(q)* with the wave vector *q* at temperature $T=0.020$ with a range of solute–residue interaction $f_w=0.0$–6.0. Insets: top right—variation of the linear length (r) or wavelength with the wave vector (*q*) and bottom left—variation *S(q)* with *q* in a range of *q* covering the radius of gyration at the solute–residue interaction strength $f_w=0.0$, 1.0, and 6.0 to show the changes.

residue–residue interaction (KBRR). A large ensemble protein structures in PDB is used as an input to derive such knowledge-based residue-residue contact potentials. The diversity of proteins structures in PDB and the underlying matrix (diverse range of proteins in a variety of solvents) implies that the effect of solvent is already taken into account in the estimates of KBRR interactions. Therefore, the thermal response of a free AQP1 should be treated as a protein chain in an effective solvent environment in our coarse-grain model. The competition between the residue–residue interaction and the thermal agitation controls the structure and dynamics of the protein. A number of local and global physical quantities such as residue contact and mobility profiles, radius of gyration, structure factor, etc. are analyzed. We find that the thermal response of the protein structure in its native phase, i.e., in low temperature regime with dominating residue interaction (over the thermal noise) exhibits unexpected characteristics where the radius of gyration of the protein decreases on raising the temperature. Such an unusual negative thermal response has been also recently reported for other membrane proteins (Kitjaruwankul et al., 2018).

Continuous increasing the temperature eventually leads to thermal expansion of the protein where its radius of gyration increases on increasing the temperature before saturating to its random-coil morphology in denatured phase at high temperature.

The non-monotonic thermal response is interesting as the conformational transition occurs from its native structure to denatured phase. From the contact and mobility profiles, we are able to identify active (dynamic) segments and quasi-static globular segments that anchor the stability of the proteins. For example, three regions of persistent globularization are identified, the end segments 1M-25V and 250V-269K with a narrow region around 155A-163D in the middle region of the backbone. The effective dimension (D) of the protein for the distribution of the residues is estimated from the scaling of the structure factor with the wave vector characterizing the random coil structure with $D \sim 2$ in denatured phase and a globular conformation $D \sim 3$ in native phase. Variation in structure factor reveals a systematic redistribution of self-organizing residues with globular and fibrous morphologies which depends on the length scale and the temperature.

The effect of the crowded membrane environment on the protein structure is also analyzed in a preliminary study by incorporating constituting elements (solute particles) with unique residue–solute interaction based on the hydropathy index. In such representation of effective membrane environment, each residue–solute interaction is not only unique but also the strength (f_w) of these unique interaction can be varied while maintaining their relative characteristics (as described above). In general, the solute particles move much faster than the residues except those reaching their targeted binding sites (with specific residues). As a result the structure of the protein is pinned in a conformation different from that of an AQP1 in absence of such crowded environment. The matrix-directed structure of the protein becomes more stable with the increasing number of solute particles at their targets forming an effective cage that enhances the anchoring of protein particularly at low temperatures. The matrix-contact profiles reveal that a narrow segment of the protein (around 158D–163D) has the maximum number of bound solute particles followed by terminal regions, a small segments (1M-25V) at one end while a larger segment (225L-269K) at the other end; pockets of solute-bound segments do exist throughout along the protein backbone. Scaling of the structure factor provides a valuable insight into quantifying the distribution of residues from the estimates of its effective dimension, e.g., random-coil conformation in denatured phase to globular structure in native phase and a range morphology in between.

In short, the structure of AQP1 can be controlled by temperature and the constitutive components of the matrix in which the protein is embedded in. Therefore, a desirable conformation of AQP1 can be achieved by manipulating appropriate parameters.

Even with the simplified representation of the underlying host matrix of the protein, many questions need to be addressed systematically for a better understanding of the conformational evolution of the protein. For example, how does the dynamics and concentration of solute constituents (slow-to-fast, i.e., its make-up) affect the structural relaxation? What happens if there are many AQP1 protein in the same host matrix? There are many interesting issues to be addressed with an ultimate goal to understand the structure, dynamics, and self-organizing morphology in a realistic membrane with more realistic constituents.

Acknowledgments

The authors gratefully acknowledge financial support from the Chulalongkorn University Dusadi Phipat scholarship and Ratchadaphiseksomphot Endowment Fund. RBP appreciates support for a visiting professorship by the Chulalongkorn University and warm hospitality by the Department of Chemistry.

References

Aquaporin. (2018). https://en.wikipedia.org/wiki/Aquaporin. [(Accessed 20 December 2018)].

Ash, W. L., Zlomislic, M. R., Oloo, E. O., & Tieleman, D. P. (2004). Computer simulations of membrane proteins. *Biochimica et Biophysica Acta, 1666*, 158–189.

Betancourt, M. R., & Thirumalai, D. (1999). Pair potentials for protein folding: Choice of reference states and sensitivity of predicted native states to variations in the interaction schemes. *Protein Science, 2*, 361–369.

Binder, K. (1995). *Monte Carlo and molecular dynamics simulations in polymer science*. Oxford University Press.

Boassa, D., Stamer, W. D., & Yool, A. J. (2006). Ion channel function of aquaporin-1 natively expressed in choroid plexus. *The Journal of Neuroscience, 26*, 7811–7819.

Bond, P. J., Holyoake, J., Ivetac, A., Khalid, S., & Sansom, M. S. P. (2007). Coarse-grained molecular dynamics simulations of membrane proteins and peptides. *Journal of Structural Biology, 157*, 593–605.

de Jong, D. H., Singh, G., Drew Bennett, W. F., Arnarez, C., Wassenar, T. A., et al. (2013). Improved parameters for the martini coarse-grained protein force field. *Journal of Chemical Theory and Computation, 9*, 687–697.

Ercolessi, F., & Adams, J. (1994). Interatomic potentials from first-principle calculations: The force-matching method. *Europhysics Letters, 26*, 583–588.

Fritsche, M., Pandey, R. B., Farmer, B. L., & Heermann, D. (2012). Conformational temperature-dependent behavior of a histone h2ax: A coarse-grained Monte Carlo approach via knowledge-based interaction potentials. *PLoS One, 7*, e32075-1–e32075-8.

Godzik, A., Kolinski, A., & Skolnick, J. (1996). Knowledge-based potentials for protein folding: What can we learn from protein structures? *Proteins, 4*, 363–366.

Gumbart, J., Wang, Y., Aksimentiev, A., Tajkhorshid, E., & Schulten, K. (2005). Molecular dynamics simulations of proteins in lipid bilayers. *Current Opinion in Structural Biology*, *15*, 423–431.

Haliloglu, T., & Bahar, I. (1998). Coarse-grained simulations of conformational dynamics of proteins: Application to apomyoglobin. *Proteins*, *31*, 27–281.

Hashido, M., Ikeguchi, M., & Kidera, A. (2005). Comparative simulations of aquaporin family: AQP1, AQPZ, AQP0 and GlpF. *FEBS Letters*, *579*, 5549–5552.

Hashido, M., Kidera, A., & Ikeguchi, M. (2007). Water transport in aquaporins: Osmotic permeability matrix analysis of molecular dynamics simulations. *Biophysical Journal*, *93*, 373–385.

Huang, S.-Y., & Xiaoqin, Z. (2011). Statistical mechanics-based method to extract atomic distance-dependent potentials from protein structures. *Proteins*, *79*, 2648–2661.

Jensen, M. O., & Mouritsen, O. G. (2006). Single-channel water permeabilities of Escherichia coli aquaporins AqpZ and GlpF. *Biophysical Journal*, *90*, 2270–2284.

Kitjaruwankul, S., et al. (2018). Thermal-induced folding and unfolding of a transmembrane protein (CorA). *Physica A: Statistical Mechanics and its Applications*, *506*, 987–992.

Knepper, M. A., & Nielsen, S. (2004). Peter Agre, 2003 Nobel Prize winner in chemistry, special commentary. *Journal of the American Society of Nephrology*, *15*, 1093–1095.

Kong, Y., & Ma, J. (2001). Dynamic mechanisms of the membrane water channel aquaporin-1 (AQP1). *Proceedings of the National Academy of Sciences of the United States of America*, *98*, 14345–14349.

Kumar, M., Grzelakowski, M., Zilles, J., Clark, M., & Meier, W. (2007). Highly permeable polymeric membranes based on the incorporation of the functional water channel protein aquaporin z. *Proceedings of the National Academy of Sciences of the United States of America*, *104*, 20719–20724.

Lindahl, E., & Sansom, M. S. P. (2008). Membrane proteins: Molecular dynamics simulations. *Current Opinion in Structural Biology*, *18*, 425–431.

Liwo, A., Czaplewski, C., Oldziej, S., & Scheraga, H. A. (2008). Computational techniques for efficient conformational sampling of protein. *Current Opinion in Structural Biology*, *18*, 134–139.

Lyubartsev, A. P., & Laaksonen, A. (1995). Calculation of effective interaction potential from radial distribution functions: A reverse Monte Carlo approach. *Physical Review E*, *52*, 3730–3737.

Mirau, P., Farmer, B. L., & Pandey, R. B. (2015). Structural variation of alpha-synuclein with temperature by a coarse-grained approach with knowledge-based interactions. *AIP Advances*, *5*, 092504-1–092504-10.

Miyazawa, S., & Jernigan, R. L. (1985). Estimation of effective inter residue contact energies from protein crystal structures: Quasi-chemical approximation. *Macromolecules*, *18*, 534–552.

Miyazawa, S., & Jernigan, R. L. (1996). Residue-residue potentials with a favorable contact pair term for simulation and treading. *Journal of Molecular Biology*, *256*, 623–644.

Pandey, R. B., & Farmer, B. L. (2010). Global structure of a human immunodeficiency virus-1 protease (1DIFA dimer) in an effective solvent medium by a Monte Carlo simulation. *The Journal of Chemical Physics*, *132*, 125101–125106.

Pandey, R. B., & Farmer, B. L. (2012). Random coil to globular thermal response of a protein (H3.1) with three knowledge-based coarse-grained potentials. *PLoS One*, *7*, e49352-1–e32075-9.

Pandey, R. B., & Farmer, B. L. (2013). Conformational response to solvent interaction and temperature of a protein (histone h3.1) by a multi-grained Monte Carlo simulation. *PLoS One*, *8*, e76069-1–e76069-9.

Pandey, R. B., & Farmer, B. L. (2017). Structure and dynamics of a free aquaporin (AQP1) by a coarse-grained Monte Carlo simulation. *Structural Chemistry*, *28*, 625–633.

Pandey, R. B., Farmer, B. L., & Gerstman, B. S. (2015). Self-assembly dynamics for the transition of a globular aggregate to a fibril network of lysozyme proteins via a coarse-grained Monte Carlo simulation. *AIP Advances*, *5*, 092502-1–092502-12.

Pandey, R. B., Jcobs, D. J., & Farmer, B. L. (2017). Preferential binding effects on protein structure and dynamics revealed by coarse-grained Monte Carlo simulation. *The Journal of Chemical Physics*, *146*, 195101.

Papadopoulos, M. C., & Saadoun, S. (2015). Key roles of aquaporins in tumor biology. *Biochimica et Biophysica Acta*, *1848*, 2576–2583. https://doi.org/10.1016/j.bbamem.2014.09.001.

Pelagalli, A., Squillacioti, C., Mirabella, N., & Meli, R. (2016). Aquaporins in health and disease: An overview focusing on the gut of different species. *International Journal of Molecular Sciences*, *17*, 1213.

Reith, D., Putz, M., & Muller-Plathe, F. (2003). Deriving effective mesoscale potentials from atomistic simulations. *Journal of Computational Chemistry*, *24*, 1624.

Sorensen, J., Xavier, P., Skeby, K. K., Marrink, S. J., & Schiott, B. (2011). Protofibrillar assembly towards the formation of amyloid fibrils. *Journal of Physical Chemistry Letters*, *2*, 2385–2390.

Tanaka, S., & Scheraga, H. A. (1976). Medium and long range interaction parameters between amino acids for predicting three dimensional structures of proteins. *Macromolecules*, *9*, 945–950.

To, J., & Torres, J. (2015). Can stabilization and inhibition of aquaporins contribute to future development of biomimetic membranes? *Membranes*, *5*, 352–368.

Tomita, Y., Dorward, H., Yool, A. J., Smith, E., Townsend, A. R., Price, T. J., et al. (2017). Role of aquaporin 1 signalling in cancer development and progression. *International Journal of Molecular Sciences*, *18*, 299.

van Giessen, A. E., & Straub, J. E. (2005). Mote Carlo simulations of polyalanine using a reduced model and statistics-based interaction potential. *The Journal of Chemical Physics*, *122*, 0249041–0249049.

Wang, J., Feng, L., Zhu, Z., Zheng, M., Wang, D., Chen, Z., et al. (2015). Aquaporins as diagnostic and therapeutic targets in cancer: How far we are? *Journal of Translational Medicine*, *13*, 96.

Wang, Y., & Tajkhorshid, E. (2007). Molecular mechanisms of conduction and selectivity in aquaporin water channels. *The Journal of Nutrition*, *137*, 1509S–1515S.

Werber, J. R., & Elimelech, M. (2018). Permselectivity limits of biomimetic desalination membranes. *Science Advances*, *4*, eaar8266. 29 June 2018.

Werber, J. R., Osuji, C. O., & Elimelech, M. (2016). Materials for next-generation desalination and water purification membranes. *Nature Reviews Materials*, *1*, 16018 [5 April 2016].

Yu, J., Yool, A. J., Schulten, K., & Tajkhorshid, E. (2006). Mechanism of gating and ion conductivity of a possible tetrameric pore in aquaporin-1. *Structure*, *14*, 1411–1423.

Zhou, J., Chen, S., & Jiang, S. (2003). Orientation of adsorbed antibodies on charged surfaces by computer simulation based on a united-residue model. *Langmuir*, *19*, 3472–3478.

Zhou, J., Thorpe, I. F., Izvekov, S., & Voth, G. A. (2007). Coarse-grained peptide modeling using a systematic multiscale approach. *Biophysical Journal*, *92*, 4289–4303.

CHAPTER THREE

Selectivity and transport in aquaporins from molecular simulation studies

Siladitya Padhi[a,*], U. Deva Priyakumar[b,*]
[a]TCS Innovation Labs—Hyderabad (Life Sciences Division), Tata Consultancy Services Limited, Hyderabad, India
[b]Center for Computational Natural Sciences and Bioinformatics, International Institute of Information Technology, Hyderabad, India
*Corresponding authors: e-mail address: siladitya.padhi@tcs.com; deva@iiit.ac.in

Contents

Abstract

The transport of water through aquaporins is a dynamic process that involves rapid movement of a chain of water molecules through the pore of the aquaporin. Structures of aquaporins solved using X-ray crystallography have provided some insights into how water is transported through these channels, and how certain structural features of the pore might help exclude other solutes from passing through the pore. However, such techniques provide only a static picture, and a dynamic picture of the transport and selectivity mechanism at work in aquaporins is possible with molecular dynamics (MD) simulations. In MD simulations, the forces between the different atoms in a system are computed, and the atoms are then allowed to move under the influence of these forces. This allows the sampling of different conformations of the molecule being studied, including conformations that are crucial in driving biological phenomena like water transport. Simulation studies have provided insights into a number of aspects of aquaporins, including the role of the asparagine-proline-alanine (NPA) motif and the aromatic/arginine (ar/R) constriction, water transport mechanism, mechanisms defining the selectivity of the channel, interaction with lipids, response to external electric field,

Vitamins and Hormones, Volume 112
ISSN 0083-6729
https://doi.org/10.1016/bs.vh.2019.10.001

and binding of putative drug molecules. This chapter provides a brief review of the current status of computational modeling of aquaporins using MD simulations. Initially, a brief account of force fields and MD simulations is presented followed by an account of how MD simulations have contributed to further our understanding of different aspects of aquaporins.

Computational methods have become one of the indispensable tools in the investigation of chemical and biological systems. Among a wide variety of computational methods available, molecular dynamics (MD) simulations have proven to be one of the most powerful tools to examine the structural, dynamic and thermodynamic properties of biomolecular systems. MD simulations have been successfully used to explain experimental observations, and to make predictions based on which new experiments are designed. The simulation captures the time evolution of the position and momenta of each of the nuclei in the biomolecular systems, which can further be used to calculate desirable properties related to the system. A qualitative and brief account of MD simulations is provided below followed by a summary of insights obtained on aquaporins using simulations.

MD simulations are performed by integrating Newton's equation of motion for each atom/nuclei in the system (Adcock & McCammon, 2006; Padhi & Priyakumar, 2017a). Interatomic interactions are defined by a potential energy function (in other words, molecular mechanics) along with a set of empirical parameters usually referred to as force fields. Molecular mechanics treats atoms as spheres centered at the nuclear positions connected by harmonic springs. CHARMM and AMBER are two most popular force fields of choice for biomolecular systems (Hart et al., 2012; Huang et al., 2016; Denning, Priyakumar, Nilsson, & MacKerell, 2011; Klauda et al., 2010; Maier et al., 2015). These force fields express the total energy of the system as the sum of bonded and nonbonded energy contributions. The bonded energy includes energy terms corresponding to bond stretching, angle bending and dihedral rotations, whereas the nonbonded energy is modeled using a classical Coulomb potential and a Lennard-Jones potential as given in the equation below.

$$U(r) = \sum_{bonds} k_b(b - b_0)^2 + \sum_{angles} k_\theta(\theta - \Theta_0)^2 + \sum_{dihedrals} k_\phi[1 + \cos(n\phi + \delta)]$$
$$+ \sum_{i}^{N} \sum_{j=i+1}^{N} \left[\frac{q_i q_j}{\varepsilon r_{ij}} + \epsilon_{ij} \left(\left(\frac{R_{\min i,j}}{r_{ij}} \right)^{12} - 2 \left(\frac{R_{\min i,j}}{r_{ij}} \right)^{6} \right) \right]$$

U(r) is the potential energy expressed as a function of the positions of the atoms in the molecule. The first term on the right hand side accounts for changes in energy arising from stretching of bonds. A bond between two atoms is treated as a system of two rigid spheres connected by a spring. k_b denotes the spring constant of the spring connecting the two atoms, b_0 is the equilibrium bond length, and b is the bond length at any instant. Like bond stretching, angle bending is also treated by a harmonic term. This is given by the second term in the equation, in which θ_0 is the equilibrium bond angle, θ is the instantaneous bond angle, and k_θ is the force constant. Rotations about every bond in the system are quantified by a geometric measure called the dihedral angle whose contribution to the energy is typically modeled by the third term in the equation, where ϕ is the dihedral angle, k_ϕ accounts for the height of the potential energy curve, n determines the number of maxima/minima that occur as ϕ varies from $-\pi$ to π, and δ gives the phase shift for ϕ. The fourth term in the equation is the Coulomb term for quantifying the electrostatic interactions between any two atoms (*i* and *j*) in the system except when the pair forms a bond or when they are part of the same angle. In the above equation, q_i and q_j denote the partial charge on particles *i* and *j*, respectively; ε is the permittivity in vacuum, and r_{ij} is the distance between the particles *i* and *j*. The fifth term is the LJ potential that quantifies the van der Waals energy.

The force field is used to calculate the force acting on each particle in the system due to the other particles in the system (Adcock & McCammon, 2006). The first derivative of the potential energy (force field) with respect to the position is the force:

$$F_i^\alpha = -\frac{dU}{dr_i^\alpha}$$

Here F_i^α is the force acting along the ith component for atom α, U is the potential energy, and r_i^α is the position of atom α along the ith component (a given atom α can have three components x, y, and z). The acceleration of each particle then is calculated using Newton's equation of motion:

$$a_i^\alpha = \frac{F_i^\alpha}{m^\alpha}$$

where m^α is the mass of atom α. Typically, the simulation starts with a set of nuclear positions (for example, atomic coordinates from a PDB file) of the biomolecule and solvent. In case of transmembrane simulations, a lipid model is chosen to build the lipid bilayer and aqueous layer is added to both sides of the membrane. An initial guess of the velocities for each particle in the system is obtained, using which the particles are moved. Using an integration algorithm such as the Verlet algorithm, the subsequent position is calculated from the current position, past position and the acceleration. This process is repeated until the desired simulation time length is achieved or until the computer infrastructure limits further progress! The MD trajectory (positions of each of the atoms in the system as a function of time) is collected and analyzed to learn more about the system at hand. When the simulation length scale is significantly longer than the actual event, modeling of such systems become more complicated. Advanced methods such as replica exchange MD, umbrella sampling and metadynamics can be used to study such rare events. Umbrella sampling is the most widely used method in this context. An event like glycerol permeation across an aquaporin in the membrane is an example of such a system that can be modeled using umbrella sampling. The method enables sampling of glycerol configurations along the pore axis by applying a bias on the glycerol molecule in the form of a harmonic potential (Kästner, 2011). A number of windows are defined along the pore axis, and independent simulations are performed for each window. For a window i, the biasing potential has the form

$$w_i(z) = \frac{1}{2}k(z - z_{i,0})^2$$

where k is the harmonic constant, $z_{i,0}$ is the center of the window i, and z is the position of the glycerol along the pore axis at any given instant. The potential energy function then has the form

$$E^b(r^N) = E^u(r^N) + w_i(z)$$

where r^N denotes the positions of all the atoms in the system, and $E^u(r^N)$ is the unbiased potential energy function.

The following sections systematically summarize how MD simulations have been helpful in gaining insights into several aspects of aquaporins.

1. Conformational dynamics and water transport

The solving of the crystal structure of the AQP1 tetramer (Murata et al., 2000) provided an atomistic level view of the structure of the channel. Simulation studies on this channel show a single file of hydrogen bonded water molecules in the pores of each of the four monomeric units (Zhu, Tajkhorshid, & Schulten, 2001). Water molecules also form hydrogen bonds with the asparagine residues of the two NPA motifs lining the pore. No water permeation, however, was observed in the central pore of the tetrameric structure. MD simulations by Grubmüller and coworkers on a more refined structure of AQP1 (Fu et al., 2000) revealed that another major water-protein interaction site is located at the ar/R region constituted by F56, H180, and R195 residues (de Groot & Grubmüller, 2001). Strong interactions in this region between water and the channel lead to weakening of water-water hydrogen bonds and a disruption in the water chain. In general, strong orientational control of the water dipoles is seen inside the pore, and the dipoles undergo an inversion about the NPA region. It is proposed that this dipole inversion is important in defining the selectivity of the channel toward water, since only small molecules with a large dipole moment would be able to undergo such inversion. The free energy for the transport of a water molecule along the pore axis z has been calculated using the relation

$$G(z) = -k_B T \ln n(z)$$

where k_B is the Boltzmann constant, T is the absolute temperature and $n(z)$ is the residence frequency along the pore axis z. An energy barrier of slightly more than 1 kcal/mol is observed, indicating a high water transport rate.

The above study also provided an account of water permeation through the aquaglyceroporin GlpF. There is a disruption of the water chain around the ar/R region in a manner similar to that seen in AQP1. The major difference between the two channels is seen in the vicinity of the NPA motif; while AQP1 has the bulky F24 side chain facing the pore, the equivalent residue in GlpF is the smaller L21 residue. The smaller leucine side chain in GlpF permits the passage of glycerol, which is bigger in size than a water molecule. A more comprehensive view of water transport through GlpF was provided in a study by Schulten and coworkers (Tajkhorshid et al., 2002). Fig. 1A gives a view of the arrangement of water molecules inside the pore of GlpF. The water molecules are hydrogen bonded with one another as well as with the residues of the two NPA motifs and the ar/R constriction. Two α-helices, namely, M3 and M7, have their termini facing the pore, and these charged termini also interact with water molecules. The orientation of the water molecules is opposite in the two halves of the channel, with the molecules undergoing a rotation around the NPA region. At the two ends of the pore, the waters orient themselves such that they engage in stable interactions with the helical dipoles (Fig. 1B). At the middle of the pore, the water

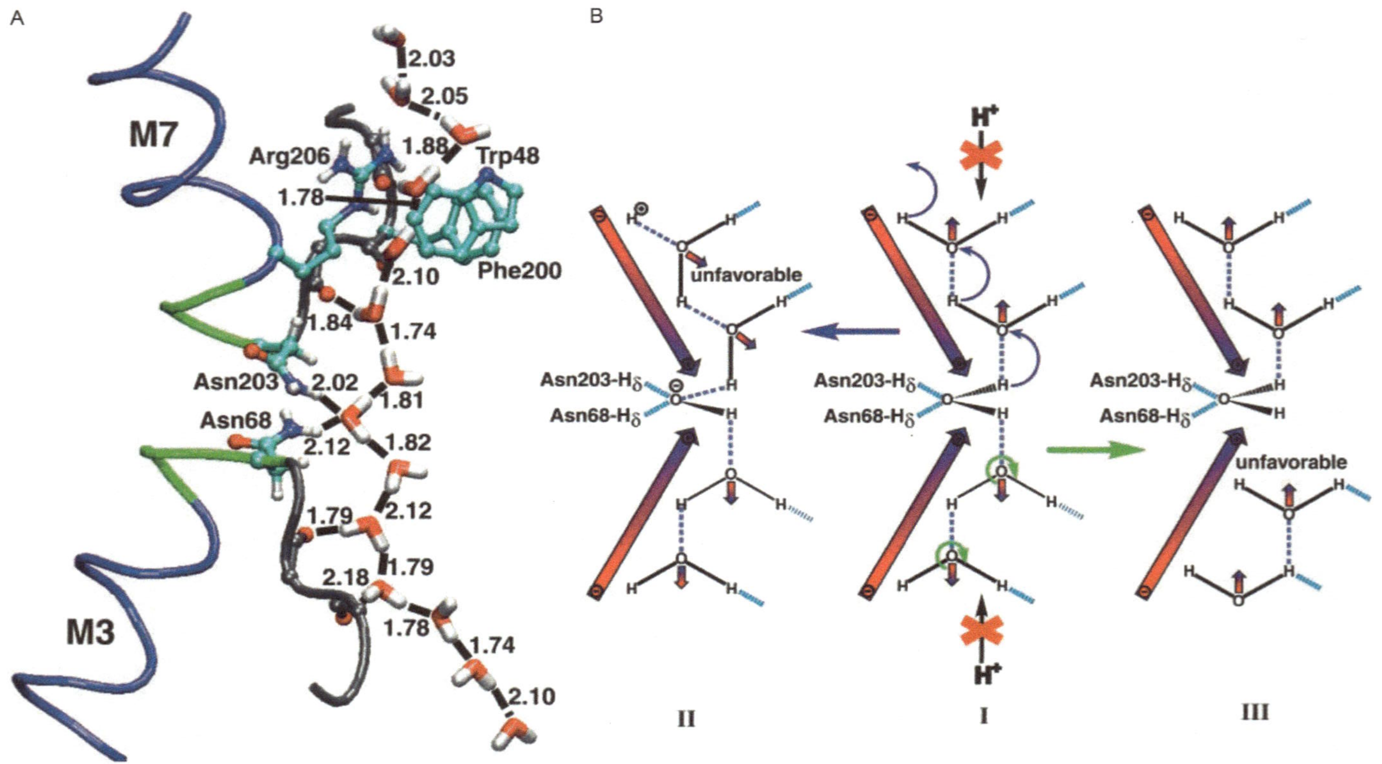

Fig. 1 (A) Arrangement of the water chain in the pore of the aquaglyceroporin GlpF. Hydrogen bond distances for water-water and water-protein interactions are shown in Angstroms. Important pore-lining residues and the helices M3 and M7 are also shown. (B) Illustration of the proton exclusion mechanism. **I** shows that protonation at either of the termini is unfavorable because of the water orientation. **II** illustrates the unfavorable orientation of water molecules with respect to the helical dipole that would arise if water protonation were to occur at one of the termini. **III** shows that if the water oxygen at the terminus were to be available for protonation, it would lead to breakage of hydrogen bonding in the water chain. *Reproduced with permission from Tajkhorshid, E., Nollert, P., Jensen, M. Ø., Miercke, L. J., O'connell, J., Stroud, R. M., et al. (2002). Control of the selectivity of the aquaporin water channel family by global orientational tuning.* Science, 296*(5567), 525–530.*

oxygen forms hydrogen bonds with the asparagine side chains on the two NPA motifs. The arrangement of the water chain inside the pore ensures that protons are not allowed to permeate. Protonation of water molecules at the two ends of the pore is not favorable because of the orientation. Furthermore, proton movement through the water chain would require reorientation of water molecules and breakage of hydrogen bonds, which is an unfavorable process.

Side chain dynamics of pore-lining residues in aquaporins is important in the movement of water molecules through the pore, as shown in simulations on a monomeric form of bovine AQP1 (Smolin, Li, Beck, & Daggett, 2008). Upon simulating the system with the protein fixed, water molecules occupied the same positions as those reported in the crystal structure, but no complete permeation events (from one end of the pore to the other) were observed. As the protein is allowed to move, the side chains of the asparagines in the two NPA motifs hydrogen bond with water molecules and assist their movement through the pore by aligning them. The pore-facing H76 and V155 residues act as a valve and modulate the flow of water through the channel.

Apart from providing insights into the water transport mechanism in aquaporins, MD simulation studies have also unraveled the gating mechanism. A study on the human aquaporin AQP5 showed that water movement is gated at two points in the channel: the cytoplasmic end and the ar/R constriction (Janosi & Ceccarelli, 2013). While the cytoplasmic end switches between open and closed states, the ar/R region switches between wide and narrow states. The closed state of the cytoplasmic end completely blocks the passage of water. When the cytoplasmic end is in an open configuration, the ar/R region can adopt either the wide state or the narrow state. The two states of the ar/R region are defined by the orientation of the H173 side chain, with the ring switching between conformations that are either parallel to (wide state) or perpendicular to (narrow state) the pore axis. Extracellular loops have also been proposed to play a role in regulating water transport, as shown in a study on AQP1 that combined MD simulations with nuclear magnetic resonance (NMR)-derived restraints (Wang et al., 2016). Simulations of a tetrameric form of the human aquaporin AQP4 show that, while the pores of each of the individual monomers possess a single file water chain, the central pore in the tetramer is impermeable to water, owing to the presence of hydrophobic residues (Cui & Bastien, 2011).

An analysis of the rotational and translational components of water molecules permeating aquaporins has provided a thermodynamic basis for the spontaneous entry of water molecules into the channel (Barati Farimani, Aluru, & Tajkhorshid, 2014). Although water entry into the pore is accompanied by a reduction in translational motion, there is a significant increase in the rotational motion of water molecules inside the pore. There is thus an entropic gain upon entry of water molecules into the pore, making the process spontaneous.

An investigation of the free energy landscape for water orientation at different points along the pore axis in the human aquaporin AQP2 provided a comprehensive view of the transport mechanism in the channel (Padhi & Priyakumar, 2017b). The free energy for an angle θ between the water dipole and the pore axis, at a position i along the pore axis, is given by

$$G(n_i, \theta) = G(n_i) - k_B T \ln \rho(\theta; n_i) + c_i$$

where n_i is the normalized number density for water molecules at position i, ρ (θ; n_i) is the probability of occurrence of angle θ for a given value of i, and C_i is a constant for a given value of i. Fig. 2 shows the free energy profile as a

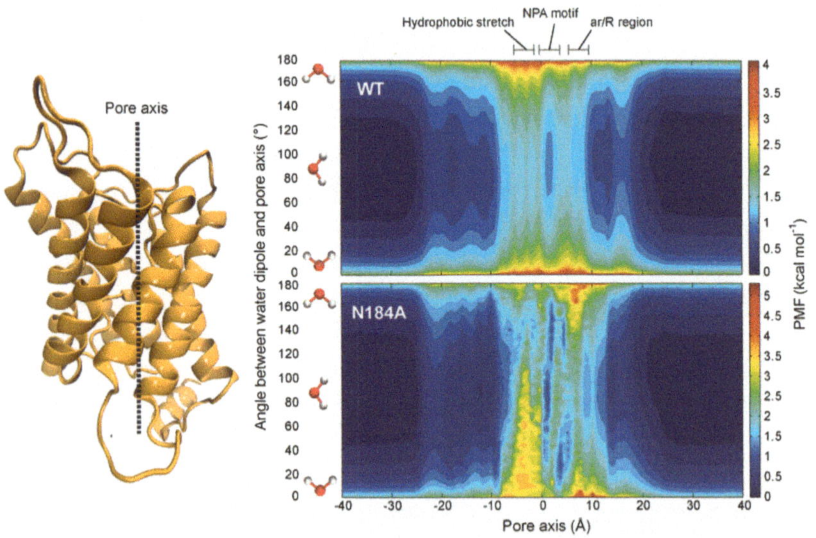

Fig. 2 Free energy profiles for water transport through aquaporin AQP2 as a function of water position along pore axis and angle between water dipole and pore axis. Orientations of water corresponding to 0°, 90° and 180° are shown alongside y-axis labels. The free energy profile on the top corresponds to the wild-type and that at the bottom corresponds to the N184A mutant. The figure on the left hand side depicts the pore axis. *Reproduced with permission from Padhi, S., & Priyakumar, U. D. (2017b). Microsecond simulation of human aquaporin 2 reveals structural determinants of water permeability and selectivity.* Biochimica et Biophysica Acta (BBA)-Biomembranes, 1859*(1), 10–16.*

function of water position along pore axis and angle between water dipole and pore axis. When the water molecules are in the peripheral regions of the channel, they are free to attain a variety of orientations, as shown by the flat free energy landscape in this region. However, as the molecules cross the NPA motif, they are oriented such that their dipole is perpendicular to the pore axis. Such an orientation allows the molecules to form hydrogen bonds with the asparagine side chain. The role of the asparagine in restricting the orientation of water was further confirmed by designing a mutant in which the asparagine was replaced with an alanine. Water molecules in the mutant had greater orientational freedom in the region corresponding to the NPA motif. It is proposed that the restriction of water orientation around the NPA motif in the wild-type is possibly a mechanism that prevents proton transport through the water chain.

2. Selectivity: Proton exclusion

An understanding of the water transport mechanism in aquaporins provided some insights into how the passage of protons through the channel might be blocked. Proton exclusion in these channels continue to be a fascinating problem, and a number of studies have further delved into the atomistic details of this phenomenon. An MD study on the aquaglyceroporin GlpF made an estimate of the barrier for proton permeation from the distribution of water oxygens in the pore (Jensen, Tajkhorshid, & Schulten, 2003). Initially, the positions of the water oxygens were approximated to be the positions of a hypothetical anion passing through the pore, and this was used to calculate the electrostatic interaction energy between the channel and an anion at different positions along the pore. The negative value of this electrostatic interaction energy for a permeating anion, in turn, gave the electrostatic interaction energy for a permeating cation. The most negative interaction energy for the anion is seen at the NPA motif, not at the ar/R selectivity filter that has a positively charged arginine. While the strong interaction at the NPA region is because of the two asparagine residues, the strong attraction between the anion and the arginine at the ar/R region is canceled by repulsive forces from nearby glutamates and the pore-lining M3 and M7 helices. The results suggest that the highest barrier to proton permeation would occur at the NPA region.

The transport of protons along the water chain in AQP1 has been modeled using quantum hopping (Q-HOP) MD simulations, which are MD simulations that allow stochastic proton hopping events to occur

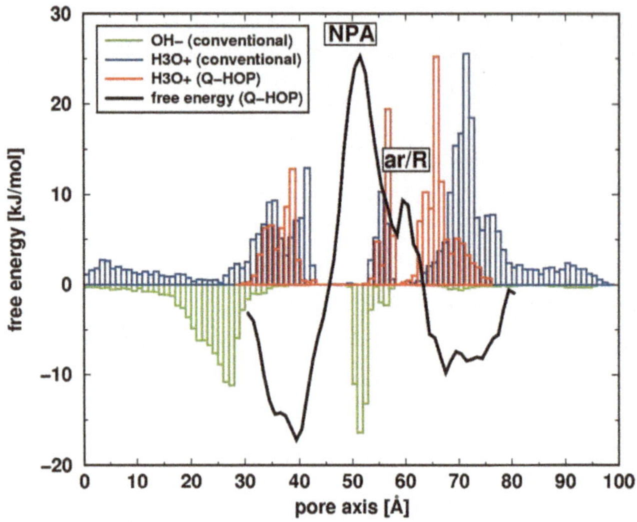

Fig. 3 Free energy for proton transport through the pore of AQP1 from Q-HOP simulations. The figure also shows the distribution of OH^- and H_3O^+ from conventional and Q-HOP simulations. *Reproduced with permission from De Groot, B. L., Frigato, T., Helms, V., & Grubmüller, H. (2003). The mechanism of proton exclusion in the aquaporin-1 water channel.* Journal of Molecular Biology, 333*(2), 279–293.*

(De Groot, Frigato, Helms, & Grubmüller, 2003). Fig. 3 shows the free energy profile for the transport of a proton across the pore of AQP1. The figure also shows the distributions of protons computed from Q-HOP simulations and the distributions of H_3O^+ and OH^- computed from conventional MD simulations. The major barrier for proton transport is seen around the NPA motif, which arises from the occurrence of the asparagine side chains of the two NPA motifs and the positively charged N-termini of two pore-facing helices. A secondary barrier is seen around the ar/R constriction, which is caused by a disruption in the water chain in this region. Thus, the study further confirmed the hypothesis that proton exclusion in aquaporins is brought about by the electrostatics of the NPA region, not by a disruption in the water chain around the ar/R constriction. Results from the study also throw light on hydroxide ion exclusion in aquaporins. Hydroxide ion exclusion is an important feature of aquaporins, since binding of hydroxide ions in the pore can block the pore, thereby rendering the channel incapable of conducting water molecules. The distribution of OH^- shown in Fig. 3 reveals two regions on either side of the NPA region where OH^- does not occur, suggesting that these are regions that are likely to pose an energy barrier to the passage of OH^-. Indeed, the simulations show that when hydroxide ions are placed in the center of the pore, they are attracted toward the NPA motif. However, when they are placed in the peripheral region of the pore, they are not drawn toward the NPA motif, but are instead expelled from the channel.

The effect of the electrostatics of the pore on proton transport has been studied by designing artificial variants of the aquaglyceroporin GlpF (Chakrabarti, Roux, & Pomès, 2004).

Fig. 4A shows the free energy profile for the transfer of an excess proton along the water chain in the different variants of the channel. Switching off the charge on arginine (system "R") does not reduce the barrier, suggesting that the positive charge on proton is not responsible for the barrier.

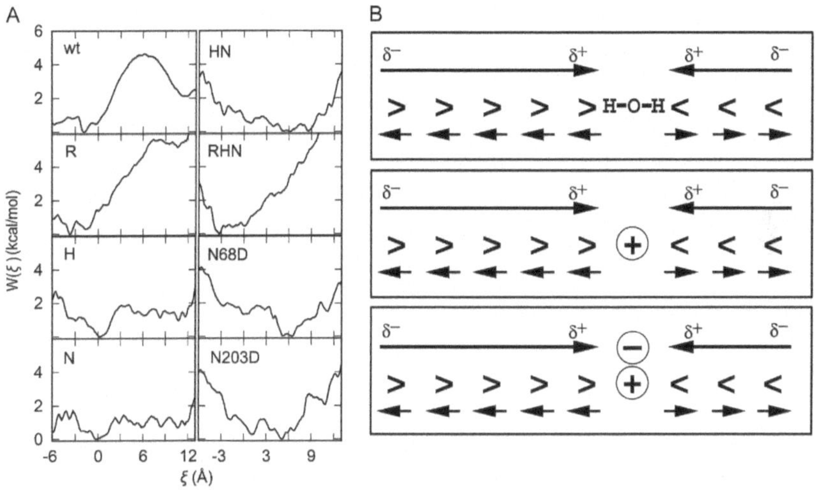

Fig. 4 (A) Free energy profiles for the transfer of a proton through the water chain in GlpF. A number of artificial variants of the channel were studied in which pairwise electrostatic interactions between the pore lumen contents and specific parts of the channel were turned off. These systems are as follows. wt: wild-type; R: electrostatic interactions with R206 turned off; H: electrostatic interactions with M3/M7 helix backbone turned off; N: electrostatic interactions with NH_2 of N68/N203 turned off; HN: both M3/M7 helix backbone and NH_2 of N68/N203 turned off; RHN: R206, M3/M7 helix backbone and NH_2 of N68/N203 turned off. Two mutants, N68D and N203D, were also studied. (B) Schematic showing how the helical dipoles and the bipolar orientation of water aid in proton exclusion. *Top panel*: The bipolar orientation of water molecules ensures favorable interactions with helical dipoles. *Middle panel*: Protonation of a water molecule in the middle leads to repulsion between the protonated water and the helical dipoles. *Bottom panel*: In the N65D and N203D mutants, the negative charge on the glutamate in the middle stabilizes the protonated water. *Reproduced with permission from Chakrabarti, N., Roux, B., & Pomès, R. (2004). Structural determinants of proton blockage in aquaporins.* Journal of Molecular Biology, 343*(2), 493–510.*

Removing the charge at the N-termini of the pore-facing M3 and M7 helices (system "H") reduces the barrier for proton permeation, as does removal of the charge on the asparagine residues in the NPA region (system "N"). Thus, it is the electrostatics of the asparagine residues and the M3/M7 helices that pose a barrier to proton permeation. Mutation of either of the two asparagines to the negatively charged aspartate also reduces the barrier. The observed trends in the free energy profiles are accounted for in Fig. 4B. The charges on the M3 and M7 helices leads to a bipolar orientation of water in the channel (Fig. 4B, top panel). Protonation of a water molecule around the NPA region does not disturb water orientation, but leads to unfavorable interactions between the protonated water and the helix dipoles (Fig. 4B, middle panel). Mutation of an asparagine to aspartate stabilizes the positive charge on the protonated water, thereby facilitating proton permeation (Fig. 4B, bottom panel). The role of the NPA region in posing a major barrier to proton transport through aquaporins has been suggested in a number of other modeling studies (Burykin & Warshel, 2004; Ilan, Tajkhorshid, Schulten, & Voth, 2004; Chakrabarti, Tajkhorshid, Roux, & Pomès, 2004). It has also been suggested that a major factor governing proton permeation through aquaporins is the electrostatic barrier encountered by the proton for passing through the pore of the channel, which is a low dielectric region (Burykin & Warshel, 2003).

3. Selectivity: Permeation of other solutes

Aquaporins also conduct solute molecules other than water, the most common example being the permeation of glycerol through aquaglyceroporins. A comparison of the pores of the pure water channel AQPZ and the aquaglyceroporin GlpF shows that the pore in AQPZ is narrower than GlpF (Wang, Schulten, & Tajkhorshid, 2005). The most significant difference occurs around the ar/R selectivity filter, where the pore in AQPZ is narrower than that in GlpF by 1 Å. AQPZ therefore poses a steric barrier to the bulky glycerol molecule, and passage of a glycerol molecule through the pore of the channel would necessitate major conformational changes in the pore-lining residues. In fact, pore-lining residues were seen to undergo major conformational changes in simulations in which the permeation of glycerol through AQPZ was modeled by application of an external force on glycerol.

The permeation of urea, glycerol, water, ammonia, carbon dioxide, and oxygen through three different systems has been studied, including

AQP1, GlpF, and a lipid bilayer of 1-palmitoyl-2-oleoyl-sn-glycero-3-phosphocholine (POPC) molecules (Hub & De Groot, 2008). A comparison of the free energy barriers for the permeation of the different solutes shows a correlation between permeability and solute hydrophobicity. For AQP1, the barrier increases with an increase in hydrophobicity, indicating an anti-correlation between permeability and solute hydrophobicity. However, urea and glycerol, being bulky molecules, encounter a high barrier in spite of their polar nature. GlpF does not exhibit any difference in selectivity arising from solute hydrophobicity. On the other hand, the POPC bilayer shows a positive correlation between permeability and solute hydrophobicity, with the free energy barrier going down with an increase in solute hydrophobicity.

The transport of glycerol and urea through the *Plasmodium falciparum* aquaglyceroporin PfAQP has been compared with their transport through GlpF from *Escherichia coli* (Aponte-Santamaría, Hub, & de Groot, 2010). The free energy profiles for solute permeation show that for both PfAQP and GlpF, the primary barrier for permeation occurs around the ar/R selectivity filter. Furthermore, PfAQP has lower energy barriers for both urea and glycerol, indicating that the channel conducts these solutes at a faster rate than GlpF. A possible explanation for the lower barrier in case of PfAQP is that the number of water-arginine hydrogen bonds broken at the ar/R region upon passage of the solute through this region is lower than that broken in case of GlpF. In fact, an analysis of the number of hydrogen bonds between water and the arginine in the ar/R constriction shows that there are higher water-arginine hydrogen bonds in GlpF compared to PfAQP.

The permeability of aquaporins toward various solute molecules that are small in size has made it a potential target for drug delivery. H_2O_2 is one of the reactive oxygen species created by cold atmospheric plasmas (CAPs), which have been shown to have anti-cancer activity (Yan, Sherman, & Keidar, 2017). It has been hypothesized that the delivery of H_2O_2 into the cell interior occurs through aquaporins, but the permeability of aquaporins toward H_2O_2 has not been demonstrated. A comparison of the free energy profiles for the permeation of H_2O_2 through AQP1 and a POPC bilayer has been performed using MD simulations (Yusupov, Yan, Cordeiro, & Bogaerts, 2018). The free energy profiles show that the energy barrier for H_2O_2 permeation through the POPC bilayer is about three times higher than the barrier posed by AQP1. This explains the selective uptake of H_2O_2 by cancer cells compared to normal cells, since cancer cells have a higher expression of aquaporins in their cell membrane.

In addition to the permeability toward various solutes exhibited by the pores from the individual monomeric units in a tetrameric aquaporin structure, the central cavity in the tetramer has also been proposed to be permeable toward certain solutes, in particular cations (Anthony et al., 2000). The ion conduction activity arises in the presence of an increased concentration of intracellular cyclic guanosine monophosphate (cGMP). In order to find out if there is any direct interaction between cGMP and aquaporins, the binding of cGMP to AQP1 was modeled by placing cGMP molecules on the cytoplasmic face of the channel (Yu, Yool, Schulten, & Tajkhorshid, 2006). The cGMP molecules spontaneously drift toward arginine residues located on a cytoplasmic loop in AQP1, and binding of cGMP to these arginines leads to an outward motion of the loop away from the central cavity. The role of cGMP and the arginines in the ion transport activity of AQP1 was confirmed by experimental measurements of ion conductance of wild-type AQP1 and a double arginine mutant (R159A + R160A) in the presence of cGMP. While the wild-type exhibited significant ion conduction activity, the double arginine mutant had negligible ion conduction activity. This provides a molecular mechanism for the conduction of ions through the central cavity of AQP1. The binding of cGMP to arginine residues on the cytoplasmic loop makes the loop move away from the central cavity, thereby enabling the passage of ions through the cavity.

4. Quantifying water permeability

The permeation of water through aquaporins has been quantified by Schulten and coworkers in terms of the diffusive permeability p_d and the osmotic permeability p_f (Zhu, Tajkhorshid, & Schulten, 2004a, 2004b). The diffusion permeability for single-file water molecules permeating a channel is given by

$$p_d = v_w q_0$$

where v_w is the volume of a single water molecule and q_0 is the number of complete permeation events in one direction across the channel in unit time. If the average number of water molecules in the lumen of the channel is N, the osmotic permeability constant can be calculated as

$$\frac{p_f}{p_d} = N + 1$$

The permeabilities of a number of aquaporins have subsequently been estimated in a number of studies. Osmotic permeabilities for the channels AQP1, AQPZ, AQP0 and GlpF follow the order GlpF ~ AqpZ > AQP1 ≫ AQP0 (Hashido, Ikeguchi, & Kidera, 2005). The pore radii of these channels, on the other hand, decrease as GlpF ≫ AQP1 > AQPZ ≫ AQP0. It is stated that, in narrow channels, the single file nature of water transport is important, and if the pore is too narrow to allow a continuous single file to exist, the rate of water permeation is greatly slowed down. As the pore radius increases, the single file nature becomes less prevalent, and there is no correlation between rate of permeation and pore radius. A later study showed that, while the p_f for GlpF is two- to three-fold greater than that for AQPZ, the p_d for GlpF is two–nine times greater than that for AQPZ (Jensen & Mouritsen, 2006). The narrower AQPZ has stronger water-protein interactions and weaker water-water interactions. This leads to lower water-water correlation and a slower rate of water transport in AQPZ. The water osmotic permeability has been shown to be affected by correlated motions of both distantly separated and neighboring water molecules (Hashido, Kidera, & Ikeguchi, 2007).

The low permeability of AQP0, an aquaporin abundant in eye lens fiber cells, has been accounted for via MD simulations on the junctional (octameric) and non-junctional (tetrameric) forms of the channel (Jensen et al., 2008). Two pore-facing tyrosine residues regulate water flow and are responsible for the low water permeation rate. It is proposed that the low permeability of the channel makes it possible to maintain the structural stability of the junction.

The effect of pH on the permeability of aquaporin AQP4 has been modeled by altering the protonation state of the pore-facing H95 residue (Kaptan et al., 2015). Under conditions of low pH, which correspond to the positively charged histidine, the permeability of the channel toward water increases. In order to validate the histidine-dependent sensitivity of the channel toward pH, *in vitro* studies were performed on a H95A mutant. The mutant was insensitive to changes in cytoplasmic pH, thereby confirming the role of the histidine in pH-mediated gating.

An aquaporin that is permeable toward both water and ammonia is TIP2;1 from wheat (Jahn et al., 2004). Interestingly, a double mutant of the channel, I185H + G194C, is permeable to water but not ammonia (Jahn et al., 2004; Kirscht et al., 2016). MD simulations on the double mutant show that the permeability of the mutant relative to the wild-type is lower by a factor of 2.5 for water and 4 for ammonia (Lindahl, Gourdon, Andersson, & Hess, 2018). There is thus an increase in the overall selectivity toward water, and this is attributed to increased water-channel hydrogen bonding around the selectivity filter. The impermeability of the double mutant toward ammonia arises from a narrowing of the pore. The permeabilities of a number of other aquaporins have been quantified using MD simulations, including the archaeal aquaporin AqpM and the human aquaporins AQP2, AQP4 and AQP5 (Alishahi & Kamali, 2018; Araya-Secchi, Garate, Holmes, & Perez-Acle, 2011; Binesh & Kamali, 2015; Wambo, Rodriguez, & Chen, 2017).

5. Interaction with lipids

Lipids play a critical role in governing the activity of membrane proteins (Phillips, Ursell, Wiggins, & Sens, 2009). Recognizing the importance of lipid-protein interactions, a number of studies have investigated the interaction of aquaporins with lipids. The electron crystallography structure of lens-specific aquaporin AQP0 revealed annular lipids, which are lipids found in the first shell around the protein (Gonen et al., 2005). To determine if the position of these lipids correspond to those of unconstrained lipids around the protein, MD simulations were performed on a tetrameric form of the channel embedded in a bilayer of dimyristoyl-phosphatidylcholine (DMPC) lipids (Aponte-Santamaría, Briones, Schenk, Walz, & de Groot, 2012). A comparison between a lipid density map obtained from the MD simulation with the position of the lipids as seen in the crystal structure shows that the lipids occupy nearly the same position in the two cases (Fig. 5A). Fig. 5B shows the contribution of different moieties on lipids to the lipid density map. The greatest contribution comes from the acyl chains rather than the headgroup, indicating that the positions of lipids around the protein is defined by the acyl chains. A later study on the same system showed that, while protein mobility affects the localization of annular lipids, immobilization of the lipids (by inducing gel phase) does not affect protein mobility (Briones, Aponte-Santamaría, & de Groot, 2017).

Extensive simulations have been employed to model interactions between the aquaporin AQP1 and gangliosides, which are a group of glycolipids that occur in the outer leaflet of cell membranes (Gu, Ingólfsson, de Vries, Marrink, & Tieleman, 2016). Results show that the ganglioside GM1 binds to specific sites on the surface of AQP1, the mode of interaction being hydrogen bonds between GM1 headgroups and polar residues on the protein.

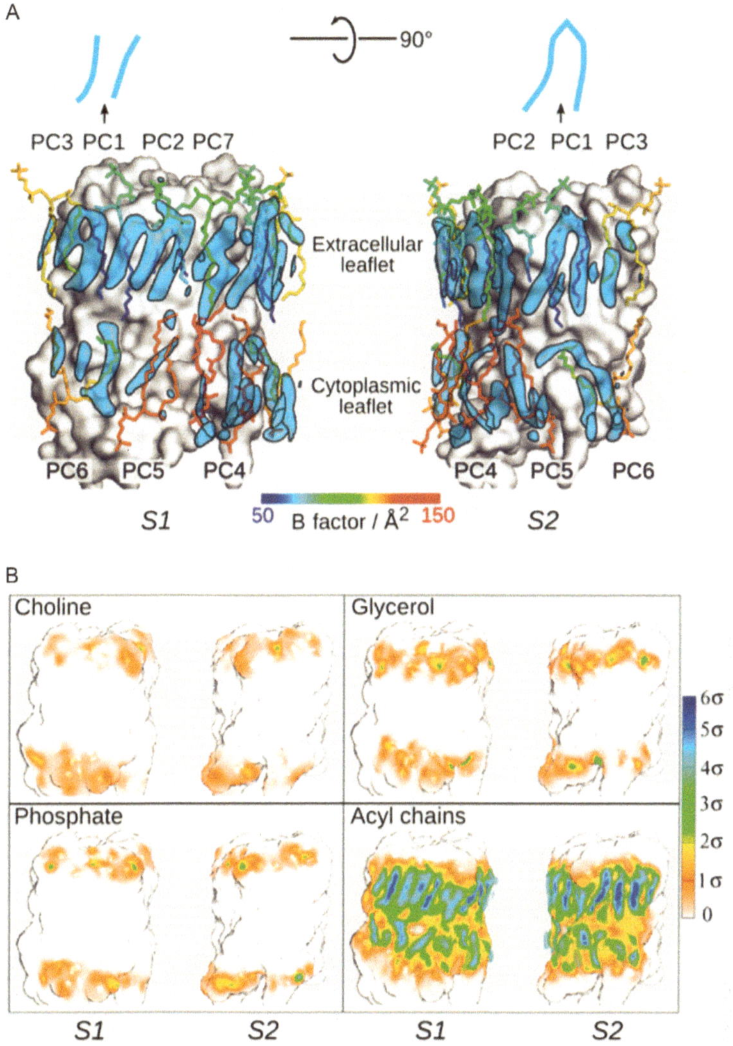

Fig. 5 (A) Comparison between the lipid positions around AQP0 in the crystal structure and the lipid density map obtained from MD simulations. The crystallographic lipids, named PC1 to PC7, are shown in stick representation and colored according to their B factor. Probable lipid positions from the MD-derived lipid density map are shown as blue regions. S1 and S2 indicate the two lipid-facing surfaces of a monomeric unit of AQP0. (B) Contribution of different lipid moieties to the lipid density map. *Reproduced with permission from Aponte-Santamaría, C., Briones, R., Schenk, A. D., Walz, T., & de Groot, B. L. (2012). Molecular driving forces defining lipid positions around aquaporin-0.* Proceedings of the National Academy of Sciences of the United States of America, 109*(25), 9887–9892.*

6. Response to electric field

There have not been many attempts at investigating the effect of an external electric field on aquaporins, possibly because none of the physiological functions associated with aquaporins have any relation with voltage regulation. The first study that investigated the effect of an electrostatic potential on an aquaporin was, in fact, a simulation study (Hub, Aponte-Santamaría, Grubmüller, & de Groot, 2010). Two AQP1 tetramers were set up in a lipid bilayer environment and then stacked on top of each other such that there was a water compartment in between them (Fig. 6A). The membrane potential was modulated by altering the number of cations and anions in the water compartments. Single channel permeabilities of the channel as a function of membrane potential are shown in Fig. 6B. While a positive membrane potential led to an increase in permeability, a negative membrane potential led to a decrease. A visual inspection of the trajectories revealed the arginine in the ar/R constriction to be the voltage gate. The arginine exhibited two distinct conformations, one corresponding to the up state or the positive potential and another corresponding to the down state or the negative potential (Fig. 6C and D).

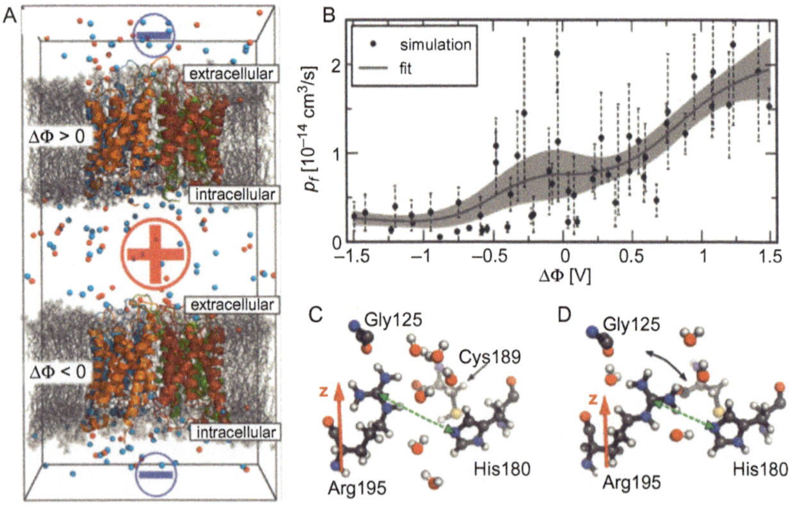

Fig. 6 (A) Simulation setup showing two lipid-embedded AQP1 tetramers stacked on top of each other with a water compartment in between. (B) Variation of single channel permeability p_f with the membrane potential $\Delta\Phi$. (C,D) The arginine from the ar/R constriction in the up (C) and down (D) states, respectively. *Reproduced with permission from Hub, J. S., Aponte-Santamaría, C., Grubmüller, H., & de Groot, B. L. (2010). Voltage-regulated water flux through aquaporin channels in silico.* Biophysical Journal, *99(12), L97-L99.*

The effect of external static electric fields on human AQP4 has been studied using MD simulations by applying pulses acting in a direction perpendicular to the pore (Marracino et al., 2016). Reduced water permeability is seen in the presence of the electric field, and this is attributed to dipolar flipping of residues at the two ends of the pore, in particular H95, C178 and S211. On the other hand, the application of a circularly polarized electric field, that acts axially along the pore, enhances the water permeability of the channel (Burnham & English, 2017).

7. *In silico* drug design

The association of aquaporins with several diseased conditions, such as edema, cancer, glaucoma and obesity, has made these channels an attractive drug target (Tradtrantip, Jin, Yao, Anderson, & Verkman, 2017). The possibility of tetraethylammonium (TEA) acting as an inhibitor for AQP1 has been investigated using three approaches: (1) docking of the ligand followed by molecular dynamics, (2) unbiased MD simulations on membrane-embedded AQP1 in an aqueous solution of 100 mM TEA, and (3) umbrella sampling (Müller, Hub, Grubmüller, & de Groot, 2008). The study identified the binding site of TEA in the pore, and this binding led to decreased water permeation. Virtual screening of human AQP1 followed by molecular dynamics helped identify three inhibitors of the channel, the chemical structures of which are depicted in Fig. 7A (Seeliger et al., 2013). The inhibitors bind in the vicinity of the ar/R constriction toward the extracellular exit of the channel, as shown in Fig. 7B. The binding involves interaction with a lysine residue, which is not conserved in the other aquaporins. Thus, the compounds identified in the study could serve as AQP1-specific inhibitors.

MD simulations on the aquaglyceroporin AQP3 in the presence of Hg^{2+} ions revealed the mechanism of inhibition of the channel (Spinello, de Almeida, Casini, & Barone, 2016). The covalent attachment of mercury to a C40 residue leads to the collapse of the ar/R constriction, leading to the blockage of the pore. AQP3 has also been shown to be inhibited by gold(III) complexes (Martins et al., 2013). The drugs sumatriptan and rizatriptan have been identified as inhibitors of AQP4 (Huber, Tsujita, & Nakada, 2009). Other inhibitors identified for the aquaporin family include bumetanide and the gold(III) complex Auphen (Martins et al., 2012; Migliati et al., 2009), while agonists include the furosemide derivative AqF026 (Yool et al., 2013).

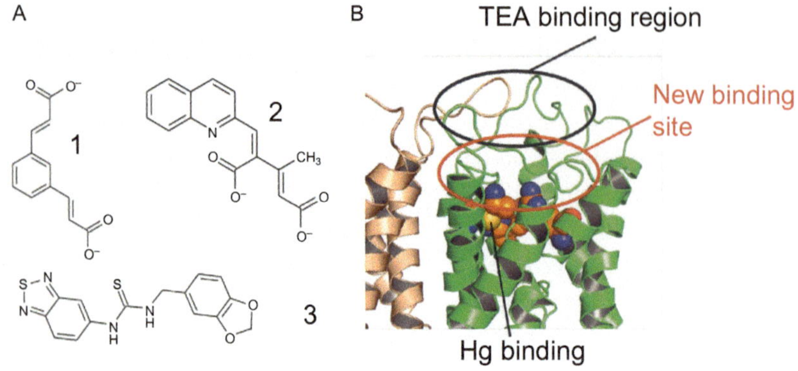

Fig. 7 (A) Structures of the three novel inhibitors of AQP1 identified by virtual screening and MD simulations. (B) Binding sites for TEA, Hg and the compounds 1 and 2. *Reproduced with permission from Seeliger, D., Zapater, C., Krenc, D., Haddoub, R., Flitsch, S., Beitz, E., et al. (2013). Discovery of novel human aquaporin-1 blockers.* ACS Chemical Biology, *8(1), 249–256.*

8. Future directions

Molecular dynamics simulations in general have been instrumental in understanding various structural, dynamic and thermodynamic properties related to aquaporins. Modeling these channels at longer length scales and longer timescales would further our knowledge. An important aspect of aquaporin simulations is the lipid composition that is being used in the bilayer. Since there are well defined interactions between lipids and proteins (Aponte-Santamaría et al., 2012; Briones et al., 2017; Gu et al., 2016), it is important to use a combination of lipids that closely resemble the cell membranes of living cells. The effect of different membrane constituents—like cholesterol, for instance—on aquaporin structure and function is worth investigating. A major goal in the future would be to model multiple aquaporin molecules in realistic cell membranes, as opposed to present day studies that focus on the functioning of a single aquaporin tetramer (or monomer) embedded in a membrane. This would provide further insights into how aquaporins function dynamically in real cells, especially in tumor cells, which have higher expression of aquaporins in their membranes. The permeability of aquaporins toward H_2O_2 has been demonstrated in simulation studies, and this offers hope for the use of CAPs in tumor cell-targeted drug delivery (Yusupov et al., 2018). Apart from being a target for cancer, aquaporins are associated with a number of other diseases (Tradtrantip et al., 2017). The development of modulators for these channels can be accelerated by using drug repurposing approaches.

Acknowledgments

UDP thanks the financial support of Department of Science and Technology-SERB and the Indian National Science Academy.

References

Adcock, S. A., & McCammon, J. A. (2006). Molecular dynamics: Survey of methods for simulating the activity of proteins. *Chemical Reviews*, *106*(5), 1589–1615.

Alishahi, M., & Kamali, R. (2018). Forced diffusion of water molecules through aquaporin-5 biomembrane; a molecular dynamics study. *Biophysics and Physicobiology*, *15*, 255–262.

Anthony, T. L., Brooks, H. L., Boassa, D., Leonov, S., Yanochko, G. M., Regan, J. W., et al. (2000). Cloned human aquaporin-1 is a cyclic GMP-gated ion channel. *Molecular Pharmacology*, *57*(3), 576–588.

Aponte-Santamaría, C., Briones, R., Schenk, A. D., Walz, T., & de Groot, B. L. (2012). Molecular driving forces defining lipid positions around aquaporin-0. *Proceedings of the National Academy of Sciences of the United States of America*, *109*(25), 9887–9892.

Aponte-Santamaría, C., Hub, J. S., & de Groot, B. L. (2010). Dynamics and energetics of solute permeation through the Plasmodium falciparum aquaglyceroporin. *Physical Chemistry Chemical Physics*, *12*(35), 10246–10254.

Araya-Secchi, R., Garate, J. A., Holmes, D. S., & Perez-Acle, T. (2011). Molecular dynamics study of the archaeal aquaporin AqpM. In *BMC genomics. Vol. 12* (p. S8). BioMed Central. No. 4.

Barati Farimani, A., Aluru, N. R., & Tajkhorshid, E. (2014). Thermodynamic insight into spontaneous hydration and rapid water permeation in aquaporins. *Applied Physics Letters*, *105*(8), 083702.

Binesh, A. R., & Kamali, R. (2015). Molecular dynamics insights into human aquaporin 2 water channel. *Biophysical Chemistry*, *207*, 107–113.

Briones, R., Aponte-Santamaría, C., & de Groot, B. L. (2017). Localization and ordering of lipids around aquaporin-0: Protein and lipid mobility effects. *Frontiers in Physiology*, *8*, 124.

Burnham, C. J., & English, N. J. (2017). Electropumping of water through human aquaporin 4 by circularly polarized electric fields: Dramatic enhancement and control revealed by non-equilibrium molecular dynamics. *The Journal of Physical Chemistry Letters*, *8*(19), 4646–4651.

Burykin, A., & Warshel, A. (2003). What really prevents proton transport through aquaporin? Charge self-energy versus proton wire proposals. *Biophysical Journal*, *85*(6), 3696–3706.

Burykin, A., & Warshel, A. (2004). On the origin of the electrostatic barrier for proton transport in aquaporin. *FEBS Letters*, *570*(1–3), 41–46.

Chakrabarti, N., Roux, B., & Pomès, R. (2004). Structural determinants of proton blockage in aquaporins. *Journal of Molecular Biology*, *343*(2), 493–510.

Chakrabarti, N., Tajkhorshid, E., Roux, B., & Pomès, R. (2004). Molecular basis of proton blockage in aquaporins. *Structure*, *12*(1), 65–74.

Cui, Y., & Bastien, D. A. (2011). Water transport in human aquaporin-4: Molecular dynamics (MD) simulations. *Biochemical and Biophysical Research Communications*, *412*(4), 654–659.

De Groot, B. L., Frigato, T., Helms, V., & Grubmüller, H. (2003). The mechanism of proton exclusion in the aquaporin-1 water channel. *Journal of Molecular Biology*, *333*(2), 279–293.

de Groot, B. L., & Grubmüller, H. (2001). Water permeation across biological membranes: Mechanism and dynamics of aquaporin-1 and GlpF. *Science*, *294*(5550), 2353–2357.

Denning, E. J., Priyakumar, U. D., Nilsson, L., & MacKerell, A. D., Jr. (2011). Impact of 2′-hydroxyl sampling on the conformational properties of RNA: Update of the CHARMM all-atom additive force field for RNA. *Journal of Computational Chemistry*, *32*(9), 1929–1943.

Fu, D., Libson, A., Miercke, L. J., Weitzman, C., Nollert, P., Krucinski, J., et al. (2000). Structure of a glycerol-conducting channel and the basis for its selectivity. *Science, 290*(5491), 481–486.

Gonen, T., Cheng, Y., Sliz, P., Hiroaki, Y., Fujiyoshi, Y., Harrison, S. C., et al. (2005). Lipid–protein interactions in double-layered two-dimensional AQP0 crystals. *Nature, 438*(7068), 633.

Gu, R. X., Ingólfsson, H. I., de Vries, A. H., Marrink, S. J., & Tieleman, D. P. (2016). Ganglioside-lipid and ganglioside-protein interactions revealed by coarse-grained and atomistic molecular dynamics simulations. *The Journal of Physical Chemistry B, 121*(15), 3262–3275.

Hart, K., Foloppe, N., Baker, C. M., Denning, E. J., Nilsson, L., & MacKerell, A. D., Jr. (2012). Optimization of the CHARMM additive force field for DNA: Improved treatment of the BI/BII conformational equilibrium. *Journal of Chemical Theory and Computation, 8*(1), 348–362.

Hashido, M., Ikeguchi, M., & Kidera, A. (2005). Comparative simulations of aquaporin family: AQP1, AQPZ, AQP0 and GlpF. *FEBS Letters, 579*(25), 5549–5552.

Hashido, M., Kidera, A., & Ikeguchi, M. (2007). Water transport in aquaporins: Osmotic permeability matrix analysis of molecular dynamics simulations. *Biophysical Journal, 93*(2), 373–385.

Huang, J., Rauscher, S., Nawrocki, G., Ran, T., Feig, M., de Groot, B. L., et al. (2016). CHARMM36m: An improved force field for folded and intrinsically disordered proteins. *Nature Methods, 14*(1), 71.

Hub, J. S., Aponte-Santamaría, C., Grubmüller, H., & de Groot, B. L. (2010). Voltage-regulated water flux through aquaporin channels in silico. *Biophysical Journal, 99*(12), L97–L99.

Hub, J. S., & De Groot, B. L. (2008). Mechanism of selectivity in aquaporins and aquaglyceroporins. *Proceedings of the National Academy of Sciences of the United States of America, 105*(4), 1198–1203.

Huber, V. J., Tsujita, M., & Nakada, T. (2009). Identification of aquaporin 4 inhibitors using in vitro and in silico methods. *Bioorganic & Medicinal Chemistry, 17*(1), 411–417.

Ilan, B., Tajkhorshid, E., Schulten, K., & Voth, G. A. (2004). The mechanism of proton exclusion in aquaporin channels. *Proteins: Structure, Function, and Bioinformatics, 55*(2), 223–228.

Jahn, T. P., Møller, A. L., Zeuthen, T., Holm, L. M., Klærke, D. A., Mohsin, B., et al. (2004). Aquaporin homologues in plants and mammals transport ammonia. *FEBS Letters, 574*(1–3), 31–36.

Janosi, L., & Ceccarelli, M. (2013). The gating mechanism of the human aquaporin 5 revealed by molecular dynamics simulations. *PLoS One, 8*(4), e59897.

Jensen, M. Ø., Dror, R. O., Xu, H., Borhani, D. W., Arkin, I. T., Eastwood, M. P., et al. (2008). Dynamic control of slow water transport by aquaporin 0: Implications for hydration and junction stability in the eye lens. *Proceedings of the National Academy of Sciences of the United States of America, 105*(38), 14430–14435.

Jensen, M. Ø., & Mouritsen, O. G. (2006). Single-channel water permeabilities of Escherichia coli aquaporins AqpZ and GlpF. *Biophysical Journal, 90*(7), 2270–2284.

Jensen, M. Ø., Tajkhorshid, E., & Schulten, K. (2003). Electrostatic tuning of permeation and selectivity in aquaporin water channels. *Biophysical Journal, 85*(5), 2884–2899.

Kaptan, S., Assentoft, M., Schneider, H. P., Fenton, R. A., Deitmer, J. W., MacAulay, N., et al. (2015). H95 is a pH-dependent gate in aquaporin 4. *Structure, 23*(12), 2309–2318.

Kästner, J. (2011). Umbrella sampling. *Wiley Interdisciplinary Reviews: Computational Molecular Science, 1*(6), 932–942.

Kirscht, A., Kaptan, S. S., Bienert, G. P., Chaumont, F., Nissen, P., de Groot, B. L., et al. (2016). Crystal structure of an ammonia-permeable aquaporin. *PLoS Biology, 14*(3), e1002411.

Klauda, J. B., Venable, R. M., Freites, J. A., O'Connor, J. W., Tobias, D. J., Mondragon-Ramirez, C., et al. (2010). Update of the CHARMM all-atom additive force field for lipids: Validation on six lipid types. *The Journal of Physical Chemistry B*, *114*(23), 7830–7843.

Lindahl, V., Gourdon, P., Andersson, M., & Hess, B. (2018). Permeability and ammonia selectivity in aquaporin TIP2; 1: Linking structure to function. *Scientific Reports*, *8*(1), 2995.

Maier, J. A., Martinez, C., Kasavajhala, K., Wickstrom, L., Hauser, K. E., & Simmerling, C. (2015). ff14SB: Improving the accuracy of protein side chain and backbone parameters from ff99SB. *Journal of Chemical Theory and Computation*, *11*(8), 3696–3713.

Marracino, P., Liberti, M., Trapani, E., Burnham, C. J., Avena, M., Garate, J. A., et al. (2016). Human aquaporin 4 gating dynamics under perpendicularly-oriented electric-field impulses: A molecular dynamics study. *International Journal of Molecular Sciences*, *17*(7), 1133.

Martins, A. P., Ciancetta, A., de Almeida, A., Marrone, A., Re, N., Soveral, G., et al. (2013). Aquaporin inhibition by gold (III) compounds: New insights. *ChemMedChem*, *8*(7), 1086–1092.

Martins, A. P., Marrone, A., Ciancetta, A., Cobo, A. G., Echevarría, M., Moura, T. F., et al. (2012). Targeting aquaporin function: Potent inhibition of aquaglyceroporin-3 by a gold-based compound. *PLoS One*, 7(5), e37435.

Migliati, E., Meurice, N., DuBois, P., Fang, J. S., Somasekharan, S., Beckett, E., et al. (2009). Inhibition of aquaporin-1 and aquaporin-4 water permeability by a derivative of the loop diuretic bumetanide acting at a internal pore-occluding binding site. *Molecular Pharmacology*, *76*(1), 105–112.

Müller, E. M., Hub, J. S., Grubmüller, H., & de Groot, B. L. (2008). Is TEA an inhibitor for human Aquaporin-1? *Pflügers Archiv/European Journal of Physiology*, *456*(4), 663–669.

Murata, K., Mitsuoka, K., Hirai, T., Walz, T., Agre, P., Heymann, J. B., et al. (2000). Structural determinants of water permeation through aquaporin-1. *Nature*, *407*(6804), 599.

Padhi, S., & Priyakumar, U. D. (2017a). Modeling complex biomolecular systems and processes using molecular mechanics force fields and molecular dynamics simulations. In D. Kumar (Ed.), *Theoretical and computational advances: From atoms to molecules to materials* (pp. 105–127). Lucknow, India: Self Publishing.

Padhi, S., & Priyakumar, U. D. (2017b). Microsecond simulation of human aquaporin 2 reveals structural determinants of water permeability and selectivity. *Biochimica et Biophysica Acta (BBA)—Biomembranes*, *1859*(1), 10–16.

Phillips, R., Ursell, T., Wiggins, P., & Sens, P. (2009). Emerging roles for lipids in shaping membrane-protein function. *Nature*, *459*(7245), 379.

Seeliger, D., Zapater, C., Krenc, D., Haddoub, R., Flitsch, S., Beitz, E., et al. (2013). Discovery of novel human aquaporin-1 blockers. *ACS Chemical Biology*, *8*(1), 249–256.

Smolin, N., Li, B., Beck, D. A., & Daggett, V. (2008). Side-chain dynamics are critical for water permeation through aquaporin-1. *Biophysical Journal*, *95*(3), 1089–1098.

Spinello, A., de Almeida, A., Casini, A., & Barone, G. (2016). The inhibition of glycerol permeation through aquaglyceroporin-3 induced by mercury (II): A molecular dynamics study. *Journal of Inorganic Biochemistry*, *160*, 78–84.

Tajkhorshid, E., Nollert, P., Jensen, M. Ø., Miercke, L. J., O'connell, J., Stroud, R. M., et al. (2002). Control of the selectivity of the aquaporin water channel family by global orientational tuning. *Science*, *296*(5567), 525–530.

Tradtrantip, L., Jin, B. J., Yao, X., Anderson, M. O., & Verkman, A. S. (2017). Aquaporin-targeted therapeutics: State-of-the-field. In B. Yang (Ed.), *Aquaporins* (pp. 239–250). Dordrecht: Springer.

Wambo, T. O., Rodriguez, R. A., & Chen, L. Y. (2017). Computing osmotic permeabilities of aquaporins AQP4, AQP5, and GlpF from near-equilibrium simulations. *Biochimica et Biophysica Acta (BBA)—Biomembranes*, *1859*(8), 1310–1316.

Wang, S., Ing, C., Emami, S., Jiang, Y., Liang, H., Pomès, R., et al. (2016). Structure and dynamics of extracellular loops in human aquaporin-1 from solid-state NMR and molecular dynamics. *The Journal of Physical Chemistry B*, *120*(37), 9887–9902.

Wang, Y., Schulten, K., & Tajkhorshid, E. (2005). What makes an aquaporin a glycerol channel? A comparative study of AqpZ and GlpF. *Structure*, *13*(8), 1107–1118.

Yan, D., Sherman, J. H., & Keidar, M. (2017). Cold atmospheric plasma, a novel promising anti-cancer treatment modality. *Oncotarget*, *8*(9), 15977.

Yool, A. J., Morelle, J., Cnops, Y., Verbavatz, J. M., Campbell, E. M., Beckett, E. A., et al. (2013). AqF026 is a pharmacologic agonist of the water channel aquaporin-1. *Journal of the American Society of Nephrology*, *24*(7), 1045–1052.

Yu, J., Yool, A. J., Schulten, K., & Tajkhorshid, E. (2006). Mechanism of gating and ion conductivity of a possible tetrameric pore in aquaporin-1. *Structure*, *14*(9), 1411–1423.

Yusupov, M., Yan, D., Cordeiro, R. M., & Bogaerts, A. (2018). Atomic scale simulation of H2O2 permeation through aquaporin: Toward the understanding of plasma cancer treatment. *Journal of Physics D: Applied Physics*, *51*(12), 125401.

Zhu, F., Tajkhorshid, E., & Schulten, K. (2001). Molecular dynamics study of aquaporin-1 water channel in a lipid bilayer. *FEBS Letters*, *504*(3), 212–218.

Zhu, F., Tajkhorshid, E., & Schulten, K. (2004a). Collective diffusion model for water permeation through microscopic channels. *Physical Review Letters*, *93*(22), 224501.

Zhu, F., Tajkhorshid, E., & Schulten, K. (2004b). Theory and simulation of water permeation in aquaporin-1. *Biophysical Journal*, *86*(1), 50–57.

CHAPTER FOUR

Aquaporin regulation in metabolic organs

Matteo Tardelli[a,b,†], **Thomas M. Stulnig**[b,*,†]
[a]Division of Gastroenterology and Hepatology, Joan & Sanford I. Weill Cornell Department of Medicine, Weill Cornell Medical College, New York, NY, United States
[b]Clinical Division of Endocrinology and Metabolism, Department of Medicine III, Medical University of Vienna, Vienna, Austria
*Corresponding author: e-mail address: thomas.stulnig@meduniwien.ac.at

Contents

Abstract

Aquaporins (AQPs) are a family of 13 small trans-membrane proteins, which facilitate shuttling of glycerol, water and urea. The peculiar role of AQPs in glycerol transport makes them attractive targets in metabolic organs since glycerol represents the backbone of triglyceride synthesis. Importantly, AQPs are known to be regulated by various nuclear receptors which in turn govern lipid and glucose metabolism as well as inflammatory cascades. Here, we review the role of AQPs regulation in metabolic organs exploring their physiological impact in health and disease.

1. Aquaporins and nuclear receptors

Cells in the animal and plant kingdom strictly depend on water for their basic metabolic needs. Water balance is maintained by aquaporins

† Current address: Third Department of Medicine, Hietzing Hospital, Vienna, Austria

Vitamins and Hormones, Volume 112
ISSN 0083-6729
https://doi.org/10.1016/bs.vh.2019.09.002

(AQPs), which are membrane channels that facilitate diffusion of water and small solutes such as urea and glycerol (Verkman, 2011). They are a family of 13 known channel proteins, which can be subdivided on the basis of their solute selectivity. On the one hand, orthodox AQPs allow the passage of water. On the other hand, aquaglyceroporins instead allow glycerol as well urea, gases such as CO_2 and ammonia (AQP3, -7, - 9, -10) (Verkman, Anderson, & Papadopoulos, 2014). Importantly, AQPs mediate osmotic water transport across cell plasma membranes and glycerol diffusion thereby regulating pivotal physiologic function such as cell proliferation and lipid metabolism. AQPs are expressed in many organs and tissues such as adipose tissue (AT), liver, kidney, muscle as well as immune cells and adipocytes, with several functions depending on their localization (Verkman, 2011) (Table 1). AQPs have 25–60% homology inprotein sequence, and are structurally organized in homotetrameric complexes where each monomer is built up of six transmembrane segments forming a water pore (Agre & Kozono, 2003). Most AQPs are localized across the lipid bilayer of the cell plasma membrane and their expression is mainly regulated at transcriptional level. They have been shown to translocate upon certain stimuli from intracellular vesicles to the plasma membrane, therefore modulating its permeability and cellular functions (Conner et al., 2012).

Different studies investigated the role of AQPs in lipid metabolism elucidating the importance of glycerol channels in many physiologic and pathologic processes such as obesity and type 2 diabetes (T2D) (Catalán et al., 2008; Lebeck, Cheema, Skowronski, Nielsen, & Praetorius, 2015; Lebeck, Søndergaard, & Nielsen, 2018). Indeed, glycerol is a prerequisite for the formation of triglycerides (TGs) backbones. TGs are the main source of energy storage for the human body, taking part in metabolic pathways such as fatty acid oxidation, biosynthesis of other lipids including polar lipids for biological membranes, and transport via lipoproteins (Reshef et al., 2003).

The regulation of TG metabolism is finely tuned by nuclear receptors (NRs), a family of transcriptional regulators involved in diverse functions of physiology (Olefsky, 2001). Notably, NRs play key roles in cell differentiation/proliferation and metabolism, being crucially connected with numerous pathologies such as liver steatosis, inflammation and cancer. The impressive network of target proteins associated with NRs action is tremendously complex and comprises many co-regulators and other micro- and long non-coding RNAs. The NR superfamily includes 48 transcription factors (Robinson-Rechavi, Escriva, & Laudet, 2003); the PPAR

Table 1 Aquaporins solute selectivity, predominant localization and disease implications.

AQP	Solute selectivity	Tissue expression	Disease implications	References
AQP0	Water	Eye	Cataract	Berry, Francis, Kaushal, Moore, and Bhattacharya (2000)
AQP1	Water	Brain, kidney	Tumor genesis, metastasis	Verkman, Hara-Chikuma, and Papadopoulos (2008)
AQP2	Water	Kidney	Nephrogenic diabetes	Deen et al. (1994)
AQP3	Water, glycerol, urea	Liver, skin, intestine, eye	Skin cancer	Qin et al. (2011)
AQP4	Water	Brain, intestine, kidney, eye, nose, muscle	Brain oedema, epilepsy, neuromyelitis optica	Fukuda and Badaut (2012), Hsu, Lee, and Binder (2007), Jarius and Wildemann (2010), and Manley, Zador, Stiver, and Wang (2009)
AQP5	Water	Lungs, eye, pancreas	Asthma, cystic fibrosis	Dong et al. (2012) and Song and Verkman (2001)
AQP6	Water	Kidney	Unknown	Yasui, Kwon, Knepper, Nielsen, and Agre (1999)
AQP7	Water, glycerol, urea	Adipose tissue, liver, kidney, heart muscle	Obesity, Type 2 diabetes	Hara-Chikuma et al. (2005), Hibuse et al. (2005), Maeda, Hibuse, and Funahashi (2009)
AQP8	Water	Intestine	Ulcerative colitis	Zahn et al. (2007)
AQP9	Water, glycerol, urea	Liver, adipose tissue	Obesity, Type 2 diabetes. NAFLD	Gena et al. (2013), Lebeck (2014), and Maeda et al. (2009)
AQP10	Water, glycerol, urea	Small intestine	Unknown	Li et al. (2005)
AQP11	Water	Liver, kidney	Polycystic kidney disease	Morishita et al. (2005)
AQP12	Unknown	Pancreas	Pancreatitis	Ohta et al. (2009)

family includes PPARα, -δ, -γ which form heterodimers with the retinoid X receptors (RXRs) to regulate gene expression. PPARα is highly expressed in the heart, liver, kidney and skeletal muscle. It senses fatty acids and promotes fatty acid oxidation (Kliewer, Lehmann, & Willson, 1999). In addition to its metabolic actions, PPARα plays an anti-inflammatory role in immune cells and can be activated by drugs such as fibrates (van Bilsen & van Nieuwenhoven, 2010). In contrast, PPARδ is ubiquitously expressed and promotes fat burning and weight loss in mice (Wang et al., 2003). PPARγ is the master regulator of lipogenesis expressed in adipose tissue, skeletal muscle and liver (Abbott, 2009). PPARγ is be activated thiazolidinediones, anti-diabetic drugs which improve insulin sensitivity by promoting FA storage in adipocytes (Jacobi, Stanya, & Lee, 2012). Farnesoid X receptor (FXR) is expressed in the intestine and liver and like PPARs forms heterodimers with RXR to trigger gene expression (Wang, Chen, Moore, & Huang, 2008; Zhang & Edwards, 2008). Importantly, FXR regulates bile acid synthesis (Goodwin et al., 2000; Watanabe et al., 2004) and controls insulin sensitivity (Massafra & van Mil, 2018). The liver X receptor (LXR) is activated by oxysterols and regulates genes involved in lipid metabolism and cholesterol uptake, efflux and transport (Kalaany & Mangelsdorf, 2006).

Therefore, NRs regulate key components of lipid including glycerol shuttling mediated by aquaglyceroporins (Tardelli, Claudel, Bruschi, & Trauner, 2018).

2. Aquaporins regulation in the adipose tissue

The adipose tissue (AT) is by far the largest TG reservoirs in the human body. AT is mainly composed by adipocytes, immune cells and extracellular matrix formed by proteins and collagen. When energy balance is positive, adipocytes increase in size and number to store excess lipids, thus dramatically adapting and changing their morphology due to increased FA uptake and TG synthesis (Scherer, 2006). Of interest, AT secretes a number of mediators, called adipokines, for endocrine and paracrine action. Adipokines allow AT to communicate with other organs coordinating biological processes such as inflammation, immunity and energy metabolism including lipid and glucose homeostasis (Choe, Huh, Hwang, Kim, & Kim, 2016; Kajimura, 2017). The ability of aquaglyceroporins to facilitate glycerol diffusion through adipocyte membranes makes them an interesting target in the treatment of metabolic disorders (Verkman et al., 2014).

Glycerol is a polyol produced from various substrates such as glucose, amino acids and glycerolipids. On the other hand, glycerol is absorbed from exogenous sources particularly after release from dietary fat released during digestion (Hibuse, Maeda, Nagasawa, & Funahashi, 2006). Notably, glycerol not only forms the TG backbone, but is also a precursor of carbohydrate, lipids and phospholipids, and takes also part in the glycolysis in form of glycerol 3 phosphate (Petersen, Vatner, & Shulman, 2017). With regards to glycerol shuttling, aquaporin 7 (AQP7) is highly expressed in white and brown AT (Sjöholm et al., 2005). AQP7 plays a key role in adipocyte metabolism, insulin resistance and lipid as well as glucose control. Interestingly, AQP7 deficient mice (AQP7 KO) have low plasma glycerol levels but normal level of free fatty acid compared to WT (Maeda et al., 2004). In vitro, epinephrine-stimulated glycerol secretion was also impaired in AQP7 knockdown adipocytes. During prolonged fasting, plasma glycerol elevated and glucose level was maintained in WT mice, while AQP7 KO mice showed a disrupted increase of plasma glycerol and rapid reduction of glucose (Maeda et al., 2004). In another study, impaired glycerol efflux in AQP7 KO mice led to accumulation of TGs in adipocytes resulting in weight gain and adipocytes hypertrophy (Hara-Chikuma et al., 2005). In line, the same authors found that AQP7 KO mice developed adult-onset obesity, with increased body weight and fat mass after 12 weeks of age. AQP7 KO mice developed obesity and insulin resistance even at younger age when challenged with high-fat/high-sucrose diet (Hibuse et al., 2005). Mechanistically, they demonstrated that AQP7 disruption elevated adipose glycerol kinase activity, also accelerating TG synthesis in adipocytes, and therefore resulting in an obese phenotype. (Hibuse et al., 2005). As a potential mode of action, AQP7 was shown to be a direct target of PPARγ in 3T3-L1 cells and in mice treated with pioglitazone (Kishida et al., 2000). A direct regulation was further confirmed by binding of PPARγ and RXRα complex to the AQP7 PPRE (Kishida et al., 2001). Further studies discovered that PPARα deficiency in mice blocked the increase in AQP7 transcript levels in AT triggered by fasting (Walker, Holness, Gibbons, & Sugden, 2007). Hormonal regulation of AQP7 by estrogens was also demonstrated in adipocytes. Accordingly, ovarectomized mice had decreased AQP7 expression in visceral fat which was restored by estradiol supplementation (Jin et al., 2017).

AQP3, another aquaglyceroporin, is also expressed in subcutaneous AT and primary adipocytes (Rodríguez, Catalán, Gómez-Ambrosi, García-Navarro, et al., 2011). Insulin stimulation significantly increases AQP3,

-7 and -9 expression in human adipocytes, whereas leptin elicits opposite effects via PI3K signaling (Rodríguez, Catalán, Gómez-Ambrosi, García-Navarro, et al., 2011). Leptin deficiency in ob/ob mice was strongly accompanied by increased AQP3 and AQP7 in WAT, without changes in hepatic AQP9. When exogenous leptin was administrated in ob/ob mice, it subsequently downregulated AQP3 and AQP7 in AT while upregulating hepatic AQP9. Thus, restoring AQPs regulation and glycerol shuttling between liver and AT might be important in preventing lipid overaccumulation in these two organs (Rodríguez, Moreno, et al., 2015). In obese Wistar rats subjected to sleeve gastrectomy, AQP7 and AQP3 expression was increased in epididymal and subcutaneous fat, whereas hepatic AQP9 remained unchanged, contributing to the reduced liver steatosis and weight loss (Méndez-Giménez et al., 2015). AQP9 was also shown to cooperate with AQP3 during lipolysis to facilitate glycerol efflux (Rodríguez, Catalán, Gómez-Ambrosi, & Frühbeck, 2011). Since PPARγ agonists repress leptin expression, molecular studies are still needed to understand the effects of PPARγ vs leptin on aquaporin regulation (De Vos et al., 1996).

Importantly, AQP10, another aquaglyceroporin, is expressed in adipocytes as an alternative pathway for glycerol efflux (Laforenza, Scaffino, & Gastaldi, 2013). AQP10 might also take part in glycerol metabolism and be connected to metabolic dysfunctions such as obesity (Laforenza et al., 2013). A recent study demonstrated that the pH decrease during lipolysis correlates with glycerol release and stimulation of AQP10 expression, suggesting that targeting of AQP10 may offer an attractive option for treating obesity (Gotfryd, Mósca, Missel, & Truelsen, 2018). However, studies on aquaporin 10 gene regulation including PPARs and adipokines such as leptin are still awaited in order to understand its role in AT glycerol transport (Laforenza et al., 2013). AQP11 was also discovered in subcutaneous and visceral human mature adipocytes with a preferential localization in proximity of lipid droplets (Madeira, Fernández-Veledo, et al., 2014). However further studies are needed to assess its function and confirm its subcellular localization.

Fig. 1 summarizes the known AQPs expressed in AT and their role in glycerol homeostasis during lipogenesis/lipolysis.

3. Aquaporins in the digestive system

AQPs play important physiological role in the digestive system. For instance AQP3 and -4 are highly expressed in the stomach and contribute

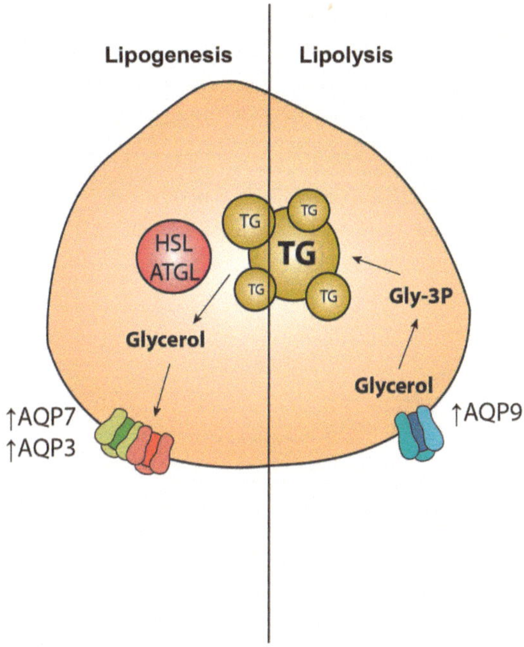

Fig. 1 AQPs expressed in AT and their role in glycerol homeostasis during lipogenesis/lipolysis. AQP9 expression is induces in lipolysis where increased amount of glycerol from peripheral tissues is diffused into the adipocyte and stored in form of TG. During lipogenesis instead TG hydrolysis by ATGL and HSL generates FA and glycerol which is then shuttled through AQP3/7 outside the adipocytes.

to water secretion, whereas AQP3, -4 and -8 are present in the large intestine, and AQP7 and -9 in the liver (Eto, Noda, Horikawa, Uchida, & Sasaki, 2010; Farinas, Kneen, Moore, & Verkman, 1997). In the pancreas, AQP1 is expressed in capillaries and ductal cells, AQP5 in the apical membrane of ductal cells and acinar cells, whereas AQP12 is expressed but with unknown cellular localization (Isokpehi, Rajnarayanan, Jeffries, Oyeleye, & Cohly, 2009; Ohta et al., 2009). In the following, we describe important AQPs and their functions in various organs of the digestive system in more detail.

3.1 Aquaporins in the liver

Aquaporin 1, -3, -7, -8, -9 are all expressed in human liver, but with different cellular distribution (Gregoire et al., 2015). AQP3 localizes in Kupffer cells, AQP7 in hepatocytes and AQP9 in cholangiocytes (Gregoire et al., 2015; Laforenza et al., 2005). AQP9 is particularly abundant in hepatocytes, where it plays a key role in hepatic glycerol and glucose metabolism

(Rojek et al., 2007). AQP9 favors glycerol influx into hepatocytes during fasting AQP9 (Carbrey et al., 2003). Therefore, it has been proposed that glycerol released from adipose tissue through AQP7 is taken up by the liver via AQP9 for gluconeogenesis (Kuriyama et al., 2002). The coordination of these two aquaporins, located in adipose tissue and liver, respectively, seems to be pivotal in glucose metabolism under physiological conditions as well as in obesity and T2DM (Kuriyama et al., 2002) (Fig. 2). Coordinated glycerol release and uptake mediated by AQPs controlled by PPARα and PPARγ in

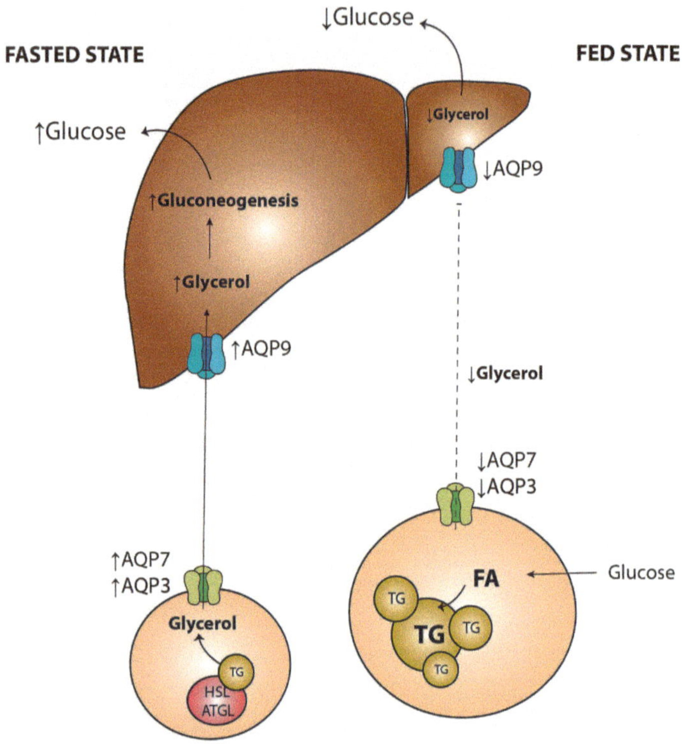

Fig. 2 Regulation of adipose AQP3/AQP7 and liver AQP9 during fasting and feeding states. Fasted state induces lipolysis in adipocytes and in turn gluconeogenesis in liver (left side), AQP7 and AQP3 expression upregulates in AT facilitating glycerol release into the blood stream. In line, AQP9 increases in the liver in order to transport glycerol in the portal vein. In the liver, glycerol is used as substrate for gluconeogenesis, increasing glucose levels in the blood. In fed state (right side), the abundance of nutrients triggers a metabolic switch from lipolysis to lipogenesis in adipocytes, suppressing glucose production from liver. Therefore, adipose AQP3 and AQP7 decrease and FA are stored in form of TG. Circulating glycerol levels are also decreased in parallel with liver AQP9 and glycerol uptake.

liver and adipose tissue, respectively, reinforces the central role of PPARs in regulating whole body lipid homeostasis, making sense of their activation by similar ligands.

In addition to lipid regulation, hepatic AQPs crucially contribute to water transport and bile formation. Bile emulsifies dietary lipids and its production is a complex process of concerted canalicular water secretion in response to osmotic gradients originated by active transport of solutes (Boyer, 2013). Bile flow is strictly dependent on water secretion in the canalicular lumen, driven by AQPs. In fact, a role of AQP8 in biliary water transport was shown in hepatocytes. Silencing of AQP8 in HepG2 cells by RNA interference results in a significant reduction of canalicular volume in AQP8, followed by reduced water transport in response to several stimuli such as inward osmotic gradient and bile secretory agonist. These data suggest that AQP8 is the rate-limiting channel aimed at water secretion in canalicular membranes during bile secretion (Larocca, Soria, Espelt, Lehmann, & Marinelli, 2009). Moreover, AQP8 canalicular expression and translocation is strongly induced by glucagon via protein kinase A increasing hepatocyte membrane water permeability (Gradilone et al., 2003). In contrast to AQP8, AQP9 is involved in the sinusoidal uptake of water (Bernardino et al., 2016).

AQPs have also been demonstrated to play diverse roles in a cell-specific manner. Hepatocytes express AQP0, -8, 9, -11 (Elkjar et al., 2000; Huebert, Splinter, Garcia, Marinelli, & Larusso, 2002) mediating water transport and glycerol shuttling. Expression of AQP9 in hepatocytes is regulated by PPARα, which preferentially directs glycerol shuttling into lipogenesis rather than gluconeogenesis (Lebeck et al., 2015). AQPs are expressed at both apical and basolateral plasma membrane in cholangiocytes. In particular, AQP1 is secretin responsive and responsible for the apical transport of water. In contrast, AQP4 expression is restricted to the basolateral membrane of cholangiocytes and non-responsive to secretin (Marinelli et al., 1999; Marinelli, Pham, Tietz, & LaRusso, 2000; Masyuk, Marinelli, & LaRusso, 2002). However, deletion of AQP1 in mice did not change bile flow and bile salts concentration, but inhibited dietary fat absorption. In contrast, AQP1 seems to be pivotal in ductal bile formation in rats (Ma et al., 2001; Splinter, Masyuk, & LaRusso, 2003).

We and others elucidated the importance of AQPs in activation, apoptosis and quiescence of hepatic stellate cells (HSC). Interestingly, only quiescent HSCs express AQP0, -1, -5, -8, -9, -12. The changes in AQP expression is associated to increased resistance to apoptosis in activated

HSCs (Lakner, Walling, McKillop, & Schrum, 2011). We demonstrated that AQP3 expression is driven by adiponectin stimulation in LX2 hepatic stellate cells (Tardelli, Moreno-Viedma, et al., 2017). This regulation was mediated by PPARα signaling (Tardelli, Claudel, Bruschi, Moreno-Viedma, & Trauner, 2017), which exerts a profound effect on glycerol uptake, lipogenesis and lipid storage in HSC. Moreover, PPARα activation also downregulates HSC activation and fibrosis markers. We further explored the role of AQP3 in primary HSCs isolated from patients with NASH, showing an up-regulation of its protein and gene expression, which was proportional to the fibrosis stage and regulated by PPARγ through the JNK pathway (Tardelli, Bruschi, et al., 2017).

3.2 Aquaporin in pancreas

The exocrine pancreas has the capacity to secrete large amounts of digestive fluids with digestive enzymes into the duodenum. AQPs are particularly important to generate the osmotic gradient of the pancreatic juice (Lee, Ohana, Park, Yang, & Muallem, 2012). AQP12 is expressed in pancreatic acinar cells, and AQP12 KO mice are less resistant to cholecystokinin-8 (CCK-8) induced pancreatitis (Ohta et al., 2009). In contrast, AQP8 KO mice display normal pancreatic function as do AQP1 and AQP5 KO probably due to redundant functions (Delporte, 2014). Interestingly, AQP7 is expressed in the β-cells of the islets of Langerhans in both rats and mice. AQP7 deficiency was found to be associated with impairment of β-cell swelling and changes in membrane potential in response to glycerol administration, thus proposing an alternative mechanism for AQP7 influencing insulin secretion (Ma et al., 2001). Recently, the pancreatic AQP7 expression was reported to be increased in HFD fed obese rats. When evaluating the effect of sleeve gastrectomy as a surgical intervention for the treatment of obesity, a further increase in pancreatic AQP7 expression was observed. Altogether, the role of pancreatic AQP7 expression in the regulation of insulin secretion is still to be resolved and further studies are needed.

3.3 Aquaporins in stomach and intestine

The stomach expresses mRNA for AQP1, -3, -4, -5, -7, -8, -10 and -11 (Laforenza, 2012). In one study, AQP4 was successfully knocked out in mice, however, without generating a clear phenotype or changes in acid and fluid secretion (Wang et al., 2000). This led to the conclusion that AQP4 is unlikely to play a major role in gastric acid production.

Moreover, AQP5 was found in rat lower stomach with localization in pyloric glands and duodenum (Parvin, Tsumura, Akamatsu, Kanamori, & Hosoi, 2002). In addition, AQP3 is expressed in the stomach of rats (Koyama et al., 1999). Importantly, AQP3 and -4 are correlated with stomach pathologies including chronic gastritis and spleen-stomach dampness-heat syndrome in a Chinese cohort indicating an impact of AQPs in human stomach disease (Mei et al., 2010).

As the intestine is the major site for digestion and absorption of nutrients, a rapid bidirectional movement of water is key for secretion and absorption in this organ. AQP9 is expressed in duodenum, jejunum, ileum, colon but not stomach. Interestingly, AQP9's expression is limited to the basolateral membrane of goblet cells which are responsible for mucus secretion (Okada, Misaka, Matsumoto, Watanabe, & Abe, 2003). Mucus in turn is of great importance to protect the intestinal surface and guarantee the proper flow of intestinal content. Also AQP7, -8, -10 and -11 are expressed in the small and large intestine localizing on the apical membrane of enterocytes (Zhu, Chen, & Jiang, 2016). Notably, AQPs in colon were recently connected to pathologies such as constipation and diarrhea (Ikarashi, Kon, & Sugiyama, 2016). For instance, AQP3 is expressed in the mucosal epithelial cells and is downregulated together with AQP8 and AQP1 in Chron's disease and ulcerative colitis patients (Zahn et al., 2007). AQP3 expression also correlates with constipation (Kon et al., 2015) and changes in water content in the stools following osmotic laxatives subministration (Ikarashi et al., 2011). In fact, the protein expression levels of AQP3 in the colon increased dramatically over time following the administration of $MgSO_4$ highlighting a mechanistic response of AQP3 to treatment (Ikarashi et al., 2011).

4. Aquaporins regulation in kidney

Water and solute trafficking/reabsorption are pivotal kidney functions. AQP1 and -7 are expressed in the proximal tubules and AQP2, -3, -4 localize to the collecting duct system (Agarwal & Gupta, 2008). AQP2 is the most studied AQP in kidney with several studies showing its regulation by vasopressin (Bouley et al., 2006; Deen et al., 1994; Wilson, Miranda, & Knepper, 2013). AQP2 KO mice do not survive to adulthood because of excessive fluid loss (Rojek, Fuchtbauer, Kwon, Frokiaer, & Nielsen, 2006). AQP2 knockout restricted to kidney connecting tubule evidenced its essential role in regulation of body water balance that cannot be compensated for by other mechanisms (Kortenoeven, Pedersen, Miller, Rojek, & Fenton, 2013; Rojek et al., 2006). In another study, authors studied -F204V mutation in mice (which is due to thymine to guanine (T–G) transversion, leading to a substitution of a valine with phenylalanine at amino acid 204 of the protein) and showed improper AQP2 translocation in connection to the pathogenesis of nephrogenic diabetes insipidus (Lloyd, Hall, Tarantino, & Gekakis, 2005). AQP6 is expressed in collecting ducts, however, localizing on the intracellular vesicles but not the plasma membrane (Yasui et al., 1999). AQP7 and -8 localize to the proximal tubule brush border, whereas AQP9 is found only at gene but not protein level in the kidney (Nielsen et al., 2002). Notably, AQP7 is not involved in water reabsorption, however it plays an important role in glycerol reabsorption in the kidney. Consequently, AQP7 KO mice suffer marked loss of glycerol in the urine (glyceroluria) (Sohara et al., 2005; Sohara, Rai, Sasaki, & Uchida, 2006). AQP11 is localized intracellularly in the endoplasmic reticulum of proximal tubule cells. Its importance is highlighted by the fact that AQP11 KO mice develop polycystic kidney disease (Okada et al., 2008).

5. Aquaporins in muscle

AQP7 expression was found in skeletal and cardiac muscle. In skeletal muscle, AQP7 and AQP4 are localized to the sarcolemma and are responsible for water influx into the muscle during intense activity (Crosbie et al., 2002; Wakayama et al., 2004). In obese leptin deficient mice, AQP7, but not AQP9, expression in skeletal muscle is increased compared to WT at RNA and protein level (Wakayama, Hirako, Ogawa, Jimi, & Shioda, 2014). In line, a recent study confirmed these preclinical data in obese patients with T2D, showing that also in obese men AQP7 expression is increased in skeletal muscle contributing to the excess lipid accumulation (Lebeck et al., 2018). Of interested, the heart is known to require FA and glycerol as energy substrates and AQP7 seems to be the most abundant AQP in cardiomyocytes (Hibuse et al., 2009). AQP7 KO mice exhibit a significant reduction in cardiac glycerol consumption and ATP content. Accordingly, knockdown of AQP7 in H9C2 cardiomyotubes significantly decreases glycerol uptake (Hibuse et al., 2009). This demonstrates the key role for AQP7 in controlling cardiac glycerol metabolism and ATP production (Gladka, El Azzouzi, De Windt, & Da Costa Martins, 2009; Hibuse et al., 2009). Moreover, when AQP7 KO are treated with isoproterenol to induce vasodilation and pressure overload (cardiac stress), mice developed severe left ventricular hypertrophy and higher mortality (Hibuse et al., 2009). Altogether, these studies support a significant role for AQP7 in controlling muscular and particularly cardiac glycerol metabolism.

6. Clinical relevance of AQPs

Clinical relevance of AQPs has been evaluated in several studies. Twenty-four genetic variants were identified for AQP4 including four novel single nucleotide polymorphism (I128T, D184E, I205L, M224T) (Sorani et al., 2008) showing that the SNP M278T results in a gain of function whereas the other four reduce water permeability (Sorani et al., 2008). Interestingly, a significant association between AQP7 and lipogenic/lipolytic gene expression in obese human AT was found, with a strong increase of AQP7 gene expression in VAT from T2D obese subjects (Miranda et al., 2010). In line, another study demonstrated a strong increase of AQP7 expression in VAT of obese patients, accompanied by a decrease of hepatic AQP9 in obese T2D (Catalán et al., 2008). Importantly, subjects with a genetic variant of AQP7 called (-953G) showed increased risk to develop T2D and/or obesity in a study with 977 Caucasian subjects (Prudente et al., 2007).

In contrast, AQP1, -8, -3 and -7 were found to be reduced in IBD patients, with expression changes in Crohn's disease and ulcerative colitis (Ricanek et al., 2015). AQPs could possibly contribute to cancer development as their expression has been shown to fluctuate during the G1 phase (Galán-Cobo, Ramírez-Lorca, & Echevarría, 2016), with AQP2 slowing the transit time through S and G2/M phases of the cell cycle (Rivarola et al., 2010). For instance, Chen et al. discovered an up-regulation of AQP3 and concomitant downregulation of AQP7/AQP9 in hepatocellular carcinoma, correlating with tumor grade and metastasis (Chen, Li, Lü, & Mei, 2016). In another report, AQP9 expression was shown to be decreased in NASH patients translating its down-regulation into a compensatory mechanism against TG accumulation (Rodríguez et al., 2014). Gender was also found to impact AQPs biology: AQP9 expression in hepatocytes was documented to be higher in males than in females (Rodríguez, Marinelli, Tesse, Frühbeck, & Calamita, 2015). In another study after 90 min exercise women had higher levels of plasma glycerol as compared to men, probably due to reduced carbohydrate and increased lipid utilization (Davis, Shavers, & Costa, 2000). This difference might be explained by higher expression of AQP7 in visceral fat in women, or by their physiologically higher fat mass compared to men. AQPs were proposed as interesting targets in oncology with special emphasis on colorectal, lung, liver and breast cancers, although their therapeutic potentials still remain questionable (Wang et al., 2015). Gold based compounds act as modulators of AQPs (Calamita, Perret, & Delporte, 2018). Auphen for instance selectively blocks AQP3′s glycerol permeability (Martins et al., 2012) and also inhibits mouse and human AQP7 (Madeira, De Almeida, et al., 2014). PPARγ agonists such as thiazolidinediones increase AQP7 expression in AT (Kishida et al., 2001) whereas metformin does alter AQP3 expression in vitro (Asai et al., 2006).

7. Conclusions and future directions

A large body of evidence connects aquaglyceroporins to liver, AT and pancreas metabolism in which they control energy and therefore glucose and lipid homeostasis. Promising AQP inhibitors have been recently identified whereas others are subject to ongoing research, making AQPs still elusive pharmacological targets (Verkman et al., 2014). However, their complex tissue specific functions have to be taken into consideration. Therapeutic activation of AQPs expression was demonstrated for rosiglitazone in kidney (Procino et al., 2015) and liver (Tardelli, Bruschi, et al., 2017; Tardelli, Claudel, et al., 2017; Tardelli, Moreno-Viedma, et al., 2017), whereas statins and fibrates were tested with regards to AQP4 expression (Pierno et al., 2006) in preclinical models. Recently, promising mathematical models were developed to better decipher AQPs function in multiple organs, which may become useful tools to understand the complex dynamics of glycerol, glucose and energy homeostasis in metabolic disorders (Gena et al., 2017). AQPs coordination in liver and AT is driven by PPARα and PPARγ, respectively (Patsouris et al., 2004), and many AQPs are regulated by both PPARs as well as other NRs such as FXR (Zhang et al., 2014) or aryl hydrocarbon receptor (Bui et al., 2016). Efforts should be made to develop selective blocking or agonistic strategies in chronic disorders such as obesity, fatty liver or skin diseases as well as diarrhea and constipation in which evidence suggests important roles in glycerol and water trafficking.

References

Abbott, B. D. (2009). Review of the expression of peroxisome proliferator-activated receptors alpha (PPAR alpha), beta (PPAR beta), and gamma (PPAR gamma) in rodent and human development. *Reproductive Toxicology*, *27*, 246–257. https://doi.org/10.1016/j.reprotox.2008.10.001.

Agarwal, S., & Gupta, A. (2008). Aquaporins: The renal water channels. *Indian Journal of Nephrology*, *18*, 95–100. https://doi.org/10.4103/0971-4065.43687.

Agre, P., & Kozono, D. (2003). Aquaporin water channels: Molecular mechanisms for human diseases. *FEBS Letters, 555*, 72–78. https://doi.org/10.1016/S0014-5793(03)01083-4.

Asai, M., Higuchi, S., Kubota, M., Iguchi, K., Usui, S., & Hirano, K. (2006). Regulators for blood glucose level affect gene expression of aquaporin 3. *Biological & Pharmaceutical Bulletin, 29*, 991–996. https://doi.org/10.1248/bpb.29.991.

Bernardino, R., Marinelli, R., Maggio, A., Gena, P., Cataldo, I., Alves, M., et al. (2016). Hepatocyte and sertoli cell aquaporins, recent advances and research trends. *International Journal of Molecular Sciences, 17*, 1096. https://doi.org/10.3390/ijms17071096.

Berry, V., Francis, P., Kaushal, S., Moore, A., & Bhattacharya, S. (2000). Missense mutations in MIP underlie autosomal dominant "polymorphic" and lamellar cataracts linked to 12q. *Nature Genetics, 25*, 15–17. https://doi.org/10.1038/75538.

Bouley, R., Hawthorn, G., Russo, L. M., Lin, H. Y., Ausiello, D. A., & Brown, D. (2006). Aquaporin 2 (AQP2) and vasopressin type 2 receptor (V2R) endocytosis in kidney epithelial cells: AQP2 is located in 'endocytosis-resistant' membrane domains after vasopressin treatment. *Biology of the Cell, 98*, 215–232. https://doi.org/10.1042/BC20040054.

Boyer, J. L. (2013). Bile formation and secretion. *Comprehensive Physiology, 3*, 1035–1078. https://doi.org/10.1002/cphy.c120027.

Bui, L. C., Tomkiewicz, C., Pierre, S., Chevallier, A., Barouki, R., & Coumoul, X. (2016). Regulation of aquaporin 3 expression by the AhR pathway is critical to cell migration. *Toxicological Sciences, 149*, 158–166. https://doi.org/10.1093/toxsci/kfv221.

Calamita, G., Perret, J., & Delporte, C. (2018). Aquaglyceroporins: Drug targets for metabolic diseases? *Frontiers in Physiology, 9*, 851. https://doi.org/10.3389/fphys.2018.00851.

Carbrey, J. M., Gorelick-Feldman, D. A., Kozono, D., Praetorius, J., Nielsen, S., & Agre, P. (2003). Aquaglyceroporin AQP9: Solute permeation and metabolic control of expression in liver. *Proceedings of the National Academy of Sciences, 100*, 2945–2950. https://doi.org/10.1073/pnas.0437994100.

Catalán, V., Gómez-Ambrosi, J., Pastor, C., Rotellar, F., Silva, C., Rodríguez, A., et al. (2008). Influence of morbid obesity and insulin resistance on gene expression levels of AQP7 in visceral adipose tissue and AQP9 in Liver. *Obesity Surgery, 18*, 695–701. https://doi.org/10.1007/s11695-008-9453-7.

Chen, X.-F., Li, C.-F., Lü, L., & Mei, Z.-C. (2016). Expression and clinical significance of aquaglyceroporins in human hepatocellular carcinoma. *Molecular Medicine Reports, 13*, 5283–5289. https://doi.org/10.3892/mmr.2016.5184.

Choe, S. S., Huh, J. Y., Hwang, I. J., Kim, J. I., & Kim, J. B. (2016). Adipose tissue remodeling: Its role in energy metabolism and metabolic disorders. *Frontiers in Endocrinology (Lausanne), 7*, 30. https://doi.org/10.3389/fendo.2016.00030.

Conner, M. T., Conner, A. C., Bland, C. E., Taylor, L. H. J., Brown, J. E. P., Parri, H. R., et al. (2012). Rapid aquaporin translocation regulates cellular water flow: Mechanism of hypotonicity-induced subcellular localization of aquaporin 1 water channel. *The Journal of Biological Chemistry, 287*, 11516–11525. https://doi.org/10.1074/jbc.M111.329219.

Crosbie, R. H., Dovico, S. A., Flanagan, J. D., Chamberlain, J. S., Ownby, C. L., & Campbell, K. P. (2002). Characterization of aquaporin-4 in muscle and muscular dystrophy. *The FASEB Journal, 16*, 943–949. https://doi.org/10.1096/fj.01-0327com.

Davis, S. N., Shavers, C., & Costa, F. (2000). Differential gender responses to hypoglycemia are due to alterations in CNS drive and not glycemic thresholds. *American Journal of Physiology. Endocrinology and Metabolism, 279*, E1054–E1063. https://doi.org/10.1152/ajpendo.2000.279.5.E1054.

De Vos, P., Lefebvre, A. M., Miller, S. G., Guerre-Millo, M., Wong, K., Saladin, R., et al. (1996). Thiazolidinediones repress ob gene expression in rodents via activation of peroxisome proliferator-activated receptor γ. *The Journal of Clinical Investigation, 98*, 1004–1009. https://doi.org/10.1172/JCI118860.

Deen, P., Verdijk, M., Knoers, N., Wieringa, B., Monnens, L., van Os, C., et al. (1994). Requirement of human renal water channel aquaporin-2 for vasopressin-dependent concentration of urine. *Science*, *264*, 92–95. https://doi.org/10.1126/science.8140421.

Delporte, C. (2014). Aquaporins in salivary glands and pancreas. *Biochimica et Biophysica Acta, General Subjects*, *1840*, 1524–1532. https://doi.org/10.1016/j.bbagen.2013.08.007.

Dong, C., Wang, G., Li, B., Xiao, K., Ma, Z., Huang, H., et al. (2012). Anti-asthmatic agents alleviate pulmonary edema by upregulating AQP1 and AQP5 expression in the lungs of mice with OVA-induced asthma. *Respiratory Physiology & Neurobiology*, *181*, 21–28. https://doi.org/10.1016/j.resp.2011.12.008.

Elkjar, M. L., Vajda, Z., Nejsum, L. N., Kwon, T. H., Jensen, U. B., Amiry-Moghaddam, M., et al. (2000). Immunolocalization of AQP9 in liver, epididymis, testis, spleen, and brain. *Biochemical and Biophysical Research Communications*, *276*, 1118–1128. https://doi.org/10.1006/bbrc.2000.3505.

Eto, K., Noda, Y., Horikawa, S., Uchida, S., & Sasaki, S. (2010). Phosphorylation of aquaporin-2 regulates its water permeability. *The Journal of Biological Chemistry*, *285*, 40777–40784. https://doi.org/10.1074/jbc.M110.151928.

Farinas, J., Kneen, M., Moore, M., & Verkman, A. S. (1997). Plasma membrane water permeability of cultured cells and epithelia measured by light microscopy with spatial filtering. *The Journal of General Physiology*, *110*, 283–296. https://doi.org/10.1085/jgp.110.3.283.

Fukuda, A. M., & Badaut, J. (2012). Aquaporin 4: A player in cerebral edema and neuroinflammation. *Journal of Neuroinflammation*, *9*, 279. https://doi.org/10.1186/1742-2094-9-279.

Galán-Cobo, A., Ramírez-Lorca, R., & Echevarría, M. (2016). Role of aquaporins in cell proliferation: What else beyond water permeability? *Channels*, *10*, 185–201. https://doi.org/10.1080/19336950.2016.1139250.

Gena, P., Del Buono, N., D'Abbicco, M., Mastrodonato, M., Berardi, M., Svelto, M., et al. (2017). Dynamical modeling of liver aquaporin-9 expression and glycerol permeability in hepatic glucose metabolism. *European Journal of Cell Biology*, *96*, 61–69. https://doi.org/10.1016/j.ejcb.2016.12.003.

Gena, P., Mastrodonato, M., Portincasa, P., Fanelli, E., Mentino, D., Rodríguez, A., et al. (2013). Liver glycerol permeability and aquaporin-9 are dysregulated in a murine model of non-alcoholic fatty liver disease. *PLoS One*, *8*, e78139. https://doi.org/10.1371/journal.pone.0078139.

Gladka, M., El Azzouzi, H., De Windt, L. J., & Da Costa Martins, P. A. (2009). Aquaporin 7: The glycerol aquaeductus in the heart. *Cardiovascular Research*, *83*, 3–4. https://doi.org/10.1093/cvr/cvp147.

Goodwin, B., Jones, S. A., Price, R. R., Watson, M. A., McKee, D. D., Moore, L. B., et al. (2000). A regulatory cascade of the nuclear receptors FXR, SHP-1, and LRH-1 represses bile acid biosynthesis. *Molecular Cell*, *6*, 517–526. https://doi.org/10.1016/S1097-2765(00)00051-4.

Gotfryd, K., Mósca, A. F., Missel, J. W., & Truelsen, S. F. (2018). Human adipose glycerol flux is regulated by a pH gate in AQP10. *Nature Communications*, *9*, 4749. https://doi.org/10.1038/s41467-018-07176-z.

Gradilone, S. A., García, F., Huebert, R. C., Tietz, P. S., Larocca, M. C., Kierbel, A., et al. (2003). Glucagon induces the plasma membrane insertion of functional aquaporin-8 water channels in isolated rat hepatocytes. *Hepatology*, *37*, 1435–1441. https://doi.org/10.1053/jhep.2003.50241.

Gregoire, F., Lucidi, V., Zerrad-Saadi, A., Virreira, M., Bolaky, N., Delforge, V., et al. (2015). Analysis of aquaporin expression in liver with a focus on hepatocytes. *Histochemistry and Cell Biology*, *144*, 347–363. https://doi.org/10.1007/s00418-015-1341-3.

Hara-Chikuma, M., Sohara, E., Rai, T., Ikawa, M., Okabe, M., Sasaki, S., et al. (2005). Progressive adipocyte hypertrophy in aquaporin-7-deficient mice: Adipocyte glycerol permeability as a novel regulator of fat accumulation. *The Journal of Biological Chemistry*, *280*, 15493–15496. https://doi.org/10.1074/jbc.C500028200.

Hibuse, T., Maeda, N., Funahashi, T., Yamamoto, K., Nagasawa, A., Mizunoya, W., et al. (2005). From the cover: Aquaporin 7 deficiency is associated with development of obesity through activation of adipose glycerol kinase. *Proceedings of the National Academy of Sciences*, *102*, 10993–10998. https://doi.org/10.1073/pnas.0503291102.

Hibuse, T., Maeda, N., Nagasawa, A., & Funahashi, T. (2006). Aquaporins and glycerol metabolism. *Biochimica et Biophysica Acta, Biomembranes*, *1758*, 1004–1011. https://doi.org/10.1016/j.bbamem.2006.01.008.

Hibuse, T., Maeda, N., Nakatsuji, H., Tochino, Y., Fujita, K., Kihara, S., et al. (2009). The heart requires glycerol as an energy substrate through aquaporin 7, a glycerol facilitator. *Cardiovascular Research*, *83*, 34–41. https://doi.org/10.1093/cvr/cvp095.

Hsu, M. S., Lee, D. J., & Binder, D. K. (2007). Potential role of the glial water channel aquaporin-4 in epilepsy. *Neuron Glia Biology*, *3*, 287–297. https://doi.org/10.1017/S1740925X08000112.

Huebert, R. C., Splinter, P. L., Garcia, F., Marinelli, R. A., & Larusso, N. F. (2002). Expression and localization of aquaporin water channels in rat hepatocytes. Evidence for a role in canalicular bile secretion. *The Journal of Biological Chemistry*, *277*, 22710–22717. https://doi.org/10.1074/jbc.M202394200.

Ikarashi, N., Kon, R., & Sugiyama, K. (2016). Aquaporins in the colon as a new therapeutic target in diarrhea and constipation. *International Journal of Molecular Sciences*, *17*, 1172. https://doi.org/10.3390/ijms17071172.

Ikarashi, N., Ushiki, T., Mochizuki, T., Toda, T., Kudo, T., Baba, K., et al. (2011). Effects of magnesium sulphate administration on aquaporin 3 in rat gastrointestinal tract. *Biological & Pharmaceutical Bulletin*, *34*, 238–242. https://doi.org/10.1248/bpb.34.238.

Isokpehi, R. D., Rajnarayanan, R. V., Jeffries, C. D., Oyeleye, T. O., & Cohly, H. H. P. (2009). Integrative sequence and tissue expression profiling of chicken and mammalian aquaporins. *BMC Genomics*, *10*, S7. https://doi.org/10.1186/1471-2164-10-S2-S7.

Jacobi, D., Stanya, K. J., & Lee, C.-H. (2012). Adipose tissue signaling by nuclear receptors in metabolic complications of obesity. *Adipocytes*, *1*, 4–12. https://doi.org/10.4161/adip.19036.

Jarius, S., & Wildemann, B. (2010). AQP4 antibodies in neuromyelitis optica: Diagnostic and pathogenetic relevance. *Nature Reviews. Neurology*, *6*, 383–392. https://doi.org/10.1038/nrneurol.2010.72.

Jin, B., Chen, X., Xing, L., Xu, W., Fu, X., Zhu, J., et al. (2017). Tissue-specific effects of estrogen on glycerol channel aquaporin 7 expression in an ovariectomized mouse model of menopause. *Climacteric*, *20*, 385–390. https://doi.org/10.1080/13697137.2017.1319920.

Kajimura, S. (2017). Adipose tissue in 2016: Advances in the understanding of adipose tissue biology. *Nature Reviews. Endocrinology*, *13*, 69–70. https://doi.org/10.1038/nrendo.2016.211.

Kalaany, N. Y., & Mangelsdorf, D. J. (2006). LXRS and FXR: The Yin and Yang of cholesterol and fat metabolism. *Annual Review of Physiology*, *68*, 159–191. https://doi.org/10.1146/annurev.physiol.68.033104.152158.

Kishida, K., Kuriyama, H., Funahashi, T., Shimomura, I., Kihara, S., Ouchi, N., et al. (2000). Aquaporin adipose, a putative glycerol channel in adipocytes. *The Journal of Biological Chemistry*, *275*, 20896–20902. https://doi.org/10.1074/jbc.M001119200.

Kishida, K., Shimomura, I., Nishizawa, H., Maeda, N., Kuriyama, H., Kondo, H., et al. (2001). Enhancement of the aquaporin adipose gene expression by a peroxisome proliferator-activated receptor gamma. *The Journal of Biological Chemistry*, *276*, 48572–48579. https://doi.org/10.1074/jbc.M108213200.

Kliewer, S. A., Lehmann, J. M., & Willson, T. M. (1999). Orphan nuclear receptors: Shifting endocrinology into reverse. *Science, 284*, 757–760. https://doi.org/10.1126/science.284.5415.757.

Kon, R., Ikarashi, N., Hayakawa, A., Haga, Y., Fueki, A., Kusunoki, Y., et al. (2015). Morphine-induced constipation develops with increased aquaporin-3 expression in the colon via increased serotonin secretion. *Toxicological Sciences, 145*, 337–347. https://doi.org/10.1093/toxsci/kfv055.

Kortenoeven, M. L. A., Pedersen, N. B., Miller, R. L., Rojek, A., & Fenton, R. A. (2013). Genetic ablation of aquaporin-2 in the mouse connecting tubules results in defective renal water handling. *The Journal of Physiology, 591*, 2205–2219. https://doi.org/10.1113/jphysiol.2012.250852.

Koyama, Y., Yamamoto, T., Tani, T., Nihei, K., Kondo, D., Funaki, H., et al. (1999). Expression and localization of aquaporins in rat gastrointestinal tract. *The American Journal of Physiology, 276*, C621–C627. 10069989.

Kuriyama, H., Shimomura, I., Kishida, K., Kondo, H., Furuyama, N., Nishizawa, H., et al. (2002). Coordinated regulation of fat-specific and liver-specific glycerol channels, aquaporin adipose and aquaporin 9. *Diabetes, 51*, 2915–2921. https://doi.org/10.2337/diabetes.51.10.2915.

Laforenza, U. (2012). Water channel proteins in the gastrointestinal tract. *Molecular Aspects of Medicine, 33*, 642–650. https://doi.org/10.1016/j.mam.2012.03.001.

Laforenza, U., Gastaldi, G., Grazioli, M., Cova, E., Tritto, S., Faelli, A., et al. (2005). Expression and immunolocalization of aquaporin-7 in rat gastrointestinal tract. *Biology of the Cell, 97*, 605–613. https://doi.org/10.1042/BC20040090.

Laforenza, U., Scaffino, M. F., & Gastaldi, G. (2013). Aquaporin-10 represents an alternative pathway for glycerol efflux from human adipocytes. *PLoS One, 8*. e54474. https://doi.org/10.1371/journal.pone.0054474.

Lakner, A. M., Walling, T. L., McKillop, I. H., & Schrum, L. W. (2011). Altered aquaporin expression and role in apoptosis during hepatic stellate cell activation. *Liver International, 31*, 42–51. https://doi.org/10.1111/j.1478-3231.2010.02356.x.

Larocca, M. C., Soria, L. R., Espelt, M. V., Lehmann, G. L., & Marinelli, R. A. (2009). Knockdown of hepatocyte aquaporin-8 by RNA interference induces defective bile canalicular water transport. *American Journal of Physiology. Gastrointestinal and Liver Physiology, 296*, G93–100. https://doi.org/10.1152/ajpgi.90410.2008.

Lebeck, J. (2014). Metabolic impact of the glycerol channels AQP7 and AQP9 in adipose tissue and liver. *Journal of Molecular Endocrinology, 52*, R165–R178. https://doi.org/10.1530/JME-13-0268.

Lebeck, J., Cheema, M. U., Skowronski, M. T., Nielsen, S., & Praetorius, J. (2015). Hepatic AQP9 expression in male rats is reduced in response to PPARα agonist treatment. *American Journal of Physiology. Gastrointestinal and Liver Physiology, 308*, G198–G205. https://doi.org/10.1152/ajpgi.00407.2013.

Lebeck, J., Søndergaard, E., & Nielsen, S. (2018). Increased AQP7 abundance in skeletal muscle from obese men with type 2 diabetes. *American Journal of Physiology. Endocrinology and Metabolism, 315*, E367–E373. https://doi.org/10.1152/ajpendo.00468.2017.

Lee, M. G., Ohana, E., Park, H. W., Yang, D., & Muallem, S. (2012). Molecular mechanism of pancreatic and salivary gland fluid and HCOFormula secretion. *Physiological Reviews, 92*, 39–74. https://doi.org/10.1152/physrev.00011.2011.

Li, H., Kamiie, J., Morishita, Y., Yoshida, Y., Yaoita, E., Ishibashi, K., et al. (2005). Expression and localization of two isoforms of AQP10 in human small intestine. *Biology of the Cell, 97*, 823–829. https://doi.org/10.1042/BC20040091.

Lloyd, D. J., Hall, F. W., Tarantino, L. M., & Gekakis, N. (2005). Diabetes insipidus in mice with a mutation in aquaporin-2. *PLoS Genetics, 1*, e20. https://doi.org/10.1371/journal.pgen.0010020.

Ma, T., Jayaraman, S., Wang, K. S., Song, Y., Yang, B., Li, J., et al. (2001). Defective dietary fat processing in transgenic mice lacking aquaporin-1 water channels. *American Journal of Physiology. Cell Physiology*, *280*, C126–C134.

Madeira, A., De Almeida, A., De Graaf, C., Camps, M., Zorzano, A., Moura, T. F., et al. (2014). A gold coordination compound as a chemical probe to unravel aquaporin-7 function. *Chembiochem*, *15*, 1487–1494. https://doi.org/10.1002/cbic.201402103.

Madeira, A., Fernández-Veledo, S., Camps, M., Zorzano, A., Moura, T. F., Ceperuelo-Mallafré, V., et al. (2014). Human aquaporin-11 is a water and glycerol channel and localizes in the vicinity of lipid droplets in human adipocytes. *Obesity*, *22*, 2010–2017. https://doi.org/10.1002/oby.20792.

Maeda, N., Funahashi, T., Hibuse, T., Nagasawa, A., Kishida, K., Kuriyama, H., et al. (2004). Adaptation to fasting by glycerol transport through aquaporin 7 in adipose tissue. *Proceedings of the National Academy of Sciences of the United States of America*, *101*, 17801–17806. https://doi.org/10.1073/pnas.0406230101.

Maeda, N., Hibuse, T., & Funahashi, T. (2009). Role of aquaporin-7 and aquaporin-9 in glycerol metabolism; involvement in obesity. *Handbook of Experimental Pharmacology*, *190*, 233–249. https://doi.org/10.1007/978-3-540-79885-9_12.

Manley, G. T., Zador, Z., Stiver, S., & Wang, V. (2009). Role of aquaporin-4 in cerebral edema and stroke. *Handbook of Experimental Pharmacology*, *190*, 159–170. https://doi.org/10.1007/978-3-540-79885-9_7.

Marinelli, R. A., Pham, L. D., Tietz, P. S., & LaRusso, N. F. (2000). Expression of aquaporin-4 water channels in rat cholangiocytes. *Hepatology*, *31*, 1313–1317. https://doi.org/10.1053/jhep.2000.7986.

Marinelli, R. A., Tietz, P. S., Pham, L. D., Rueckert, L., Agre, P., & LaRusso, N. F. (1999). Secretin induces the apical insertion of aquaporin-1 water channels in rat cholangiocytes. *The American Journal of Physiology*, *276*, G280–G286.

Martins, A. P., Marrone, A., Ciancetta, A., Cobo, A. G., Echevarría, M., Moura, T. F., et al. (2012). Targeting aquaporin function: Potent inhibition of aquaglyceroporin-3 by a gold-based compound. *PLoS One*, *7*, e37435. https://doi.org/10.1371/journal.pone.0037435.

Massafra, V., & van Mil, S. W. C. (2018). Farnesoid X receptor: A "homeostat" for hepatic nutrient metabolism. *Biochimica et Biophysica Acta, Molecular Basis of Disease*, *1864*, 45–59. https://doi.org/10.1016/j.bbadis.2017.10.003.

Masyuk, A. I., Marinelli, R. A., & LaRusso, N. F. (2002). Water transport by epithelia of the digestive tract. *Gastroenterology*, *122*, 545–562. https://doi.org/10.1053/gast.2002.31035.

Mei, W. X., Lao, S. X., Yu, N., Zhou, Z., Huang, L. P., & Hu, B. (2010). Relationship between gene expressions of aquaporin 3 and 4 and various degrees of spleen-stomach dampness-heat syndrome in chronic superficial gastritis. *Chinese Journal of Integrative Medicine*, *8*, 111–115. https://doi.org/10.3736/jcim20100203.

Méndez-Giménez, L., Becerril, S., Moncada, R., Valentí, V., Ramírez, B., Lancha, A., et al. (2015). Sleeve gastrectomy reduces hepatic steatosis by improving the coordinated regulation of aquaglyceroporins in adipose tissue and liver in obese rats. *Obesity Surgery*, *25*, 1723–1734. https://doi.org/10.1007/s11695-015-1612-z.

Miranda, M., Escoté, X., Ceperuelo-Mallafré, V., Alcaide, M. J., Simón, I., Vilarrasa, N., et al. (2010). Paired subcutaneous and visceral adipose tissue aquaporin-7 expression in human obesity and type 2 diabetes: Differences and similarities between depots. *The Journal of Clinical Endocrinology and Metabolism*, *95*, 3470–3479. https://doi.org/10.1210/jc.2009-2655.

Morishita, Y., Matsuzaki, T., Hara-chikuma, M., Andoo, A., Shimono, M., Matsuki, A., et al. (2005). Disruption of aquaporin-11 produces polycystic kidneys following vacuolization of the proximal tubule. *Molecular and Cellular Biology*, *25*, 7770–7779. https://doi.org/10.1128/MCB.25.17.7770-7779.2005.

Nielsen, S., Frøkiær, J., Marples, D., Kwon, T.-H., Agre, P., & Knepper, M. A. (2002). Aquaporins in the kidney: From molecules to medicine. *Physiological Reviews*, *82*, 205–244. https://doi.org/10.1152/physrev.00024.2001.

Ohta, E., Itoh, T., Nemoto, T., Kumagai, J., Ko, S. B. H., Ishibashi, K., et al. (2009). Pancreas-specific aquaporin 12 null mice showed increased susceptibility to caerulein-induced acute pancreatitis. *American Journal of Physiology. Cell Physiology*, *297*, C1368–C1378. https://doi.org/10.1152/ajpcell.00117.2009.

Okada, S., Misaka, T., Matsumoto, I., Watanabe, H., & Abe, K. (2003). Aquaporin-9 is expressed in a mucus-secreting goblet cell subset in the small intestine. *FEBS Letters*, *540*, 157–162. https://doi.org/10.1016/S0014-5793(03)00256-4.

Okada, S., Misaka, T., Tanaka, Y., Matsumoto, I., Ishibashi, K., Sasaki, S., et al. (2008). Aquaporin-11 knockout mice and polycystic kidney disease animals share a common mechanism of cyst formation. *The FASEB Journal*, *22*, 3672–3684. https://doi.org/10.1096/fj.08-111872.

Olefsky, J. M. (2001). Nuclear receptor minireview series. *The Journal of Biological Chemistry*, *276*, 36863–36864. https://doi.org/10.1074/jbc.R100047200.

Parvin, M. N., Tsumura, K., Akamatsu, T., Kanamori, N., & Hosoi, K. (2002). Expression and localization of AQP5 in the stomach and duodenum of the rat. *Biochimica et Biophysica Acta, Molecular Cell Research*, *1542*, 116–124. https://doi.org/10.1016/S0167-4889(01)00172-0.

Patsouris, D., Mandard, S., Voshol, P. J., Escher, P., Tan, N. S., Havekes, L. M., et al. (2004). PPARα governs glycerol metabolism. *The Journal of Clinical Investigation*, *114*, 94–103. https://doi.org/10.1172/JCI200420468.

Petersen, M. C., Vatner, D. F., & Shulman, G. I. (2017). Regulation of hepatic glucose metabolism in health and disease. *Nature Reviews. Endocrinology*, *13*, 572–587. https://doi.org/10.1038/nrendo.2017.80.

Pierno, S., Didonna, M. P., Cippone, V., De Luca, A., Pisoni, M., Frigeri, A., et al. (2006). Effects of chronic treatment with statins and fenofibrate on rat skeletal muscle: A biochemical, histological and electrophysiological study. *British Journal of Pharmacology*, *149*, 909–919. https://doi.org/10.1038/sj.bjp.0706917.

Procino, G., Gerbino, A., Milano, S., Nicoletti, M. C., Mastrofrancesco, L., Carmosino, M., et al. (2015). Rosiglitazone promotes AQP2 plasma membrane expression in renal cells via a ca-dependent/cAMP-independent mechanism. *Cellular Physiology and Biochemistry*, *35*, 1070–1085. https://doi.org/10.1159/000373933.

Prudente, S., Flex, E., Morini, E., Turchi, F., Capponi, D., De Cosmo, S., et al. (2007). A functional variant of the adipocyte glycerol channel aquaporin 7 gene is associated with obesity and related metabolic abnormalities. *Diabetes*, *56*, 1468–1474. https://doi.org/10.2337/db06-1389.

Qin, H., Zheng, X., Zhong, X., Shetty, A. K., Elias, P. M., & Bollag, W. B. (2011). Aquaporin-3 in keratinocytes and skin: Its role and interaction with phospholipase D2. *Archives of Biochemistry and Biophysics*, *508*, 138–143. https://doi.org/10.1016/j.abb.2011.01.014.

Reshef, L., Olswang, Y., Cassuto, H., Blum, B., Croniger, C. M., Kalhan, S. C., et al. (2003). Glyceroneogenesis and the triglyceride/fatty acid cycle. *The Journal of Biological Chemistry*, *278*, 30413–30416. https://doi.org/10.1074/jbc.R300017200.

Ricanek, P., Lunde, L. K., Frye, S. A., Støen, M., Nygård, S., Morth, J. P., et al. (2015). Reduced expression of aquaporins in human intestinal mucosa in early stage inflammatory bowel disease. *Clinical and Experimental Gastroenterology*, *8*, 49–67. https://doi.org/10.2147/CEG.S70119.

Rivarola, V., Flamenco, P., Melamud, L., Galizia, L., Ford, P., & Capurro, C. (2010). Adaptation to alkalosis induces cell cycle delay and apoptosis in cortical collecting duct cells: Role of aquaporin-2. *Journal of Cellular Physiology*, *224*, 405–413. https://doi.org/10.1002/jcp.22136.

Robinson-Rechavi, M., Escriva, H., & Laudet, V. (2003). The nuclear receptor superfamily. *Journal of Cell Science*, *116*, 585–586. https://doi.org/10.1242/jcs.00247.

Rodríguez, A., Catalán, V., Gómez-Ambrosi, J., & Frühbeck, G. (2011). Aquaglyceroporins serve as metabolic gateways in adiposity and insulin resistance control. *Cell Cycle*, *10*, 1548–1556. https://doi.org/10.4161/cc.10.10.15672.

Rodríguez, A., Catalán, V., Gómez-Ambrosi, J., García-Navarro, S., Rotellar, F., Valentí, V., et al. (2011). Insulin- and leptin-mediated control of aquaglyceroporins in human adipocytes and hepatocytes is mediated via the PI3K/Akt/mTOR signaling cascade. *The Journal of Clinical Endocrinology and Metabolism*, *96*, E586–E597. https://doi.org/10.1210/jc.2010-1408.

Rodríguez, A., Gena, P., Méndez-Giménez, L., Rosito, A., Valentí, V., Rotellar, F., et al. (2014). Reduced hepatic aquaporin-9 and glycerol permeability are related to insulin resistance in non-alcoholic fatty liver disease. *International Journal of Obesity*, *38*, 1213–1220. https://doi.org/10.1038/ijo.2013.234.

Rodríguez, A., Marinelli, R. A., Tesse, A., Frühbeck, G., & Calamita, G. (2015). Sexual dimorphism of adipose and hepatic aquaglyceroporins in health and metabolic disorders. *Frontiers in Endocrinology (Lausanne)*, *6*, 171. https://doi.org/10.3389/fendo.2015.00171.

Rodríguez, A., Moreno, N. R., Balaguer, I., Méndez-Giménez, L., Becerril, S., Catalán, V., et al. (2015). Leptin administration restores the altered adipose and hepatic expression of aquaglyceroporins improving the non-alcoholic fatty liver of ob/ob mice. *Scientific Reports*, *5*, 12067. https://doi.org/10.1038/srep12067.

Rojek, A., Fuchtbauer, E.-M., Kwon, T.-H., Frokiaer, J., & Nielsen, S. (2006). Severe urinary concentrating defect in renal collecting duct-selective AQP2 conditional-knockout mice. *Proceedings of the National Academy of Sciences*, *103*, 6037–6042. https://doi.org/10.1073/pnas.0511324103.

Rojek, A. M., Skowronski, M. T., Fuchtbauer, E.-M., Fuchtbauer, A. C., Fenton, R. A., Agre, P., et al. (2007). Defective glycerol metabolism in aquaporin 9 (AQP9) knockout mice. *Proceedings of the National Academy of Sciences*, *104*, 3609–3614. https://doi.org/10.1073/pnas.0610894104.

Scherer, P. E. (2006). Adipose tissue: From lipid storage compartment to endocrine organ. *Diabetes*, *55*, 1537–1545. https://doi.org/10.2337/db06-0263.

Sjöholm, K., Palming, J., Olofsson, L. E., Gummesson, A., Svensson, P. A., Lystig, T. C., et al. (2005). A microarray search for genes predominantly expressed in human omental adipocytes: Adipose tissue as a major production site of serum amyloid A. *The Journal of Clinical Endocrinology and Metabolism*, *90*, 2233–2239. https://doi.org/10.1210/jc.2004-1830.

Sohara, E., Rai, T., Miyazaki, J., Verkman, A. S., Sasaki, S., & Uchida, S. (2005). Defective water and glycerol transport in the proximal tubules of AQP7 knockout mice. *American Journal of Physiology. Cell Physiology*, *289*, F1195–F1200. https://doi.org/10.1152/ajprenal.00133.2005.

Sohara, E., Rai, T., Sasaki, S., & Uchida, S. (2006). Physiological roles of AQP7 in the kidney: Lessons from AQP7 knockout mice. *Biochimica et Biophysica Acta, Biomembranes*, *1758*, 1106–1110. https://doi.org/10.1016/j.bbamem.2006.04.002.

Song, Y., & Verkman, A. S. (2001). Aquaporin-5 dependent fluid secretion in airway submucosal glands. *The Journal of Biological Chemistry*, *276*, 41288–41292. https://doi.org/10.1074/jbc.M107257200.

Sorani, M. D., Zado, Z., Hurowitz, E., Yan, D., Giacomini, K. M., & Manley, G. T. (2008). Novel variants in human aquaporin-4 reduce cellular water permeability. *Human Molecular Genetics*, *17*, 2379–2389. https://doi.org/10.1093/hmg/ddn138.

Splinter, P. L., Masyuk, A. I., & LaRusso, N. F. (2003). Specific inhibition of AQP1 water channels in isolated rat intrahepatic bile duct units by small interfering RNAs. *The Journal of Biological Chemistry*, *278*, 6268–6274. https://doi.org/10.1074/jbc.M212079200.

Tardelli, M., Bruschi, F. V., Claudel, T., Moreno-Viedma, V., Halilbasic, E., Marra, F., et al. (2017). AQP3 is regulated by PPARγ and JNK in hepatic stellate cells carrying PNPLA3 I148M. *Scientific Reports*, 7, 14661. https://doi.org/10.1038/s41598-017-14557-9.

Tardelli, M., Claudel, T., Bruschi, F. V., Moreno-Viedma, V., & Trauner, M. (2017). Adiponectin regulates AQP3 via PPARα in human hepatic stellate cells. *Biochemical and Biophysical Research Communications*, *490*, 51–54. https://doi.org/10.1016/j.bbrc.2017.06.009.

Tardelli, M., Claudel, T., Bruschi, F. V., & Trauner, M. (2018). Nuclear receptor regulation of aquaglyceroporins in metabolic organs. *International Journal of Molecular Sciences*, *19*, 1777. https://doi.org/10.3390/ijms19061777.

Tardelli, M., Moreno-Viedma, V., Zeyda, M., Itariu, B. K., Langer, F. B., Prager, G., et al. (2017). Adiponectin regulates aquaglyceroporin expression in hepatic stellate cells altering their functional state. *Journal of Gastroenterology and Hepatology*, *32*, 253–260. https://doi.org/10.1111/jgh.13415.

van Bilsen, M., & van Nieuwenhoven, F. A. (2010). PPARs as therapeutic targets in cardiovascular disease. *Expert Opinion on Therapeutic Targets*, *14*, 1029–1045. https://doi.org/10.1517/14728222.2010.512917.

Verkman, A. S. (2011). Aquaporins at a glance. *Journal of Cell Science*, *124*, 2107–2112. https://doi.org/10.1242/jcs.079467.

Verkman, A. S., Anderson, M. O., & Papadopoulos, M. C. (2014). Aquaporins: Important but elusive drug targets. *Nature Reviews. Drug Discovery*, *13*, 259–277. https://doi.org/10.1038/nrd4226.

Verkman, A. S., Hara-Chikuma, M., & Papadopoulos, M. C. (2008). Aquaporins—New players in cancer biology. *Journal of Molecular Medicine (Berlin, Germany)*, *86*, 523–529. https://doi.org/10.1007/s00109-008-0303-9.

Wakayama, Y., Hirako, S., Ogawa, T., Jimi, T., & Shioda, S. (2014). Upregulated expression of AQP 7 in the skeletal muscles of obese ob/ob mice. *Acta Histochemica et Cytochemica*, *47*, 27–33. https://doi.org/10.1267/ahc.13032.

Wakayama, Y., Inoue, M., Kojima, H., Jimi, T., Shibuya, S., Hara, H., et al. (2004). Expression and localization of aquaporin 7 in normal skeletal myofiber. *Cell and Tissue Research*, *316*, 123–129. https://doi.org/10.1007/s00441-004-0857-y.

Walker, C. G., Holness, M. J., Gibbons, G. F., & Sugden, M. C. (2007). Fasting-induced increases in aquaporin 7 and adipose triglyceride lipase mRNA expression in adipose tissue are attenuated by peroxisome proliferator-activated receptor alpha deficiency. *International Journal of Obesity*, *31*, 1165–1171. https://doi.org/10.1038/sj.ijo.0803555.

Wang, Y. D., Chen, W. D., Moore, D. D., & Huang, W. (2008). FXR: A metabolic regulator and cell protector. *Cell Research*, *18*, 1087–1095. https://doi.org/10.1038/cr.2008.289.

Wang, J., Feng, L., Zhu, Z., Zheng, M., Wang, D., Chen, Z., et al. (2015). Aquaporins as diagnostic and therapeutic targets in cancer: How far we are? *Journal of Translational Medicine*, *13*, 96. https://doi.org/10.1186/s12967-015-0439-7.

Wang, K. S., Komar, A. R., Ma, T., Filiz, F., McLeroy, J., Hoda, K., et al. (2000). Gastric acid secretion in aquaporin-4 knockout mice. *American Journal of Physiology. Gastrointestinal and Liver Physiology*, *279*, G448–G453. https://doi.org/10.1016/j.humov.2008.02.015. Changes.

Wang, Y. X., Lee, C. H., Tiep, S., Yu, R. T., Ham, J., Kang, H., et al. (2003). Peroxisome proliferator activated receptor [delta] activates fat metabolism to prevent obesity. *Cell*, *113*, 159–170. https://doi.org/10.1016/S0092-8674(03)00269-1.

Watanabe, M., Houten, S. M., Wang, L., Moschetta, A., Mangelsdorf, D. J., Heyman, R. A., et al. (2004). Bile acids lower triglyceride levels via a pathway involving FXR, SHP, and SREBP-1c. *The Journal of Clinical Investigation*, *113*, 1408–1418. https://doi.org/10.1172/JCI200421025.

Wilson, J. L. L., Miranda, C. A., & Knepper, M. A. (2013). Vasopressin and the regulation of aquaporin-2. *Clinical and Experimental Nephrology*, *17*, 751–764. https://doi.org/10.1007/s10157-013-0789-5.

Yasui, M., Kwon, T.-H., Knepper, M. A., Nielsen, S., & Agre, P. (1999). Aquaporin-6: An intracellular vesicle water channel protein in renal epithelia. *Proceedings of the National Academy of Sciences*, *96*, 5808–5813. https://doi.org/10.1073/pnas.96.10.5808.

Zahn, A., Moehle, C., Langmann, T., Ehehalt, R., Autschbach, F., Stremmel, W., et al. (2007). Aquaporin-8 expression is reduced in ileum and induced in colon of patients with ulcerative colitis. *World Journal of Gastroenterology*, *13*, 1687–1695. https://doi.org/10.1055/s-2005-919916.

Zhang, Y., & Edwards, P. A. (2008). FXR signaling in metabolic disease. *FEBS Letters*, *582*, 10–18. https://doi.org/10.1016/j.febslet.2007.11.015.

Zhang, X., Huang, S., Gao, M., Liu, J., Jia, X., Han, Q., et al. (2014). Farnesoid X receptor (FXR) gene deficiency impairs urine concentration in mice. *Proceedings of the National Academy of Sciences*, *111*, 2277–2282. https://doi.org/10.1073/pnas.1323977111.

Zhu, C., Chen, Z., & Jiang, Z. (2016). Expression, distribution and role of aquaporin water channels in human and animal stomach and intestines. *International Journal of Molecular Sciences*, *17*, 1399. https://doi.org/10.3390/ijms17091399.

CHAPTER FIVE

Phosphorylation of human AQP2 and its role in trafficking

Susanna Törnroth-Horsefield*
Department of Biochemistry and Structural Biology, Lund University, Lund, Sweden
*Corresponding author: e-mail address: susanna.horsefield@biochemistry.lu.se

Contents

Abstract

Human Aquaporin 2 (AQP2) is a membrane-bound water channel found in the kidney collecting duct whose regulation by trafficking plays a key role in regulating urine volume. AQP2 trafficking is tightly controlled by the pituitary hormone arginine vasopressin (AVP), which stimulates translocation of AQP2 residing in storage vesicles to the apical membrane. The AVP-dependent translocation of AQP2 to and from the apical membrane is controlled by multiple phosphorylation sites in the AQP2 C-terminus, the phosphorylation of which alters its affinity to proteins within the cellular membrane protein trafficking machinery. The aim of this chapter is to provide a summary of what is currently known about AVP-mediated AQP2 trafficking, dissecting the roles of individual phosphorylation sites, kinases and phosphatases and interacting proteins. From this, the picture of an immensely complex process emerges, of which many structural and molecular details remains to be elucidated.

Vitamins and Hormones, Volume 112
ISSN 0083-6729
https://doi.org/10.1016/bs.vh.2019.08.002

1. The role of AQP2 in regulating urine volume

The human adult kidney produces approximately 180 L of primary urine per day through the process of glomerular filtration in the renal corpuscle. As the primary urine passes through the nephron, 99% of this volume is reabsorbed, before the final urine volume reaches the ureter (Kwon et al., 2009). This impressive feat is carried out by aquaporins (AQP), membrane-bound water channels that facilitate water transport across cellular membranes along osmotic gradients. The majority (90%) of the primary urine is constitutively reabsorbed by AQP1, and to a lesser extent by AQP7, in the proximal tubule and descending thin limb of Henle's loop (Fig. 1A). The remaining part reaches the collecting duct, where the regulated action of AQP2 is responsible for fine-tuning the final urine volume in order to meet the current demands of the body. This is controlled by the pituitary hormone arginine vasopressin (AVP), triggering the rapid translocation of AQP2 from sub-apical storage vesicles to the apical membrane (within 5–30 min) as well as regulating AQP2 gene transcription in the long term. Once at the apical membrane, AQP2 starts to reabsorb water from urine, which is released into the blood stream through AQP3 and AQP4 located in the basolateral membrane (Fig. 1B) (Kortenoeven & Fenton, 2014; Kwon et al., 2009; Moeller, Fuglsang, & Fenton, 2016).

2. Trafficking of AQP2

The AVP-dependent trafficking of AQP2 is a critical component of the human body water balance and the major mechanism by which water reabsorption from urine can be controlled. Upon dehydration AVP is released from the pituitary gland into the blood stream and binds to the vasopressin type 2 receptor (V_2R) in the basolateral membrane of the collecting duct principal cells, resulting in increased intracellular cAMP production by adenylyl cyclase. This signaling cascade stimulates PKA-mediated phosphorylation of AQP2 residing in storage vesicles, triggering its translocation to the apical membrane via regulated exocytosis (Fig. 1B) (van Balkom et al., 2002). Once the AVP-level drops, AQP2 is removed from the apical membrane via endocytosis thereby restoring its water permeability to basal levels. Following its retrieval, AQP2 may be targeted to lysosomal degradation, redirected to storage vesicles for another round of AVP-stimulated apical membrane targeting or released into the urine as

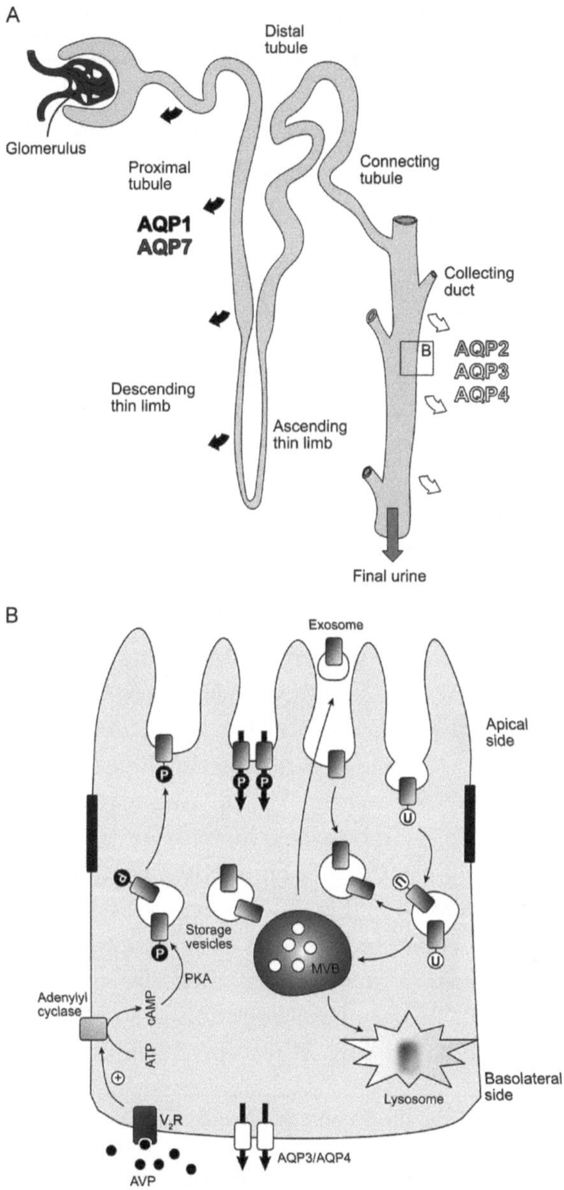

Fig. 1 Role of AQP2 in urine volume regulation. (A) Schematic picture of the nephron showing the localization of renal AQPs. The majority of the primary urine produced by glomerular filtration is constitutively reabsorbed in the proximal tubule and descending thin limb of Henle's loop (black arrows) through AQP1 and AQP7. The final urine volume is regulated in the collecting duct where AQP2 takes up water from urine into the cell in an AVP-dependent manner. The water is then released into the blood stream via AQP3

(Continued)

exosomes (Kamsteeg et al., 2006; Katsura, Ausiello, & Brown, 1996; Pisitkun, Shen, & Knepper, 2004). Dysfunctional AQP2 trafficking caused by mutations in the genes for either V_2R or AQP2 leads to Nephrogenic Diabetes Insipidus (NDI), a water balance disorder that manifests itself by excessive urine volumes and severe dehydration that, if left untreated, can lead to mental retardation (Moeller, Rittig, & Fenton, 2013).

2.1 AQP2 trafficking is controlled by post-translational modifications

The process of distributing membrane proteins between cellular compartments is known as trafficking and involves a generic mechanism whereby the membrane proteins move between cellular membranes in vesicles that bud off one membrane and fuse with the next (Kirchhausen, 2000). The recruitment of proteins into vesicles is determined by the presence of sorting signals; most often short linear amino acid sequences or post-translational modifications, that direct the protein to a particular membrane at the appropriate time (Cosson, Perrin, & Bonifacino, 2013; Derby & Gleeson, 2007). These sorting signals are believed to govern the trafficking process by controlling the affinity toward proteins within the cellular vesicle trafficking machinery.

For AQP2, several post-translational modification sites located within the C-terminus serve as sorting signals. These include four phosphorylation sites at Ser 256, Ser 261, Ser 264 and Thr 269 (serine in mouse), all of which have been shown to be phosphorylated in vivo, and a ubiquitination site at Lys 270 (Hoffert, Pisitkun, Wang, Shen, & Knepper, 2006; Kamsteeg et al., 2006; van Balkom et al., 2002) (Fig. 2). Whereas phosphorylation is essential for apical membrane targeting as well as its retention therein (see below), short-chain ubiquitination mediates its internalization and subsequent degradation (Fig. 1B). AQP2 phosphorylation and ubiquitination are thus opposing processes in controlling AQP2 apical membrane abundance.

Fig. 1—Cont'd and AQP4. (B) Schematic picture of a collecting duct principle cell showing the AQP2 trafficking mechanism. AVP-binding to the V_2R receptor in the basolateral membrane stimulates cAMP production by adenylyl cyclase. The increased cAMP levels results in PKA-mediated phosphorylation (P) of AQP2 residing in storage vesicles, triggering translocation of AQP2 to the apical membrane where it starts to reabsorb water from urine. The reabsorbed water is released into the blood stream via AQP3 and AQP4 in the basolateral membrane. When the AVP-signal ceases, dephosphorylation and ubiquitination (U) promotes AQP2 internalization via endocytosis after which AQP2 may be stored again in storage vesicles, targeted to MVBs for subsequent degradation in lysosomes or released into the urine as exosomes.

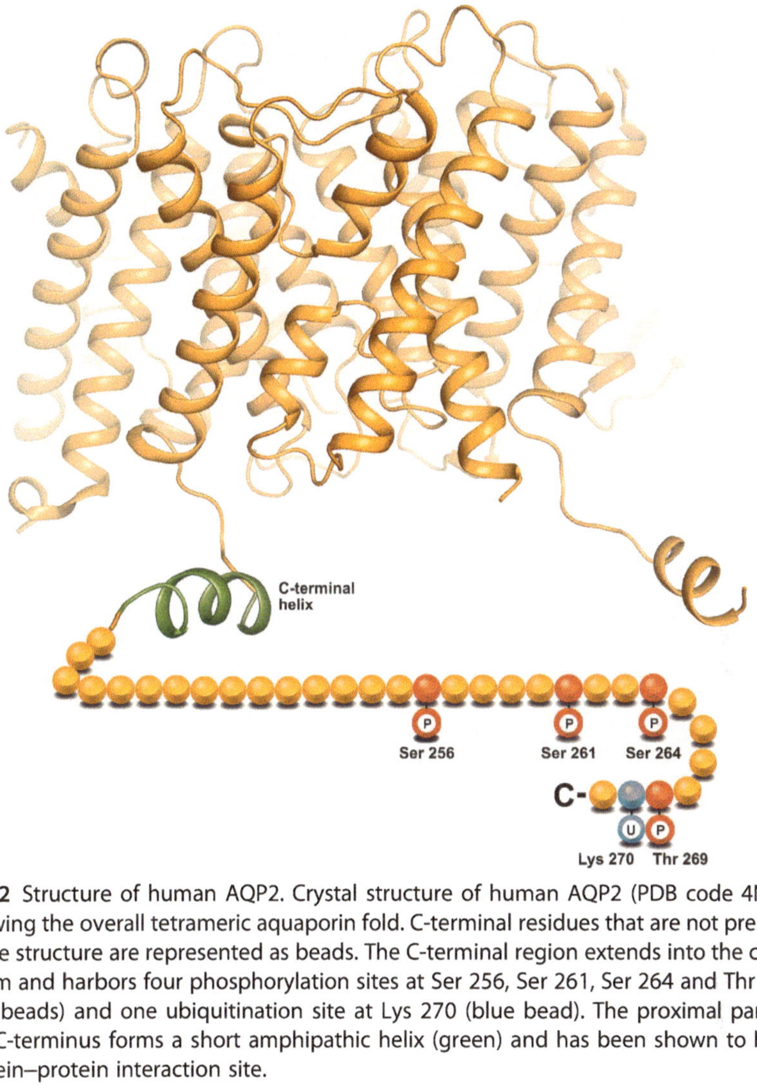

Fig. 2 Structure of human AQP2. Crystal structure of human AQP2 (PDB code 4NEF) showing the overall tetrameric aquaporin fold. C-terminal residues that are not present in the structure are represented as beads. The C-terminal region extends into the cytoplasm and harbors four phosphorylation sites at Ser 256, Ser 261, Ser 264 and Thr 269 (red beads) and one ubiquitination site at Lys 270 (blue bead). The proximal part of the C-terminus forms a short amphipathic helix (green) and has been shown to be a protein–protein interaction site.

3. AQP2 phosphorylation and its role in trafficking

All four AQP2 phosphorylation sites have been shown to be phosphorylated in an AVP-dependent manner (Hoffert et al., 2006). Ser 256 plays the most prominent role in AQP2 trafficking, as AVP-stimulated phosphorylation of this residue by PKA is essential and sufficient for translocation of AQP2 from storage vesicles to the apical membrane (Fig. 3)

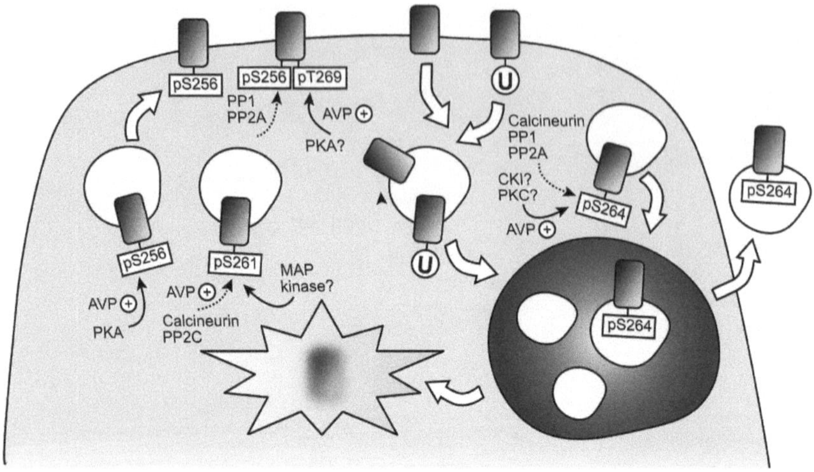

Fig. 3 Role of phosphorylation sites in AVP-dependent AQP2 trafficking. Schematic picture showing the proposed role of phosphorylation sites in AQP2 trafficking. Kinases and phosphatases are indicated by solid and dotted arrows, respectively. White block arrows illustrate translocation events. Under unstimulated conditions, AQP2 resides in storage vesicles and is phosphorylated on Ser 261. AVP stimulates dephosphorylation of Ser 261 and phosphorylation of Ser 256 by PKA, which leads to fusion of storage vesicles with the apical membrane where subsequent phosphorylation of Thr 269 increases its retention time. Dephosphorylation and ubiquitination promotes AQP2 endocytosis after which AQP2 may be transferred to storage vesicles or sorted in to MVB inner vesicles for lysosomal degradation. AVP also stimulates phosphorylation of Ser 264, which has been suggested to be involved in excretion of AQP2 in exosomes, via MVB inner vesicles.

(Katsura, Gustafson, Ausiello, & Brown, 1997; van Balkom et al., 2002). Studies in Xenopus oocytes have shown that this depends on the stoichiometry of phosphorylated and non-phosphorylated monomers within the AQP2 tetramer, with at least three out of four monomers needing to be phosphorylated at Ser 256 before translocation occurs (Kamsteeg, Heijnen, van Os, & Deen, 2000). Once at the apical membrane, additional phosphorylation at Thr 269 prolongs its retention time by decreasing AQP2 internalization. The demonstration that Thr 269 cannot be phosphorylated when Ser 256 is mutated to alanine or leucine suggests that Ser 256 phosphorylation is a pre-requisite for phosphorylation of Thr 269 to occur (Hoffert et al., 2008). Whereas AVP stimulates phosphorylation of Ser 256 and Thr 269, it has the opposite effect on Ser 261, with this residue becoming dephosphorylated when AVP-levels increase. Phosphorylation of Ser 261 coincide with AQP2 residing in storage vesicles under unstimulated conditions (Fig. 3) and its dephosphorylation occurs after phosphorylation of both Ser 256 and Thr 269 (Moeller, MacAulay, Knepper, & Fenton, 2009;

Tamma, Robben, Trimpert, Boone, & Deen, 2011). Moreover, phosphorylation of Ser 256 is able to override the effect of Ser 261 phosphorylation on sub-cellular localization and phosphorylation of Ser 261 seems to stabilize the ubiquitinated AQP2 (Tamma et al., 2011). AVP induces phosphorylation of Ser 264, the role of which is not entirely clear. Short-term AVP treatment ($<$15 min) resulted in AQP2 phosphorylated on Ser 264 being found in both the apical and basolateral membranes whereas prolonged treatment ($>$60 min) caused a relocalization to the apical membrane and early endosomes, but not to lysosomes (Fenton et al., 2008). It was further suggested that Ser 264 phosphorylation may be associated AQP2 localization in exosomes, extracellular vesicles that originates from the inner vesicles of multivesicular bodies and are released into urine (Fig. 3) (Fenton et al., 2008; Pisitkun et al., 2004). To summarize, AQP2 trafficking is governed by an intricate cross-talk between the four phosphorylation sites and the ubiquitination site, allowing AVP to dynamically control water reabsorption from urine.

3.1 Protein kinases and phosphatases in AQP2 trafficking

The role of PKA in phosphorylating AQP2 at Ser 256 is well established and, since cAMP stimulates the activity of PKA, provides a clear link between AVP-binding to the V_2R and the increased apical membrane abundance of AQP2 (Fig. 1B) (Katsura et al., 1996; van Balkom et al., 2002). Interestingly, Ser 256 has also been suggested to be phosphorylated by PKG, resulting in AVP-independent trafficking of AQP2 by an alternative mechanism involving nitric oxide and cGMP (Bouley et al., 2000). For the other phosphorylation sites, the kinases involved remains to be unambiguously identified. It has been suggested that PKA can also phosphorylate Thr 269, despite the fact that this does not sit in a canonical PKA-consensus site (RXS/T^{269} instead of RRXS/T) (Brown, Hasler, Nunes, Bouley, & Lu, 2008; Hoffert et al., 2006). However, this remains to be demonstrated experimentally. Ser 261 sits in a mitogen activator protein (MAP) kinase site and the reduced phosphorylation of Ser 261 in response to AVP has been suggested to involve inhibition of the MAP kinase p38 by cAMP. Several MAP-kinases, including p38, was shown to be able to phosphorylate an AQP2 C-terminal peptide, further supporting their involvement Ser 261 phosphorylation (Rinschen et al., 2010). Very little is known about the kinases responsible for phosphorylation of Ser 264, but sequence analysis have suggested that it may be phosphorylated by casein kinase type I or PKC (Brown et al., 2008; Hoffert et al., 2006).

A few studies have addressed the question of which specific phosphatases are involved in AQP2 trafficking. Cheung et al. showed that AVP stimulates dephosphorylation of Ser 261 by protein phosphatase 2C (PP2C) and that this is independent of the phosphorylation state of Ser 256. Inhibition of PP2C alone did not alter AVP-induced translocation to the apical membrane supporting that phosphorylation of Ser 256 rather than dephosphorylation of Ser 261 plays the most important role in AQP2 apical membrane targeting (Cheung, Ueberdiek, Day, Bouley, & Brown, 2017). In rat inner medullas, inhibition of protein phosphatase 2B (PP2B, calcineurin) increased the levels of Ser 261 and Ser 264 phosphorylation while inhibition of protein phosphatase 1 and 2A (PP1 and PP2A) increased the levels of phosphorylation at Ser 256 and Ser 264 (Jo et al., 2001; Ren et al., 2016). The role of calcineurin is further supported by the demonstration that calcineurin-null mice display normal AQP2 expression but decreased levels of AQP2 phosphorylation (Gooch, Guler, Barnes, & Toro, 2006) as well as studies of PKA-phosphorylated endosomes isolated from rat kidney collecting duct, which showed that calcineurin was able to dephosphorylate AQP2 in vitro (Jo et al., 2001). Taken together, these studies emphasize that both AQP2 phosphorylation and dephosphorylation play important roles during AQP2 trafficking. Further studies will be needed to pinpoint the specific kinases and phosphatases involved in the different trafficking events.

4. Phosphorylation-dependent protein–protein interactions in AQP2 trafficking

As mentioned above, AQP2 trafficking is controlled by interactions with proteins within the cellular trafficking machinery. A number of such interaction partners have been identified (Table 1), several of which have been shown to interact with AQP2 C-terminus where the phosphorylation sites are located (Fig. 2). Most of these interaction partners have been shown to bind AQP2 in a phosphorylation-dependent manner, supporting the notion that AQP2 phosphorylation governs trafficking by altering the affinity toward proteins in the trafficking machinery. Further details about some of the best-characterized phosphorylation-dependent interactions and their roles in trafficking are given in Table 1.

4.1 Heat shock proteins

The 70 kDa heat shock protein family (Hsp70s) are ubiquitously expressed proteins that are found in the cytoplasm where they assist in protein folding

Table 1 AQP2 trafficking interaction partners and their role in trafficking.

Interaction partner	Proposed role in trafficking	Phosphorylation-dependent	Reference
Hsc70	Endocytosis	Yes	Lu et al. (2007), Moeller, Praetorius, Rutzler, and Fenton (2010), and Zwang et al. (2009)
Hsp70	Apical membrane targeting	Yes	Lu et al. (2007) and Moeller et al. (2010)
BiP (Hsp70-5)	Increased apical membrane abundance	Yes	Zwang et al. (2009)
Actin	Vesicle translocation	Yes	Noda, Horikawa, Katayama, and Sasaki (2004), Noda, Horikawa, Katayama, and Sasaki (2005), Noda et al. (2008), and Moeller et al. (2010)
Tropomyosin 5b	Vesicle translocation	Yes	Noda et al. (2005, 2008)
Annexin II	Apical membrane fusion	Yes	Zwang et al. (2009), Tamma, Procino, Mola, Svelto, and Valenti (2008), Moeller et al. (2010), and Noda et al. (2005)
Clathrin	Endocytosis	Not known	Lu et al. (2007)
Dynamin	Endocytosis	Not known	Lu et al. (2007)
Spa-1	Apical membrane targeting	Not known	Noda, Horikawa, Furukawa, et al. (2004)
AP2	Endocytosis	Not known	Lu et al. (2007)
Ezrin	Endocytosis	No	Li et al. (2017)
MAL	Increased apical membrane retention	Yes	Kamsteeg et al. (2007)
14-3-3θ	Decreased apical membrane retention	Yes	Moeller, Slengerik-Hansen, et al. (2016)
14-3-3ζ	Prevent ubiquitination	Yes	Moeller, Slengerik-Hansen, et al. (2016)
Ndfip1	Ubiquitination	Not known	Trimpert et al. (2017)
LIP5	Targeting for degradation	Yes	Roche et al. (2017) and van Balkom et al. (2009)

and trafficking. Several members of this family have been shown to interact with AQP2 in a phosphorylation-dependent manner and affect its trafficking, including heat shock cognate protein 70 (Hsc70, also known as Hsp73), Heat shock protein 70-1 (Hsp70, also known as Hsp72, Hsp72 or Hsp70i) as well as Heat shock protein 70-5 (BiP, also known as Grp78) (Lu et al., 2007; Park et al., 2013; Zwang et al., 2009). Hsc70 plays an important role in clathrin-mediated endocytosis where it, together with auxillin, mediates ATPase-dependent uncoating of clathrin-coated vesicles (Chang et al., 2002; Chappell et al., 1986; Morgan, Prasad, Jin, Augustine, & Lafer, 2001; Newmyer & Schmid, 2001). Hsc70 have also been shown to bind directly to membrane channel proteins, including AQP2, mediating their trafficking at various steps (Bronk et al., 2001; Clay & Kuzirian, 2002; Lu et al., 2007; Meacham, Patterson, Zhang, Younger, & Cyr, 2001; Zhang et al., 2001, 2002). For AQP2, Hsc70 has been shown to interact with the C-terminus and promote its endocytosis (Fig. 4). Co-immunoprecipitation studies (co-IP) using LLC cells expressing rat AQP2 showed an increased

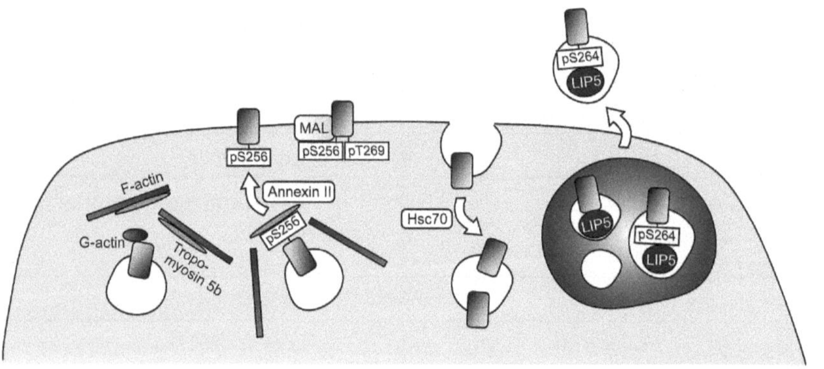

Fig. 4 Protein–protein interactions in AQP2 trafficking. Schematic picture showing the roles of some of the known interaction partners in AQP2 trafficking. Non-phosphorylated AQP2 interacts with G-actin and is inhibited from reaching the apical membrane by F-actin forming a barrier. F-actin is stabilized by Tropomyosin 5b. Upon Ser 256 phosphorylation, AQP2 interacts with Tropomyosin 5b which leads to destabilization of F-actin and, with the help of Annexin II, fusion of the AQP2 storage vesicles with the apical membrane. Once at the apical membrane, the interaction with MAL increases AQP2 retention time by interacting with phosphorylated AQP2 and preventing its endocytosis. The interaction between non-phosphorylated and Hsc70 promotes endocytosis whereas its interaction with LIP5 is involved in sorting into MVB inner vesicles for subsequent degradation. Phosphorylation of Ser 264 also promotes insertion into MVB inner vesicles from where AQP2 may be excreted as exosomes in the urine.

interaction signal, however, only after 30 min of AVP treatment (Lu et al., 2007). Since this time frame coincides with down-regulation of the AVP-signal and falling cAMP levels (van Balkom et al., 2002) this supports the role of Hsc70 in AQP2 endocytosis. Moreover, the interaction was greatly reduced when introducing a phospho-mimicking mutation at Ser 256 and Ser 269 (Thr 269 in human) (Lu et al., 2007; Moeller et al., 2010) and a pull-down study using the AQP2 C-terminal peptide with and without phosphorylation at Ser 256 showed that Hsc70 preferentially interacted with the non-phosphorylated peptide or a peptide phosphorylated at Ser 261 (Lu et al., 2007; Zwang et al., 2009). Taken together, and in line with the known role of these phosphorylation sites and Hsc70 in AQP2 trafficking and endocytosis, respectively, this suggests that phosphorylation of Ser 256 and Thr 269 prevents the interaction with Hsc70.

In contrast to Hsc70, which is constitutively expressed, Hsp70 expression is induced by cellular stress. The two proteins are highly similar in terms of sequence and have been proposed to be functionally indistinguishable, with Hsp70 replacing Hsc70 during stressful conditions (Angelidis, Lazaridis, & Pagoulatos, 1999; Gebauer, Zeiner, & Gehring, 1997; Takayama, Xie, & Reed, 1999). Nevertheless, Hsc70 and Hsp70 have been shown to play distinct roles during membrane protein trafficking and many functions associated with Hsc70 have not been demonstrated for Hsp70 (Gething & Sambrook, 1992; Goldfarb et al., 2006). Hsp70 has been shown to directly interact with the AQP2 C-terminus and studies using C-terminal peptides as well as AQP2 phospho-mimicking mutants showed that similarly as for Hsc70, phosphorylation of AQP2 at Ser 256 or Thr 269 reduces the interaction with Hsp70 (Lu et al., 2007; Moeller et al., 2010). The exact role of Hsp70 in AQP2 trafficking is not clear but it may well be different from that of Hsc70, as shown for the epithelial sodium channel ENac for which Hsc70 and Hsp70 had antagonistic effects (Goldfarb et al., 2006). This is supported by the observation that while Hsc70 seems to co-localize with AQP2 in the plasma membrane after AVP treatment, Hsp70 remains evenly distributed in the cytoplasm (Lu et al., 2007).

BiP is the only member of the Hsp70 protein family that is located in the endoplasmic reticulum where it is believed to be involved in the ER quality control system. However, several studies have shown that BiP can also be found in the cytoplasm or at the plasma membrane in a variety of cells (Altmeyer et al., 1996; Boilard et al., 2004; Davidson et al., 2005; Delpino & Castelli, 2002; Shani et al., 2008; Wiest et al., 1997; Xiao, Chung, Pyun, Fine, & Johnson, 1999), including those in the collecting duct

(Zwang et al., 2009). A direct interaction between BiP and AQP2 has been demonstrated using an AQP2 C-terminal peptide as well as in co-IP studies from native rat inner medulla collecting duct cells. In contrast to Hsc70 and Hsp70, BiP was shown to preferentially interact with AQP2 phosphorylated at Ser 256, indicating a different role in AQP2 trafficking, which may counteract that of Hsc70 and Hsp70 (Zwang et al., 2009). Alternatively, BiP-binding to AQP2 phosphorylated at Ser 256 was suggested to promote recruitment of kinases responsible for phosphorylation of other sites, thereby providing a plausible molecular mechanism for how phosphorylation of Thr 269 depends on prior phosphorylation of Ser 256.

4.2 Actin cytoskeleton

The actin cytoskeleton plays fundamental roles in eukaryotic cells, including providing mechanical support, controlling the movement of intracellular vesicles and driving cellular movements (Pollard, 2016; Pollard & Cooper, 2009). These functions are given by the ability of actin monomers (G-actin) to polymerize into filaments (F-actin) and the dynamic control of de- and re-polymerization by actin-binding proteins. One group of such proteins are the tropomyosins which are known to stabilize the filaments as well as controlling the binding of other actin-binding proteins (Robaszkiewicz, Ostrowska, Marchlewicz, & Moraczewska, 2016).

During trafficking of AQP2, the actin filaments are believed to function as a barrier, preventing AQP2-containing vesicles from fusing with the apical membrane. Upon AVP-stimulation in collecting duct cells, actin has been shown to depolymerize (Simon, Gao, Franki, & Hays, 1993) and this is suggested to clear the way for AQP2 apical membrane translocation. The actin barrier model and the facilitating role of actin depolymerization in AQP2 trafficking is supported by the demonstration that actin depolymerization agents as well as inhibition of actin polymerization promote AQP2 accumulation in the apical membrane (Klussmann et al., 2001; Tamma et al., 2001).

AQP2 has been shown to bind directly to actin and tropomyosin 5b individually (Moeller et al., 2010; Noda et al., 2008; Noda, Horikawa, Katayama, et al., 2004), as well as in a multiprotein complex consisting of actin, tropomyosin 5b and other actin-binding proteins that was proposed to function as a "motor" driving the AQP2 movement (Noda et al., 2005). Noda et al. showed that AQP2 reconstituted in liposomes binds to G-actin but not F-actin and that the interaction is strongest when AQP2 is non-phosphorylated at Ser 256 (Fig. 4) (Noda et al., 2008). This was further supported by studies in Madin-Darby canine kidney (MDCK) cells which demonstrated that mutation of Ser 256 to alanine promoted the interaction while cAMP-stimulation in cells expressing wild-type AQP2 released it (Moeller et al., 2010). This suggests that phosphorylation of Ser 256 abolishes the interaction between G-actin and AQP2. In contrast to G-actin, tropomyosin 5b preferentially interacts with AQP2 phosphorylated at Ser 256 and it was suggested that this sequesters tropomyosin 5b from its interaction with F-actin, thereby abolishing its filament stabilizing effect (Noda et al., 2008). Based on these results, a model for how phosphorylation-dependent interactions with the actin cytoskeleton controls AQP2 delivery to the apical membrane was proposed: (1) AQP2 binds G-actin under basal conditions, (2) phosphorylation of Ser 256 releases this interaction, (3) AQP2 phosphorylated at Ser 256 binds tropomyosin 5b which destabilizes the actin filament network and allows translocation of AQP2 to the apical membrane (Fig. 4).

4.3 Annexin II

Annexins are Ca^{2+}-dependent proteins that in their Ca^{2+}-bound form are able to bind to negatively charged groups in the plasma membrane. As such, they are able to link Ca^{2+}-regulated processes with membrane functions. Annexins are involved in a range of membrane-related events, including organization of membrane domains, linking the membrane to the cytoskeleton, vesicle transport and, in epithelial cells, maintaining cell polarity (Gerke, Creutz, & Moss, 2005; Markoff & Gerke, 2005). Several studies have demonstrated the direct interaction between one of the members of the annexin family, Annexin II, and AQP2 (Moeller et al., 2010; Noda et al., 2005; Tamma, Procino, Mola, Svelto, & Valenti, 2008; Zwang et al., 2009). Annexin II is expressed in the kidney (Markoff & Gerke, 2005) and has been shown to co-localize with AQP2 in intracellular vesicles (Barile et al., 2005). Since Annexin II is known to bind actin (Hayes, Rescher, Gerke, & Moss, 2004), it has been proposed that it functions as a link between AQP2 in vesicles and the actin cytoskeleton during vesicle translocation. Using an in vitro vesicle fusion assay, Tamma et al. showed that Annexin II promotes fusion of AQP2-bearing vesicles with the plasma membrane (Fig. 4). Similar results were obtained in vivo where the inhibition of Annexin II reduced cAMP-dependent accumulation of AQP2 in the plasma membrane, leading to reduced water permeability

(Tamma et al., 2008). Binding studies using AQP2 C-terminal peptides as well as co-IP-studies have shown that Annexin II have strongest affinity for non-phosphorylated AQP2 (Moeller et al., 2010; Zwang et al., 2009). This may seem counterintuitive given the role of phosphorylation in AQP2 apical membrane targeting. However, Annexin II has been identified as being one part of the multiprotein "motor" complex that control AQP2 trafficking together with cytoskeletal proteins such as actin and tropomyosin 5b (Noda et al., 2005) and it may be that the phosphorylation-dependence of the interaction with other partners within this complex plays a more dominant role.

4.4 Myelin and lymphocyte associated protein (MAL)

Myelin and lymphocyte associated protein (MAL) is a membrane-bound protein found in the apical membrane of epithelial cells, including the collecting duct (Frank, van der Haar, Schaeren-Wiemers, & Schwab, 1998; Kim, Fiedler, Madison, Krueger, & Pfeiffer, 1995). Because of its association with glycosphingolipid-enriched membranes that are believed to be important for apical membrane targeting of membrane proteins, MAL has been suggested to be involved in sorting membrane proteins into vesicles for apical membrane delivery (Cheong, Zacchetti, Schneeberger, & Simons, 1999; Simons & Ikonen, 1997). However, Kamsteeg et al. showed that, MAL is not involved in apical membrane delivery of AQP2, instead it increases the AQP2 apical membrane abundance by slowing down internalization (Fig. 4) (Kamsteeg et al., 2007). Co-IP studies demonstrated that MAL binds AQP2 directly with a preference for its Ser 256-phosphorylated form. Forskolin-stimulation, leading to increased cAMP levels, did not result in an additional increase in AQP2 apical membrane abundance or the amount of MAL that co-immunoprecipitated with AQP2. This suggests that MAL increases the steady state phosphorylation levels of AQP2 and that this coincides with increased apical membrane abundance. Since the apical membrane delivery of newly synthesized AQP2 did not increase, it was concluded that the elevated apical membrane AQP2 levels was a result of reduced internalization. Based on this, it was proposed that MAL stabilizes phosphorylated AQP2 in the apical membrane, which would, in turn, decrease AQP2 dephosphorylation thereby reducing endocytosis.

4.5 14-3-3 proteins

The 14-3-3 proteins constitute a family of ubiquitously expressed adaptor proteins that are involved in a wide range of phosphorylation-dependent cellular processes. They exert their function by binding to target proteins that have been phosphorylated at specific sites, thereby forcing a conformational change or affecting protein–protein interactions, and have been shown to play important roles in signal transduction pathways and protein trafficking (Dougherty & Morrison, 2004; Johnson et al., 2010; Mackintosh, 2004).

In mammals, seven different 14-3-3 proteins have been identified, all of which bind their target proteins via selective motifs (Mackintosh, 2004). Two of these, 14-3-3θ and 14-3-3ζ, have been shown to co-localize with AQP2 in intracellular vesicles in the collecting duct (Barile et al., 2005). Binding studies using AQP2 C-terminal peptide as well as co-IP studies in HEK293 cells showed that both isoforms are able to bind AQP2 and that this was dependent on phosphorylation Ser 256 and to a lesser degree on phosphorylation of Ser 264 and Ser 269 (Thr 269 in human). Despite these similarities, these two isoforms were shown to have distinct and opposite effects on AQP2 apical membrane abundance in a mouse kidney collecting duct cell line ($mpkCCD_{14}$) (Moeller, Slengerik-Hansen, et al., 2016). In this study, knock-down of 14-3-3ζ resulted in an increase in AQP2 ubiquitination and decreased AQP2 half-life, thereby drastically reducing its apical membrane abundance. This suggests that 14-3-3ζ plays a role in stabilizing AQP2 in the apical membrane by preventing its ubiquitination. This fits well with the fact that the interaction between AQP2 and 14-3-3ζ was shown to increase upon phosphorylation of Ser256 and Ser 269, as these two phosphorylation sites are known to promote AQP2 apical membrane localization.

In contrast, knock-down of 14-3-3θ showed the reverse effect: the AQP2 apical membrane abundance increased, suggesting a role of 14-3-3θ in removing AQP2 from the apical membrane. This could not be correlated to increased ubiquitination levels, suggesting a non-ubiquitin-dependent pathway. Since the phosphorylation of Ser 269 was significantly increased in cells lacking 14-3-3θ, this points at this site being responsible for the effect (Moeller, Slengerik-Hansen, et al., 2016). A plausible scenario could be that 14-3-3θ is responsible for recruiting a phosphatase that specifically targets Ser 269, reducing the half-life of AQP2 in the apical membrane. An attractive target for such as phosphatase would be PP1 and PP2A, which are known to be able to dephosphorylate Ser 269 (Moeller, Aroankins, Slengerik-Hansen, Pisitkun, & Fenton, 2014) as well as being able to bind 14-3-3 proteins (Angrand et al., 2006; Liang et al., 2009).

4.6 Lysosomal trafficking regulator interacting protein 5 (LIP5)

The lysosomal trafficking regulator interacting protein 5 (LIP5) plays an important role in sorting internalized membrane proteins into multivesicular bodies. This involves coordinating the actions of the endosomal sorting complex required for transport III (ESCRT-III) and the Vps4 ATPase, resulting in the transfer of sorted membrane proteins from the MVB outer limiting membrane to internal vesicles, from where they may be transferred to lysosomes for degradation (Skalicky et al., 2012; Vild, Li, Guo, Liu, & Xu, 2015; Yang & Hurley, 2010).

In addition to interacting with both ESCRT-III and Vps4, LIP5 has also been shown to interact with the sorted membrane proteins themselves. This includes AQP2, for which the interaction with LIP5 has been shown to facilitate its lysosomal degradation presumably by targeting it to MVBs (Fig. 4) (van Balkom et al., 2009). Co-IP studies as well as in vitro binding studies show that the interaction between AQP2 and LIP5 is mediated by the AQP2 C-terminus (Roche et al., 2017; van Balkom et al., 2009). This region contains a short amphipathic helix that is exposed to the cytoplasm and harbors a known LIP5-interacting motif (Fig. 2) (Vild et al., 2015). Interestingly, this C-terminal helix is known to be involved in regulatory protein interactions also in other AQP isoforms (Reichow et al., 2013). In vitro interaction studies using full-length AQP2 showed that the interaction with LIP5 was dependent on site-specific phosphorylation of the AQP2 C-terminus, despite the fact that the phosphorylation sites lies outside the LIP5 binding motif (Fig. 2) (Roche et al., 2017). Specifically, the AQP2 phospho-mimicking mutants S256E, S261E, T269E and S256E/T269E all bound LIP5 with lower affinity than non-phosphorylated AQP2. This corresponds well to the known roles of these phosphorylation sites in AQP2 trafficking, with Ser 256 and Thr 269 being involved in apical membrane localization and Ser 261 being mainly attributed to storage in intracellular vesicles, neither of which correlates with MVB targeting (Fig. 3). In contrast, S264E has similar affinity as non-phosphorylated AQP2, indicating that these two forms AQP2 phosphorylated at Ser 264 as well as the non-phosphorylated form are likely to end up in MVB inner vesicles. Since exosomes originates from MVB inner vesicles (Pisitkun et al., 2004), this would agree with a proposed role of Ser 264 phosphorylation in AQP2 exosome excretion (Fig. 4) (Fenton et al., 2008). Removal of all the AQP2 phosphorylation sites through truncation resulted in a construct that had the by far lowest affinity toward LIP5, despite the fact that the

LIP5-binding motif was retained. Since a peptide corresponding to the truncated region could not interact with LIP5 on its own (Roche et al., 2017), this suggests that phosphorylation of AQP2 allosterically controls its interaction with LIP5 by altering the conformation of the C-terminus or modulating a secondary interaction site.

5. Concluding remarks

AQP2 is one of the best-characterized membrane proteins in terms of its trafficking and often serves as the canonical example for hormone-induced regulated exocytosis. Although much is known about this process at the cellular level, the molecular mechanism behind how AQP2 moves between different sub-cellular compartments remains poorly understood. In order to understand this, the details of how AQP2 interacts with the proteins in the trafficking machinery and how this is controlled by phosphorylation must be elucidated. This can only be achieved through a cross-disciplinary approach where studies in cells are combined with in interaction and structural studies in vitro. Specifically, structural information for the transient complexes involved in AQP2 trafficking will be crucial in understanding how the molecular mechanism behind how sorting signals such as phosphorylation selectively recruits AQP2 into specific vesicles and how these vesicles reach the target destination at the right time. Such structural information would not only shed light on AQP2 trafficking *per se*, but, due to the generality of the trafficking machinery, also be tremendously important for understanding membrane protein trafficking in general.

References

Altmeyer, A., Maki, R. G., Feldweg, A. M., Heike, M., Protopopov, V. P., Masur, S. K., et al. (1996). Tumor-specific cell surface expression of the-KDEL containing, endoplasmic reticular heat shock protein gp96. *International Journal of Cancer, 69*(4), 340–349. https://doi.org/10.1002/(SICI)1097-0215(19960822)69:4<340::AID-IJC18>3.0.CO;2-9.

Angelidis, C. E., Lazaridis, I., & Pagoulatos, G. N. (1999). Aggregation of hsp70 and hsc70 in vivo is distinct and temperature-dependent and their chaperone function is directly related to non-aggregated forms. *European Journal of Biochemistry, 259*(1–2), 505–512.

Angrand, P. O., Segura, I., Volkel, P., Ghidelli, S., Terry, R., Brajenovic, M., et al. (2006). Transgenic mouse proteomics identifies new 14-3-3-associated proteins involved in cytoskeletal rearrangements and cell signaling. *Molecular & Cellular Proteomics, 5*(12), 2211–2227. https://doi.org/10.1074/mcp.M600147-MCP200.

Barile, M., Pisitkun, T., Yu, M. J., Chou, C. L., Verbalis, M. J., Shen, R. F., et al. (2005). Large scale protein identification in intracellular aquaporin-2 vesicles from renal inner medullary collecting duct. *Molecular & Cellular Proteomics*, *4*(8), 1095–1106. https://doi.org/10.1074/mcp.M500049-MCP200.

Boilard, M., Reyes-Moreno, C., Lachance, C., Massicotte, L., Bailey, J. L., Sirard, M. A., et al. (2004). Localization of the chaperone proteins GRP78 and HSP60 on the luminal surface of bovine oviduct epithelial cells and their association with spermatozoa. *Biology of Reproduction*, *71*(6), 1879–1889. https://doi.org/10.1095/biolreprod.103.026849.

Bouley, R., Breton, S., Sun, T., McLaughlin, M., Nsumu, N. N., Lin, H. Y., et al. (2000). Nitric oxide and atrial natriuretic factor stimulate cGMP-dependent membrane insertion of aquaporin 2 in renal epithelial cells. *The Journal of Clinical Investigation*, *106*(9), 1115–1126. https://doi.org/10.1172/JCI9594.

Bronk, P., Wenniger, J. J., Dawson-Scully, K., Guo, X., Hong, S., Atwood, H. L., et al. (2001). Drosophila Hsc70-4 is critical for neurotransmitter exocytosis in vivo. *Neuron*, *30*(2), 475–488.

Brown, D., Hasler, U., Nunes, P., Bouley, R., & Lu, H. A. (2008). Phosphorylation events and the modulation of aquaporin 2 cell surface expression. *Current Opinion in Nephrology and Hypertension*, *17*(5), 491–498. https://doi.org/10.1097/MNH.0b013e3283094eb1.

Chang, H. C., Newmyer, S. L., Hull, M. J., Ebersold, M., Schmid, S. L., & Mellman, I. (2002). Hsc70 is required for endocytosis and clathrin function in *Drosophila*. *The Journal of Cell Biology*, *159*(3), 477–487. https://doi.org/10.1083/jcb.200205086.

Chappell, T. G., Welch, W. J., Schlossman, D. M., Palter, K. B., Schlesinger, M. J., & Rothman, J. E. (1986). Uncoating ATPase is a member of the 70 kilodalton family of stress proteins. *Cell*, *45*(1), 3–13.

Cheong, K. H., Zacchetti, D., Schneeberger, E. E., & Simons, K. (1999). VIP17/MAL, a lipid raft-associated protein, is involved in apical transport in MDCK cells. *Proceedings of the National Academy of Sciences of the United States of America*, *96*(11), 6241–6248.

Cheung, P. W., Ueberdiek, L., Day, J., Bouley, R., & Brown, D. (2017). Protein phosphatase 2C is responsible for VP-induced dephosphorylation of AQP2 serine 261. *American Journal of Physiology. Renal Physiology*, *313*(2), F404–F413. https://doi.org/10.1152/ajprenal.00004.2017.

Clay, J. R., & Kuzirian, A. (2002). Trafficking of axonal K+ channels: Potential role of Hsc70. *Journal of Neuroscience Research*, *67*(6), 745–752. https://doi.org/10.1002/jnr.10182.

Cosson, P., Perrin, J., & Bonifacino, J. S. (2013). Anchors aweigh: Protein localization and transport mediated by transmembrane domains. *Trends in Cell Biology*, *23*(10), 511–517. https://doi.org/10.1016/j.tcb.2013.05.005.

Davidson, D. J., Haskell, C., Majest, S., Kherzai, A., Egan, D. A., Walter, K. A., et al. (2005). Kringle 5 of human plasminogen induces apoptosis of endothelial and tumor cells through surface-expressed glucose-regulated protein 78. *Cancer Research*, *65*(11), 4663–4672. https://doi.org/10.1158/0008-5472.CAN-04-3426.

Delpino, A., & Castelli, M. (2002). The 78 kDa glucose-regulated protein (GRP78/BIP) is expressed on the cell membrane, is released into cell culture medium and is also present in human peripheral circulation. *Bioscience Reports*, *22*(3–4), 407–420.

Derby, M. C., & Gleeson, P. A. (2007). New insights into membrane trafficking and protein sorting. *International Review of Cytology*, *261*, 47–116. https://doi.org/10.1016/S0074-7696(07)61002-X.

Dougherty, M. K., & Morrison, D. K. (2004). Unlocking the code of 14-3-3. *Journal of Cell Science*, *117*(Pt. 10), 1875–1884. https://doi.org/10.1242/jcs.01171.

Fenton, R. A., Moeller, H. B., Hoffert, J. D., Yu, M. J., Nielsen, S., & Knepper, M. A. (2008). Acute regulation of aquaporin-2 phosphorylation at Ser-264 by vasopressin. *Proceedings of the National Academy of Sciences of the United States of America*, *105*(8), 3134–3139. https://doi.org/10.1073/pnas.0712338105.

Frank, M., van der Haar, M. E., Schaeren-Wiemers, N., & Schwab, M. E. (1998). rMAL is a glycosphingolipid-associated protein of myelin and apical membranes of epithelial cells in kidney and stomach. *The Journal of Neuroscience*, *18*(13), 4901–4913.

Gebauer, M., Zeiner, M., & Gehring, U. (1997). Proteins interacting with the molecular chaperone hsp70/hsc70: Physical associations and effects on refolding activity. *FEBS Letters*, *417*(1), 109–113.

Gerke, V., Creutz, C. E., & Moss, S. E. (2005). Annexins: Linking Ca2+ signalling to membrane dynamics. *Nature Reviews. Molecular Cell Biology*, *6*(6), 449–461. https://doi.org/10.1038/nrm1661.

Gething, M. J., & Sambrook, J. (1992). Protein folding in the cell. *Nature*, *355*(6355), 33–45. https://doi.org/10.1038/355033a0.

Goldfarb, S. B., Kashlan, O. B., Watkins, J. N., Suaud, L., Yan, W., Kleyman, T. R., et al. (2006). Differential effects of Hsc70 and Hsp70 on the intracellular trafficking and functional expression of epithelial sodium channels. *Proceedings of the National Academy of Sciences of the United States of America*, *103*(15), 5817–5822. https://doi.org/10.1073/pnas.0507903103.

Gooch, J. L., Guler, R. L., Barnes, J. L., & Toro, J. J. (2006). Loss of calcineurin Aalpha results in altered trafficking of AQP2 and in nephrogenic diabetes insipidus. *Journal of Cell Science*, *119*(Pt. 12), 2468–2476. https://doi.org/10.1242/jcs.02971.

Hayes, M. J., Rescher, U., Gerke, V., & Moss, S. E. (2004). Annexin-actin interactions. *Traffic*, *5*(8), 571–576. https://doi.org/10.1111/j.1600-0854.2004.00210.x.

Hoffert, J. D., Fenton, R. A., Moeller, H. B., Simons, B., Tchapyjnikov, D., McDill, B. W., et al. (2008). Vasopressin-stimulated increase in phosphorylation at Ser269 potentiates plasma membrane retention of aquaporin-2. *The Journal of Biological Chemistry*, *283*(36), 24617–24627. https://doi.org/10.1074/jbc.M803074200.

Hoffert, J. D., Pisitkun, T., Wang, G., Shen, R. F., & Knepper, M. A. (2006). Quantitative phosphoproteomics of vasopressin-sensitive renal cells: Regulation of aquaporin-2 phosphorylation at two sites. *Proceedings of the National Academy of Sciences of the United States of America*, *103*(18), 7159–7164. https://doi.org/10.1073/pnas.0600895103.

Jo, I., Ward, D. T., Baum, M. A., Scott, J. D., Coghlan, V. M., Hammond, T. G., et al. (2001). AQP2 is a substrate for endogenous PP2B activity within an inner medullary AKAP-signaling complex. *American Journal of Physiology. Renal Physiology*, *281*(5), F958–F965. https://doi.org/10.1152/ajprenal.2001.281.5.F958.

Johnson, C., Crowther, S., Stafford, M. J., Campbell, D. G., Toth, R., & MacKintosh, C. (2010). Bioinformatic and experimental survey of 14-3-3-binding sites. *The Biochemical Journal*, *427*(1), 69–78. https://doi.org/10.1042/BJ20091834.

Kamsteeg, E. J., Duffield, A. S., Konings, I. B., Spencer, J., Pagel, P., Deen, P. M., et al. (2007). MAL decreases the internalization of the aquaporin-2 water channel. *Proceedings of the National Academy of Sciences of the United States of America*, *104*(42), 16696–16701. https://doi.org/10.1073/pnas.0708023104.

Kamsteeg, E. J., Heijnen, I., van Os, C. H., & Deen, P. M. (2000). The subcellular localization of an aquaporin-2 tetramer depends on the stoichiometry of phosphorylated and nonphosphorylated monomers. *The Journal of Cell Biology*, *151*(4), 919–930.

Kamsteeg, E. J., Hendriks, G., Boone, M., Konings, I. B., Oorschot, V., van der Sluijs, P., et al. (2006). Short-chain ubiquitination mediates the regulated endocytosis of the aquaporin-2 water channel. *Proceedings of the National Academy of Sciences of the United States of America*, *103*(48), 18344–18349. https://doi.org/10.1073/pnas.0604073103.

Katsura, T., Ausiello, D. A., & Brown, D. (1996). Direct demonstration of aquaporin-2 water channel recycling in stably transfected LLC-PK1 epithelial cells. *The American Journal of Physiology*, *270*(3 Pt. 2), F548–F553. https://doi.org/10.1152/ajprenal.1996.270.3.F548.

Katsura, T., Gustafson, C. E., Ausiello, D. A., & Brown, D. (1997). Protein kinase a phosphorylation is involved in regulated exocytosis of aquaporin-2 in transfected LLC-PK1 cells. *The American Journal of Physiology*, *272*(6 Pt. 2), F817–F822.

Kim, T., Fiedler, K., Madison, D. L., Krueger, W. H., & Pfeiffer, S. E. (1995). Cloning and characterization of MVP17: A developmentally regulated myelin protein in oligodendrocytes. *Journal of Neuroscience Research*, *42*(3), 413–422. https://doi.org/10.1002/jnr.490420316.

Kirchhausen, T. (2000). Three ways to make a vesicle. *Nature Reviews. Molecular Cell Biology*, *1*(3), 187–198. https://doi.org/10.1038/35043117.

Klussmann, E., Tamma, G., Lorenz, D., Wiesner, B., Maric, K., Hofmann, F., et al. (2001). An inhibitory role of Rho in the vasopressin-mediated translocation of aquaporin-2 into cell membranes of renal principal cells. *The Journal of Biological Chemistry*, *276*(23), 20451–20457. https://doi.org/10.1074/jbc.M010270200.

Kortenoeven, M. L., & Fenton, R. A. (2014). Renal aquaporins and water balance disorders. *Biochimica et Biophysica Acta*, *1840*(5), 1533–1549. https://doi.org/10.1016/j.bbagen.2013.12.002.

Kwon, T. H., Nielsen, J., Moller, H. B., Fenton, R. A., Nielsen, S., & Frokiaer, J. (2009). Aquaporins in the kidney. In *Handbook of experimental pharmacology Vol. 190.* (pp. 95–132). Springer Nature. https://doi.org/10.1007/978-3-540-79885-9_5.

Li, W., Jin, W. W., Tsuji, K., Chen, Y., Nomura, N., Su, L., et al. (2017). Ezrin directly interacts with AQP2 and promotes its endocytosis. *Journal of Cell Science*, *130*(17), 2914–2925. https://doi.org/10.1242/jcs.204842.

Liang, S., Yu, Y., Yang, P., Gu, S., Xue, Y., & Chen, X. (2009). Analysis of the protein complex associated with 14-3-3 epsilon by a deuterated-leucine labeling quantitative proteomics strategy. *Journal of Chromatography. B, Analytical Technologies in the Biomedical and Life Sciences*, *877*(7), 627–634. https://doi.org/10.1016/j.jchromb.2009.01.023.

Lu, H. A., Sun, T. X., Matsuzaki, T., Yi, X. H., Eswara, J., Bouley, R., et al. (2007). Heat shock protein 70 interacts with aquaporin-2 and regulates its trafficking. *The Journal of Biological Chemistry*, *282*(39), 28721–28732.

Mackintosh, C. (2004). Dynamic interactions between 14-3-3 proteins and phosphoproteins regulate diverse cellular processes. *The Biochemical Journal*, *381*(Pt. 2), 329–342. https://doi.org/10.1042/BJ20031332.

Markoff, A., & Gerke, V. (2005). Expression and functions of annexins in the kidney. *American Journal of Physiology. Renal Physiology*, *289*(5), F949–F956. https://doi.org/10.1152/ajprenal.00089.2005.

Meacham, G. C., Patterson, C., Zhang, W., Younger, J. M., & Cyr, D. M. (2001). The Hsc70 co-chaperone CHIP targets immature CFTR for proteasomal degradation. *Nature Cell Biology*, *3*(1), 100–105. https://doi.org/10.1038/35050509.

Moeller, H. B., Aroankins, T. S., Slengerik-Hansen, J., Pisitkun, T., & Fenton, R. A. (2014). Phosphorylation and ubiquitylation are opposing processes that regulate endocytosis of the water channel aquaporin-2. *Journal of Cell Science*, *127*(Pt. 14), 3174–3183. https://doi.org/10.1242/jcs.150680.

Moeller, H. B., Fuglsang, C. H., & Fenton, R. A. (2016). Renal aquaporins and water balance disorders. *Best Practice & Research. Clinical Endocrinology & Metabolism*, *30*(2), 277–288. https://doi.org/10.1016/j.beem.2016.02.012.

Moeller, H. B., MacAulay, N., Knepper, M. A., & Fenton, R. A. (2009). Role of multiple phosphorylation sites in the COOH-terminal tail of aquaporin-2 for water transport: Evidence against channel gating. *American Journal of Physiology. Renal Physiology*, *296*(3), F649–F657. https://doi.org/10.1152/ajprenal.90682.2008.

Moeller, H. B., Praetorius, J., Rutzler, M. R., & Fenton, R. A. (2010). Phosphorylation of aquaporin-2 regulates its endocytosis and protein-protein interactions. *Proceedings of the National Academy of Sciences of the United States of America*, *107*(1), 424–429.

Moeller, H. B., Rittig, S., & Fenton, R. A. (2013). Nephrogenic diabetes insipidus: Essential insights into the molecular background and potential therapies for treatment. *Endocrine Reviews*, *34*, 278–301. https://doi.org/10.1210/er.2012-1044.

Moeller, H. B., Slengerik-Hansen, J., Aroankins, T., Assentoft, M., MacAulay, N., Moestrup, S. K., et al. (2016). Regulation of the water channel aquaporin-2 via 14-3-3theta and -zeta. *The Journal of Biological Chemistry*, *291*(5), 2469–2484. https://doi.org/10.1074/jbc.M115.691121.

Morgan, J. R., Prasad, K., Jin, S., Augustine, G. J., & Lafer, E. M. (2001). Uncoating of clathrin-coated vesicles in presynaptic terminals: Roles for Hsc70 and auxilin. *Neuron*, *32*(2), 289–300.

Newmyer, S. L., & Schmid, S. L. (2001). Dominant-interfering Hsc70 mutants disrupt multiple stages of the clathrin-coated vesicle cycle in vivo. *The Journal of Cell Biology*, *152*(3), 607–620.

Noda, Y., Horikawa, S., Furukawa, T., Hirai, K., Katayama, Y., Asai, T., et al. (2004). Aquaporin-2 trafficking is regulated by PDZ-domain containing protein SPA-1. *FEBS Letters*, *568*(1–3), 139–145. https://doi.org/10.1016/j.febslet.2004.05.021.

Noda, Y., Horikawa, S., Kanda, E., Yamashita, M., Meng, H., Eto, K., et al. (2008). Reciprocal interaction with G-actin and tropomyosin is essential for aquaporin-2 trafficking. *The Journal of Cell Biology*, *182*(3), 587–601. https://doi.org/10.1083/jcb.200709177.

Noda, Y., Horikawa, S., Katayama, Y., & Sasaki, S. (2004). Water channel aquaporin-2 directly binds to actin. *Biochemical and Biophysical Research Communications*, *322*(3), 740–745. https://doi.org/10.1016/j.bbrc.2004.07.195.

Noda, Y., Horikawa, S., Katayama, Y., & Sasaki, S. (2005). Identification of a multiprotein "motor" complex binding to water channel aquaporin-2. *Biochemical and Biophysical Research Communications*, *330*(4), 1041–1047. https://doi.org/10.1016/j.bbrc.2005.03.079.

Park, E. J., Lim, J. S., Jung, H. J., Kim, E., Han, K. H., & Kwon, T. H. (2013). The role of 70-kDa heat shock protein in dDAVP-induced AQP2 trafficking in kidney collecting duct cells. *American Journal of Physiology. Renal Physiology*, *304*(7), F958–F971. https://doi.org/10.1152/ajprenal.00469.2012.

Pisitkun, T., Shen, R. F., & Knepper, M. A. (2004). Identification and proteomic profiling of exosomes in human urine. *Proceedings of the National Academy of Sciences of the United States of America*, *101*(36), 13368–13373. https://doi.org/10.1073/pnas.0403453101.

Pollard, T. D. (2016). Actin and actin-binding proteins. *Cold Spring Harbor Perspectives in Biology*, *8*(8), 1–17. https://doi.org/10.1101/cshperspect.a018226.

Pollard, T. D., & Cooper, J. A. (2009). Actin, a central player in cell shape and movement. *Science*, *326*(5957), 1208–1212. https://doi.org/10.1126/science.1175862.

Reichow, S. L., Clemens, D. M., Freites, J. A., Nemeth-Cahalan, K. L., Heyden, M., Tobias, D. J., et al. (2013). Allosteric mechanism of water-channel gating by Ca^{2+}–calmodulin. *Nature Structural & Molecular Biology*, *20*(9), 1085–1092. https://doi.org/10.1038/nsmb.2630.

Ren, H., Yang, B., Ruiz, J. A., Efe, O., Ilori, T. O., Sands, J. M., et al. (2016). Phosphatase inhibition increases AQP2 accumulation in the rat IMCD apical plasma membrane. *American Journal of Physiology. Renal Physiology*, *311*(6), F1189–F1197. https://doi.org/10.1152/ajprenal.00150.2016.

Rinschen, M. M., Yu, M. J., Wang, G., Boja, E. S., Hoffert, J. D., Pisitkun, T., et al. (2010). Quantitative phosphoproteomic analysis reveals vasopressin V2-receptor-dependent signaling pathways in renal collecting duct cells. *Proceedings of the National Academy of Sciences of the United States of America*, *107*(8), 3882–3887. https://doi.org/10.1073/pnas.0910646107.

Robaszkiewicz, K., Ostrowska, Z., Marchlewicz, K., & Moraczewska, J. (2016). Tropomyosin isoforms differentially modulate the regulation of actin filament polymerization and depolymerization by cofilins. *The FEBS Journal*, *283*(4), 723–737. https://doi.org/10.1111/febs.13626.

Roche, J. V., Survery, S., Kreida, S., Nesverova, V., Ampah-Korsah, H., Gourdon, M., et al. (2017). Phosphorylation of human aquaporin 2 (AQP2) allosterically controls its interaction with the lysosomal trafficking protein LIP5. *The Journal of Biological Chemistry, 292*, 14636–14648. https://doi.org/10.1074/jbc.M117.788364.

Shani, G., Fischer, W. H., Justice, N. J., Kelber, J. A., Vale, W., & Gray, P. C. (2008). GRP78 and Cripto form a complex at the cell surface and collaborate to inhibit transforming growth factor beta signaling and enhance cell growth. *Molecular and Cellular Biology, 28*(2), 666–677. https://doi.org/10.1128/MCB.01716-07.

Simon, H., Gao, Y., Franki, N., & Hays, R. M. (1993). Vasopressin depolymerizes apical F-actin in rat inner medullary collecting duct. *The American Journal of Physiology, 265*(3 Pt. 1), C757–C762. https://doi.org/10.1152/ajpcell.1993.265.3.C757.

Simons, K., & Ikonen, E. (1997). Functional rafts in cell membranes. *Nature, 387*(6633), 569–572. https://doi.org/10.1038/42408.

Skalicky, J. J., Arii, J., Wenzel, D. M., Stubblefield, W. M., Katsuyama, A., Uter, N. T., et al. (2012). Interactions of the human LIP5 regulatory protein with endosomal sorting complexes required for transport. *The Journal of Biological Chemistry, 287*(52), 43910–43926. https://doi.org/10.1074/jbc.M112.417899.

Takayama, S., Xie, Z., & Reed, J. C. (1999). An evolutionarily conserved family of Hsp70/Hsc70 molecular chaperone regulators. *The Journal of Biological Chemistry, 274*(2), 781–786.

Tamma, G., Klussmann, E., Maric, K., Aktories, K., Svelto, M., Rosenthal, W., et al. (2001). Rho inhibits cAMP-induced translocation of aquaporin-2 into the apical membrane of renal cells. *American Journal of Physiology. Renal Physiology, 281*(6), F1092–F1101. https://doi.org/10.1152/ajprenal.0091.2001.

Tamma, G., Procino, G., Mola, M. G., Svelto, M., & Valenti, G. (2008). Functional involvement of annexin-2 in cAMP induced AQP2 trafficking. *Pflügers Archiv—European Journal of Physiology, 456*(4), 729–736. https://doi.org/10.1007/s00424-008-0453-1.

Tamma, G., Robben, J. H., Trimpert, C., Boone, M., & Deen, P. M. (2011). Regulation of AQP2 localization by S256 and S261 phosphorylation and ubiquitination. *American Journal of Physiology. Cell Physiology, 300*(3), C636–C646. https://doi.org/10.1152/ajpcell.00433.2009.

Trimpert, C., Wesche, D., de Groot, T., Pimentel Rodriguez, M. M., Wong, V., van den Berg, D. T. M., et al. (2017). NDFIP allows NEDD4/NEDD4L-induced AQP2 ubiquitination and degradation. *PLoS One, 12*(9). e0183774https://doi.org/10.1371/journal.pone.0183774.

van Balkom, B. W., Boone, M., Hendriks, G., Kamsteeg, E. J., Robben, J. H., Stronks, H. C., et al. (2009). LIP5 interacts with aquaporin 2 and facilitates its lysosomal degradation. *Journal of the American Society of Nephrology, 20*(5), 990–1001.

van Balkom, B. W., Savelkoul, P. J., Markovich, D., Hofman, E., Nielsen, S., van der Sluijs, P., et al. (2002). The role of putative phosphorylation sites in the targeting and shuttling of the aquaporin-2 water channel. *The Journal of Biological Chemistry, 277*(44), 41473–41479. https://doi.org/10.1074/jbc.M207525200.

Vild, C. J., Li, Y., Guo, E. Z., Liu, Y., & Xu, Z. (2015). A novel mechanism of regulating the ATPase VPS4 by its cofactor LIP5 and the endosomal sorting complex required for transport (ESCRT)-III protein CHMP5. *The Journal of Biological Chemistry, 290*(11), 7291–7303. https://doi.org/10.1074/jbc.M114.616730.

Wiest, D. L., Bhandoola, A., Punt, J., Kreibich, G., McKean, D., & Singer, A. (1997). Incomplete endoplasmic reticulum (ER) retention in immature thymocytes as revealed by surface expression of "ER-resident" molecular chaperones. *Proceedings of the National Academy of Sciences of the United States of America, 94*(5), 1884–1889.

Xiao, G., Chung, T. F., Pyun, H. Y., Fine, R. E., & Johnson, R. J. (1999). KDEL proteins are found on the surface of NG108-15 cells. *Brain Research. Molecular Brain Research*, *72*(2), 121–128.

Yang, D., & Hurley, J. H. (2010). Structural role of the Vps4-Vta1 interface in ESCRT-III recycling. *Structure*, *18*(8), 976–984. https://doi.org/10.1016/j.str.2010.04.014.

Zhang, Y., Nijbroek, G., Sullivan, M. L., McCracken, A. A., Watkins, S. C., Michaelis, S., et al. (2001). Hsp70 molecular chaperone facilitates endoplasmic reticulum-associated protein degradation of cystic fibrosis transmembrane conductance regulator in yeast. *Molecular Biology of the Cell*, *12*(5), 1303–1314. https://doi.org/10.1091/mbc.12.5.1303.

Zhang, H., Peters, K. W., Sun, F., Marino, C. R., Lang, J., Burgoyne, R. D., et al. (2002). Cysteine string protein interacts with and modulates the maturation of the cystic fibrosis transmembrane conductance regulator. *The Journal of Biological Chemistry*, *277*(32), 28948–28958. https://doi.org/10.1074/jbc.M111706200.

Zwang, N. A., Hoffert, J. D., Pisitkun, T., Moeller, H. B., Fenton, R. A., & Knepper, M. A. (2009). Identification of phosphorylation-dependent binding partners of aquaporin-2 using protein mass spectrometry. *Journal of Proteome Research*, *8*(3), 1540–1554. https://doi.org/10.1021/pr800894p.

CHAPTER SIX

Regulation of aquaporin-2 by RNA interference

Hyun Jun Jung[a,b], Eui-Jung Park[b], Hyo-Jung Choi[b], Tae-Hwan Kwon[b,*]

[a]Division of Nephrology, Department of Medicine, Johns Hopkins University School of Medicine, Baltimore, MD, United States

[b]Department of Biochemistry and Cell Biology, School of Medicine, Kyungpook National University, Taegu, South Korea

*Corresponding author: e-mail address: thkwon@knu.ac.kr

Contents

Abstract

MicroRNAs (miRNAs) are short non-coding RNAs that interact with 3′-untranslated regions of mRNAs. miRNAs modulate gene expression by regulating mRNA translation and provide novel insights into the complex regulation of protein expression and function. In this chapter, we describe the general features of RNA interference, identification of miRNAs, and interaction of miRNAs with their target genes. In particular, we have shown that several miRNAs are responsive to arginine vasopressin or aldosterone

Vitamins and Hormones, Volume 112
ISSN 0083-6729
https://doi.org/10.1016/bs.vh.2019.08.003

stimulation in mouse cortical collecting duct mpkCCD cells. Moreover, we identified both miR-32 and miR-137 as AQP2-targeting miRNAs using *in silico* analysis and also identified several target genes of miR-32 and miR-137. As the target sequences of miR-32 and miR-137 are commonly found in mRNAs of vasopressin-regulated genes, further studies regarding the interaction of miRNAs with their target genes are required to obtain comprehensive understanding of miRNA-regulated AQP2 expression in kidney collecting duct cells.

1. General features of RNA interference

1.1 RNA interference by small RNAs

RNA interference (RNAi), also known as post-transcriptional gene silencing (PTGS), is a mechanism for regulation of gene expression that has evolved in eukaryotic cells. RNAi is mediated by "microRNAs (miRNAs)" and "small interfering RNAs (siRNAs)" derived from endogenous transcripts (Ambros, 2001; Bartel, 2004; Grivna, Beyret, Wang, & Lin, 2006; McManus & Sharp, 2002; Piatek & Werner, 2014). Endogenous transcripts of primary miRNAs (pri-miRNAs), which have a stem-loop structure, are coded from genomic regions that do not code for proteins, such as intronic regions of protein-coding genes and intergenic regions. Bioinformatic analyses have revealed that a significant fraction of primary transcripts of intergenic miRNAs is 3–4 kilobases (kb) in length with clearly defined 5′- and 3′-ends (Saini, Griffiths-Jones, & Enright, 2007). Fig. 1 depicts miRNA biogenesis. Pri-miRNAs formed from the primary transcript contain one or more miRNAs in stem-loop structures (60–120 nt long) (Lund, Guttinger, Calado, Dahlberg, & Kutay, 2004). The miRNAs that are generated from the same primary transcript form a cluster of miRNAs, which are regulated in the same direction way. Pri-miRNAs are processed to ~70-nt long pre-miRNAs by a class II RNase-III enzyme Drosha, a dsRNA-specific ribonuclease, in the nucleus. Pre-miRNAs are exported to the cytoplasm by Exportin proteins, and are converted to small double-stranded RNAs (dsRNAs, 21–25 nt), the so-called mature miRNAs by another RNase III-like enzyme, Dicer. For siRNA biogenesis, Dicer acts on long-dsRNAs in the cytoplasm and produces small dsRNAs similar to mature miRNAs. Thereafter, mature miRNAs and siRNAs are then incorporated into endonuclease (RNase H) Argonaute proteins, a component of the RNA-induced silencing complex (RISC) effector complex, which separates dsRNA into single-stranded RNA (ssRNA) by its helicase activity. Argonaute-associated miRNAs or siRNAs are separated into two single

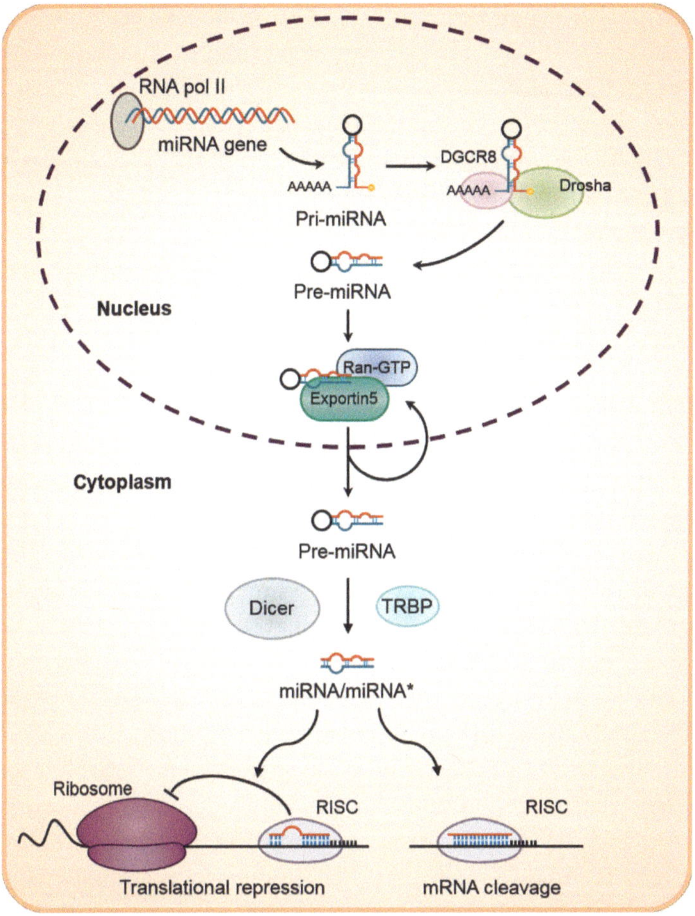

Fig. 1 Mechanism of miRNA biogenesis. miRNAs are transcribed by RNA polymerase II (RNA pol II), then, primary miRNAs (pri-miRNAs) are excised as the stem loop by Drosha and cofactor (DGCR8, DiGeorge syndrome critical region gene 8) to produce precursor miRNAs (Pre-miRNAs). The pre-miRNAs are translocated to cytoplasm by interacting with exportin 5 in the presence of Ran-GTP. Dicer (RNase III family) and TAR-RNA binding protein (TRBP) cleave the loop of pre-miRNAs to produce miRNA/miRNA* duplex. miRNA/miRNA* duplexes are transferred to Argonaute protein to assemble RNA-induced silencing complex (RISC). The passenger strand in the duplex, denoted with an asterisk (*), is relatively unstable and is normally degraded; the remaining strand, termed the guide strand, interacts with 3′ untranslated region (UTR) of target mRNA and represses the translation of mRNAs or degrades mRNAs.

strands, namely, the guide strand and passenger strand. The guide strand of miRNAs interacts with the 3′-untranslated region (UTRs) of the target mRNA based on sequence specificity; however, miRNAs and siRNAs act via different pathways leading to the degradation of target mRNAs

(Lam, Chow, Zhang, & Leung, 2015). The sequences of siRNAs are fully complementary to the 3′-UTR of their target mRNA, leading to the activation of the endonuclease activity of RISC, and the subsequent degradation of a specific target mRNA. Unlike the sequence specificity of siRNAs to their target mRNA, the sequences of miRNAs are partially complementary to the 3′-UTR of their target mRNA, which result in recruitment of other modulator complexes, including the decapping enzyme and the depolyadenylation enzyme (Nishihara, Zekri, Braun, & Izaurralde, 2013; Treiber, Treiber, & Meister, 2019).

1.2 Post-transcriptional regulation by miRNAs in the kidney

Mature miRNAs incorporated into RISC impair the translation of their target mRNAs by interacting with the target mRNAs based on sequence homology (Eichhorn et al., 2014); however, miRNAs show less specificity to their target mRNA sequence than that shown by siRNAs. Therefore, RNAi mediated by miRNAs is considered a major post-transcriptional process for regulating gene expression. Distinct subsets of genes are active in different cell types, coding for proteins that represent the features of the specific cell type including both structure and function. Approximately 200 distinct cell types exist in the human body and mechanisms for the regulation of transcription also differ with respect to cellular signaling and downstream factors such as transcription factors and coregulators. In addition to the intricacy of transcriptional regulation, gene silencing mediated by endogenous miRNAs increases the complexity of cell-type and tissue-type specific expression of genes. Development and differentiation of pluripotent stem cells or progenitor cells into specific cell types result from multiple and sequential processes associated with a subset of genes that are modulated in each process. Destabilization of mRNAs mediated by miRNAs is a critical process in cellular transformation (Cordes & Srivastava, 2009; Kumar, Lu, Mercer, Golub, & Jacks, 2007; Lu et al., 2008).

miRNA-mediated modulation of gene expression is widely been known to be involved in specification of different cell lineages during renal development (Ho & Kreidberg, 2013; Lu et al., 2005; Yu, 2014). In particular, the development of the ureteric bud epithelium lineage in the metanephric kidney involves several sequential steps during the formation of the collecting duct system. The renal cortical collecting duct is the renal tubular segment in which water and acid-base balance is regulated through membrane channels and transporter proteins expressed in heterogeneous cell types, such as principal cells (regulating reabsorption of water and sodium and secretion of potassium) and two types of intercalated cells (type A-intercalated cells which secrete hydrogen ions and type B-intercalated cells which secrete bicarbonate ions) (Agre et al., 2002; Chen et al., 2017; Kim, Kim, Cha, Tisher, & Madsen, 1999; Kim et al., 2002; Knepper, Kwon, & Nielsen, 2015). Different subsets of genes in each cell type are involved in imparting specific functions to specific cell types, and this facilitates complex regulation of water and electrolyte homeostasis in the collecting duct segment (Chen et al. 29089413). miRNAs mainly regulate the processes associated with differentiation of collecting duct cells from the ureteric bud trunks (Yu, 2014).

miRNA-mediated regulation of target genes and hence protein expression in the kidney is relevant to renal homeostasis and pathogenesis (Chandrasekaran et al., 2012). Prior studies have proposed an important physiological role of miRNAs in renal tubules based on the results of suppressing the expression of major components of the miRNA biogenesis pathway (Hajarnis et al., 2018; Nagalakshmi et al., 2011). miRNA suppression of key enzymes in the miRNA biogenesis pathway, such as DICER1, DGCR8, and the entire Argonaute family (AGO1, 2, 3, and 4), via conditional collecting duct-specific disturbance, downregulates expression of miRNAs and subsequently induces partial epithelial-to-mesenchymal transition (EMT) associated with the progression of tubulointerstitial fibrosis (Hajarnis et al., 2018).

1.3 Post-transcriptional regulation of renal aquaporins mediated by miRNAs

Aquaporins (AQPs) are water channel proteins involved in the regulation of cellular fluid volume and body water homeostasis; they are expressed in various tissues and organs (Agre et al., 2002; Borgnia, Nielsen, Engel, & Agre, 1999). In the kidney, several tubular segments of nephrons express different AQPs, representing the heterogeneity of cell types along the nephron (Kwon et al., 2001; Nielsen, Frokiaer, et al., 2002). For example, AQP1 is mainly expressed in the proximal tubule and in the long descending thin limb of the loop of Henle. AQP7 is specifically localized at the apical plasma membrane of cells in the straight segment of the proximal tubule. AQP2 is localized in the connecting tubule and the collecting duct in the renal cortex and medulla where vasopressin-responsive reabsorption of water takes place (Coleman, Wu, Liu, & Wade, 2000; Rojek, Fuchtbauer, Kwon, Frokiaer, & Nielsen, 2006). Differential expression levels of AQPs in distinct cell types and changes in physiological environment could be affected under the miRNA-associated transcriptional and post-transcriptional regulation (Gomes, da Silva, Rodrigues, Castro, & Soveral, 2018).

2. Identification of small RNA molecules

2.1 Identification of small RNA molecule expression

miRNAs were discovered in 1993 by Lee and colleagues in the nematode *Caenorhabditis elegans* (Lee, Feinbaum, & Ambros, 1993). This finding led to the discovery of a number of miRNAs expressed in vertebrates, including humans (Lagos-Quintana, Rauhut, Lendeckel, & Tuschl, 2001; Lagos-Quintana, Rauhut, Meyer, Borkhardt, & Tuschl, 2003). In addition to genome sequencing projects, currently, 38,589 hairpin precursor miRNAs and 48,860 different mature miRNA sequences have been annotated from 271 organisms (Kozomara, Birgaoanu, & Griffiths-Jones, 2019). In the human genome, 1917 hairpin precursors and 2654 mature sequences were annotated (Kozomara et al., 2019). As a large number of miRNAs have been identified in vertebrate genomes, a method for large-scale miRNA analysis is necessary for comprehensive evaluation of expression of thousands of miRNAs in cells or tissues.

The DNA microarray assay (Castoldi et al., 2006; Davison, Johnson, & Andruss, 2006; Nelson et al., 2004; Thomson, Parker, Perou, & Hammond, 2004) has widely been used in large-scale transcriptome profiling for miRNAs as well as mRNAs in specific cells and tissues. A set of oligonucleotide sequences complementary to all known miRNAs in an array platform allows the screening of novel miRNAs and comprehensive changes in expression of all known miRNAs. This method has provided new insights into integrative regulation of a large set of miRNAs associated with development and pathophysiological conditions (Barad et al., 2004; Harfe, 2005; Ichii & Horino, 2018; Miska et al., 2004; Tahiri, Aure, & Kristensen, 2018).

In the next-generation sequencing era, transcriptome profiling using RNA sequencing (RNA-seq) based on next-generation sequencing facilitates genome-wide identification of small non-coding RNAs, including miRNAs (Hunt, Broyles, Head, & Deo, 2015). Genome annotation using various bioinformatic analyses contributes to our knowledge of novel miRNAs and miRNA clusters. Although details of mechanisms underlying miRNA clustering are still unclear, a previous study provided evidence that supports evolutionary conservation and coordination between clustered miRNAs (Wang, Luo, Zhang, & Lu, 2016). Comprehensive expression analysis of miRNAs using extended genome annotation facilitates investigation of cell type- or tissue-specific miRNA expression associated with their functions (Halushka, Fromm, Peterson, & McCall, 2018). Furthermore, additional studies on the potential functions of miRNAs associated with various pathophysiological conditions could provide direction for translational miRNA research (Witwer & Halushka, 2016).

2.2 Identification of small RNA molecules-target mRNA interaction

In miRNA-mediated mRNA regulation, complementarity between the seed sequence of the miRNA (2–7-mer) to the 3′-UTR of the target mRNA is the biological principle of interaction between the miRNA and the target mRNA. Several bioinformatic tools based on algorithms that exploit sequence-based prediction were invented to identify miRNA-mRNA interaction (Table 1). Moreover, recent bioinformatic approaches provide more information about miRNA-mRNA interaction networks derived from integrative data analyses using multiple databases and high-throughput datasets generated from microarray and next-generation sequencing analysis (Table 1).

Approximately 2000 miRNAs are annotated in the human genome and approximately 20,000 protein-coding genes are expressed in each human cell (Regev et al., 2017). Regarding the number of interactions between miRNAs and protein-coding genes and the expression levels depending on the specific cell types and tissues, it needs a detailed explanation for interactive regulation between miRNAs and target mRNAs. For sequence-based computational prediction of miRNA targets, 6–8 nucleotide-long region at the 5′-end of miRNAs, a seed sequence, is used to identify sequences complementary to the 3′-UTR of the target mRNA (Lewis, Burge, & Bartel, 2005). Moreover, additional regions (3′-bases, nucleotides 12–17 of miRNAs) involved in supplementary paring are required for enhanced site efficacy for miRNA-mediated gene repression (Grimson et al., 2007). However, a few bases within these seed sequence regions only endow greater specificity of interaction between miRNAs and mRNAs. In fact, *in silico* miRNA target prediction and experimental findings showed that a single miRNA targets tens to hundreds of mRNAs. Some miRNAs widely regulate an entire regulatory network of a specific signaling pathway through interaction with multiple target mRNAs within a network. On the other hand, there is experimental evidence that reveals cross-interactions of multiple miRNAs, such as clustered miRNAs, to a single target mRNA (Bonci et al., 2008; Kim et al., 2009; Peter, 2010). Knowledge of cell type- or tissue-dependent expression level of each miRNA and mRNA from high-throughput gene expression analyses in various cells and tissues provides direct evidence to narrow down the diversity of miRNA-mRNA interaction.

Table 1 Examples of bioinformatics tools for identification of miRNA-mRNA interaction.

Program	Type	Algorithm feature	Availability
miRGate	Web-based	Sequence match Experimental validation	http://mirgate.bioinfo.cnio.es/
miRWalk (3.0)	Web-based	Sequence match Experimental validation	http://mirwalk.umm.uni-heidelberg.de/
TargetScan	Web-based	Sequence match	http://www.targetscan.org/
DIANA-Tarbase	Web-based	Sequence match Experimental validation	http://diana.imis.athena-innovation.gr
DIANA-microT-CDS	Web-based	Sequence match	http://diana.imis.athena-innovation.gr
DIANA-MirExTra	Web-based	Sequence match Experimental validation	http://diana.imis.athena-innovation.gr
Magia2	Web-based	Sequence match Expression data	http://gencomp.bio.unipd.it/magia2
BioVLAB-MMIA-NGS (MMIA)	Web-based	Sequence match Expression data	http://epigenomics.snu.ac.kr/biovlab_mmia_ngs/
miRBase	Web-based	Sequence match	http://www.mirbase.org/
PicTar	Web-based	Sequence match	https://pictar.mdc-berlin.de/
miRTarBase	Web-based	Experimental validation	http://mirtarbase.mbc.nctu.edu.tw
miRDB	Web-based	Sequence match Data mining	http://mirdb.org
RNA22	Web-based	Sequence match	https://cm.jefferson.edu/rna22/
RNA hybrid	Web-based	Sequence match	https://bibiserv.cebitec.uni-bielefeld.de/rnahybrid
miRNAtap	R package	Sequence match	https://bioconductor.org/packages/release/bioc/html/miRNAtap.html
RmiR	R package	Sequence match	https://www.bioconductor.org/packages/release/bioc/html/RmiR.html

3. Small RNA expression in response to vasopressin or aldosterone in the kidney

3.1 Vasopressin-responsive miRNAs and aquaporin 2 (AQP2)-targeting miRNAs in kidney inner medulla

miRNAs, important modulators of gene expression that act via regulation of the mRNA translation, provide novel insights in the intricate regulation of protein expression and function. Previously, we identified the vasopressin-responsive miRNAs in rat kidney inner medullary collecting duct (IMCD) cells using microarray chip assay (Kim, Jung, Lee, & Kwon, 2015). Moreover, target genes of the selected miRNAs and their biological function and processes were predicted by *in silico* analysis. Furthermore, miRNAs targeting AQP2 expression (miR-32 and miR-137) were identified, which provide a novel mechanism for regulation of AQP2 protein expression (Kim et al., 2015; Ranieri et al., 2018).

In the kidney, the collecting duct is a tubular segment that reabsorbs water under the control of arginine vasopressin (AVP) and vasopressin receptor type 2 (AVPR2) (Jung & Kwon, 2016; Knepper et al., 2015; Kwon et al., 2009; Kwon, Frokiaer, & Nielsen, 2013; Nielsen, Frokiaer, et al., 2002). In addition, sodium reabsorption and potassium secretion are regulated by aldosterone (Stanton BA, PMID: 3317634; Wang XY, PMID: 11457874). The antidiuretic hormone, AVP, is synthesized in neurons in the hypothalamus and stored in the posterior pituitary gland. The Verney receptor in the hypothalamus senses the osmotic stimuli and AVP is released into the systemic circulation when plasma osmolality is higher than the physiological threshold (290–295 mOsm/KgH_2O) (Schrier, Berl, & Anderson, 1979). AVP is also secreted in response to non-osmotic stimuli through different pathways, such as the parasympathetic afferent pathways (Schrier et al., 1979). After secretion into the systemic circulation and delivery to the kidney, AVP binds to AVPR2 and induces osmotic water permeability in the connecting tubule and collecting duct (Jung & Kwon, 2016; Knepper et al., 2015; Kwon et al., 2009, 2013; Nielsen, Frokiaer, et al., 2002). Specifically, AVP stimulates the intracellular trafficking of AQP2-expressing vesicles to the apical plasma membrane, thereby increasing the osmotic water permeability of collecting duct cells. Moreover, AVP increases AQP2 protein expression via induction of transcription of the *Aqp2* gene. Numerous studies have demonstrated that dysregulation of AQPs expressed in the renal tubular epithelial cells leads to severe disturbances in water balance (Knepper et al., 2015;

Nielsen, Frokiaer, et al., 2002). In particular, dysregulation of AVP-mediated water reabsorption in the collecting duct (i.e., dysregulation of AQP2) is a primary pathophysiological mechanism underlying disturbances in body water balance, such as systemic water retaining or water losing states (Knepper et al., 2015; Kwon, Frokiaer, Knepper, & Nielsen, 1998; Kwon et al., 2009; Nielsen, Frokiaer, et al., 2002; Nielsen, Kwon, Christensen, Frokiaer, & Nielsen, 2008; Rojek et al., 2006).

We performed microarray chip assay in the IMCD tubule suspensions from rat kidneys in the absence or the presence of desmopressin (dDAVP) stimulation (10^{-9}M, 2h) (Kim et al., 2015). The results demonstrated that 19 miRNAs, including both precursor- and mature-miRNAs, exhibited >1.3-fold change in the expression after dDAVP stimulation. To predict the interaction between miRNAs and their target genes, target mRNAs of the selected nine miRNAs (miR-127, miR-1, miR-873, miR-16, miR-206, miR-678, miR-496, miR-298, and miR-463), which showed >1.3-fold-change, were identified by TargetScan (Release 6.2). Target gene prediction analysis of miRNAs suggested the interaction of three miRNAs (miR-1, miR-16, and miR-206) with their 76 target genes. Consistent with this, a previous study showed that protein expression of 17 genes (Twf1, Ust, Arcn1, Hnrnpu, Cd2ap, Gpd2, C77370, Ash1l, Phf19, Atxn2, Med26, Usp14, Pnpla6, Snrk, Usp25, Ddx3x, and Coro1c) among these 76 predicted target genes was detected in mpkCCD (clone 11) cells stimulated with dDAVP (Yu et al. mpkCCD proteome database: http://dir.nhlbi.nih.gov/papers/lkem/mpkccdproteome/). Therefore, these miRNAs could play a role in the vasopressin signaling pathway in kidney collecting duct cells via inhibition of target gene expression. Among them, several predicted target genes were known to be involved in the vasopressin signaling in collecting duct cells. For example, E26 avian leukemia oncogene 1 (Ets1) has been suggested to be a transcription factor associated with AQP2 expression in mpkCCD (clone 11) cells (Nielsen, Frokiaer, et al., 2002). Myosin light chain kinase (Mylk), the predicted miR-16 target gene, is the calmodulin-dependent protein kinase involved in the calcium-dependent vasopressin signaling (Chou et al., 2004; Knepper, 2012). Mylk protein abundance was significantly decreased in rat kidney IMCD cells treated with dDAVP (Kim et al., 2015), consistent with a previous finding of Mylk expression in dDAVP-stimulated mpkCCD cells (http://helixweb.nih.gov/ESBL/Database/ProteinHalfLives/GeneSymbolATR.html).

Our *in silico* analysis also predicted miR-32 and miR-137 to be the AQP2-targeting miRNAs in rat and mouse kidney collecting ducts cells

(Kim et al., 2015). This finding was supported by *in situ* hybridization demonstrating the expression of these two miRNAs in the rat kidney IMCD cells. Interestingly, miR-32 and miR-137 were not identified as vasopressin-responsive miRNAs by microarray analysis, and this was further confirmed in the IMCD cells treated with dDAVP or in a rat model of altered water intake (Kim et al., 2015). Direct binding of miR-32 or miR-137 to 3′-UTR of AQP2 mRNA was demonstrated in the luciferase assay using the seed region of 3′-UTR of AQP2 mRNA, which has a sequence complementary with miR-32 or miR-137. Transfection of miR-32- or miR-137-mimic decreased the luminescence in the vector, indicating that miR-32 or miR-137 directly interacted with 3′-UTR of AQP2 mRNA. Consistent with this, quantitative real-time PCR and immunoblotting data showed that expression levels of AQP2 mRNA and protein were significantly decreased when miR-32 or miR-137 were overexpressed in mpkCCDc14 cells, indicating that both miR-32 and miR-137 target AQP2 mRNA (Kim et al., 2015). Expression of miR-32 and miR-137 was not changed by dDAVP stimulation; thus, this represents the vasopressin-independent regulation of AQP2.

A recent study also demonstrated that miR-137 plays a role in down-regulation of AQP2 (Ranieri et al., 2018). Pendrin/NCC double-knockout (dKO) mice model showed significant urinary calcium wasting and extracellular fluid volume depletion despite high vasopressin level (Ranieri et al., 2018). High concentration of calcium in urine activates calcium-sensing receptor (CaSR), which impairs vasopressin-induced trafficking and expression of AQP2 (Procino et al., 2012; Riccardi et al., 1995). Valenti and coworkers found that dKO mice demonstrated significantly increased miRNA-137 expression in the kidney inner medulla (Ranieri et al., 2018), which could contribute to the decreased expression of AQP2. In addition, a recent study showed that miR-132 regulates the hypothalamic AVP mRNA expression (Bijkerk et al., 2018). Methyl-CpG-binding protein-2 is one of the targets of a miR-132 and it inhibits hypothalamic AVP synthesis by binding to its enhancer region (Bijkerk et al., 2018).

Several genes regulate intracellular trafficking and protein expression of AQP2 (Hoorn, Pisitkun, Yu, & Knepper, 2008; Jung & Kwon, 2016; Nielsen, Frokiaer, et al., 2002; Sandoval et al., 2016; Wilson, Miranda, & Knepper, 2013). In particular, eight genes (Avpr2, Prkar1a, Gnas, Calm1, Pik3ca, Akt1, Tbc1d4, and Arhgap5) are important in the regulation of AQP2 (Knepper, 2012). We have predicted the miRNAs that target these genes. According to rat miRNA database, the dDAVP-responsive miRNAs target rat Calm1 (miR-1, miR-206, miR-16) and rat Arhgap5 (miR-16) (Table 2). Per mouse miRNA database, mouse Avpr2 (miR-298), mouse Gnas (miR-678, miR-298), mouse Calm1 (miR-206, miR-16), and mouse Arhgap5 (miR-16) were predicted to be miRNA targets. Moreover, miRNAs targeting AQP2 gene were predicted to target other genes as well, including rat Prkar1a (miR-32), rat Pik3ca (miR-32 and miR-137), mouse Prkar1a (miR-32), mouse Pik3ca (miR-32 and miR-137), and mouse Akt1 (miR-32). mRNAs of other target genes also contained the identical core sequences that are targeted by miR-32 (3′-...ACGUUA...-5′) or miR-137 (3′-...CGUUAU...-5′); however, the degree of complementarity of these miRNAs for the recognition of mRNAs might be variable. We examined whether miR-32 and miR-137 could regulate the mRNA levels of several selected target genes (Sgk3, Tgif1, and Traf3 for miR-32; Gpr137b, Mapkapk2, and Myo1c for miR-137). However, expression of none of these genes was significantly altered, at least at the mRNA level. As the target sequences of miR-32 and miR-137 are present in the mRNAs of vasopressin-regulated genes, further study on target genes of these miRNAs could provide information about miRNA-regulated AQP2 expression in kidney collecting duct cells.

3.2 Regulation of small RNAs in response to osmotic stress in the kidney

Osmotic response element binding protein (OREBP), also known as tonicity-responsive element binding protein (TonEBP) or nuclear factor of the activated T cells-5 (NFAT5), is a Rel-like transcription factor that responds to cellular osmotic stress in the kidney medulla (Miyakawa, Woo, Dahl, Handler, & Kwon, 1999). Hypertonicity enhanced transcription, improved mRNA stability, and increased translation of OREBP; miRNAs are thought to play a role in these processes (Huang et al., 2011). A previous study demonstrated that expression of miRNAs, analyzed by microarray assay, was significantly changed when mouse IMCD3 cells were cultured in hypertonic medium (with NaCl, 550 mOsm/KgH_2O), with miR-200b and miR-717 being most significantly down-regulated (Huang et al., 2011). Consistent with this, overexpression of miR-200b and miR-717 was associated with significantly decreased mRNA and protein levels of OREBP including its reduced transcription. Moreover, Luo et al. (2014) showed that *sfmbt2* 10th intron-hosted miR-466(a/b/c/e/p)-3p were also responsive to the changes in tonicity *in vitro* and *in vivo*. The miR-466(a/b/c/e/p)-3p were up-regulated in mIMCD3 cells in response

Table 2 Prediction of interaction between miRNAs and vasopressin signaling-related genes.

TargetScan (6.2) database	Gene symbol	Accession	Annotation	dDAVP-responsive miRNAs	AQP2-targeting miRNAs
Rat	Prkar1a	NM_021880	Protein kinase, cAMP dependent regulatory, type I, alpha	–	rno-miR-32
	Calm1	NM_009790	Calmodulin 1	rno-miR-1 rno-miR-206 rno-miR-16	–
	Pik3ca	NM_008839	Phosphatidylinositol 3-kinase, catalytic, alpha polypeptide	–	rno-miR-32 rno-miR-137
	Arhgap5	NM_009706	Rho GTPase activating protein 5	rno-miR-16	–
Mouse	Avpr2	NM_019404	Arginine vasopressin receptor 2	mmu-miR-298	–
	Prkar1a	NM_021880	Protein kinase, cAMP dependent regulatory, type I, alpha	–	mmu-miR-32
	Gnas	NM_019690	GNAS (guanine nucleotide binding protein, alpha stimulating) complex locus	mmu-miR-678 mmu-miR-298	–
	Calm1	NM_009790	Calmodulin 1	mmu-miR-206 mmu-miR-16	–
	Pik3ca	NM_008839	Phosphatidylinositol 3-kinase, catalytic, alpha polypeptide	–	mmu-miR-32 mmu-miR-137
	Akt1	NM_009652, NM_001165894	Thymoma viral proto-oncogene 1	–	mmu-miR-32
	Arhgap5	NM_009706	Rho GTPase activating protein 5	mmu-miR-16	–

to short-term (2 h) hypertonic conditions, whereas their expression returned to the normal level after long-term (8 h) exposure to hypertonic media. This finding was opposite to the persistent downregulation of miR-200b-3p in mIMCD3 cells. Further studies are warranted to elucidate the role of miRNAs in the medullary cells exposed to different extracellular microenvironments, such as very high osmolality, low pH, and fluid shear stress (Choi, Jung, & Kwon, 2015; Choi et al., 2012; Jang et al., 2011; Jung et al., 2016).

3.3 Aldosterone-induced changes of miRNA expression in the cortex of rat kidney

In the kidney, aldosterone, a major mineralocorticoid hormone, mediates sodium reabsorption and potassium secretion in the distal convoluted tubule, connecting tubule, and collecting duct (Nielsen, Kwon, et al., 2002; Rossier, Bochud, & Devuyst, 2017; Wang et al., 2001). The release of aldosterone is initiated by renin produced in the juxtaglomerular (JG) cells, which converts angiotensinogen to angiotensin I. Angiotensin I is subsequently converted to angiotensin II by the angiotensin-converting enzyme (ACE), which stimulates the release of aldosterone from the adrenal cortex. Conditional Dicer knock-out in renin-producing cells in the kidneys of mice was associated with the decrease of expression of miRNAs in JG cells as well as the number of JG cells, resulting in the significant reduction in renin secretion and blood pressure (Sequeira-Lopez et al., 2010). The mice also showed vascular abnormality and fibrosis in the kidney. The specific miRNAs related to JG cell development or renin production were not identified in the Dicer knock-out condition since the overall expression of miRNAs was altered. However, it could be concluded that miRNAs maintain the morphology and function of the kidneys.

The miRNA profile changed in response to aldosterone administration in mCCD cells. In particular, miRNA clusters miR-23a–24-2–27a (miRC11) and miR-23b–24-1–27b (miRC22) were highly up-regulated (Liu, Edinger, et al., 2017). Overexpression of miR-27 in mCCD cells increased ENaC-mediated sodium transport, whereas overexpression of miR-23 or miR-24 did not change ENaC activity (Liu, Edinger, et al., 2017). Potential miR-27 target genes were predicted through *in silico* approaches, and among them intersectin 2 (Itsn2) was particularly studied, which is a regulator of potassium channels in the distal nephron (Liu, Edinger, et al., 2017). Importantly, Itsn2-depleted mCCD cells exhibited significantly activated ENaC-mediated sodium transport, whereas Itsn2-overexpressing cells showed decreased ENaC-mediated sodium current in mCCD cells (Liu, Edinger, et al., 2017). Thus, this study revealed a potential role of miRNAs in aldosterone-induced sodium reabsorption in the kidney and hence in regulating the blood pressure. Other studies have also demonstrated the role of miRNAs in aldosterone-induced sodium reabsorption in the mouse collecting duct cells. Edinger et al. (2014) demonstrated that aldosterone treatment decreased miR-335-3p, miR-290-5p, and miR-1983 expression in mCCD cells. Notably, the decreased expression of these three miRNAs increased ENaC-mediated sodium transport in mCCD cells, even in the absence of mineralocorticoid supplementation (Edinger et al., 2014). In contrast, overexpression of three miRNAs diminished the effect of aldosterone on ENaC-mediated sodium transport. Among the three miRNAs, miR-335-3p was demonstrated to interact directly with 3′-UTR of ankyrin 3, which increases ENaC-mediated sodium transport in mCCD cells (Edinger et al., 2014). These results indicated that miR-335-3p, miR-290-5p, and miR-1983 are likely to regulate the activity ENaC by altering the expression of ankyrin 3, a novel aldosterone-regulated protein (Edinger et al., 2014).

In addition to its critical role in body fluid homeostasis, aldosterone is also associated with inflammatory injury and fibrosis in the kidney, in which renal hypertrophy, tubulointerstitial fibrosis, and glomerulosclerosis are observed (Brem et al., 2010; Brown, 2013; Remuzzi, Cattaneo, & Perico, 2008). Consistent with this, we demonstrated that aldosterone stimulation could induce the expression of the fibrosis marker proteins (fibronectin and α-smooth muscle actin) in mpkCCDc14 cells (Park et al., 2018). Thus, we profiled the aldosterone-regulated miRNAs in mpkCCDc14 cells and identified the signaling pathways in which the putative target genes of the identified miRNAs were enriched to examine the roles of the identified miRNAs and their putative target genes in aldosterone-mediated renal fibrosis (Park et al., 2018). Microarray chip assay of mpkCCDc14 cells treated with aldosterone (10^{-6} M) for 3 days revealed that 56 miRNAs, including mature and pre-mature miRNAs, were significantly changed when the criterion used was >30% of fold-change among the significantly changed miRNAs (P value < 0.05). The putative target genes of these miRNAs were further analyzed; the putative targets were components of several important signaling pathways, including actin cytoskeleton, Wnt signaling, insulin signaling, and TGF-β signaling pathway (Park et al., 2018). Among the several signaling pathways, seven up-regulated miRNAs and eight down-regulated miRNAs were found to be involved in the Wnt signaling. In particular, among these 15 identified miRNAs, miR-130b-3p (up-regulated), and miR-146a-5p and miR-34c-5p (down-regulated) were selected for further studies, since these miRNAs have previously been known to be involved in tissue fibrosis (Bertero et al., 2015; Bhatt et al., 2016; Morizane et al., 2014). Among the putative target genes of miR-130b-3p, miR-146a-5p, and miR-34c-5p, the mRNA and protein expression of CaMKIIβ, the putative miR-34c-5p target, was revealed to be significantly increased by aldosterone treatment. Transfection of CaMKIIβ-siRNA or miR-34c-5p mimic into mpkCCDc14 cells resulted in significantly reduced aldosterone-induced fibronectin expression, associated with reduced CaMKIIβ protein expression. This finding indicated that miR-34c-5p and CaMKIIβ play a role in aldosterone-induced fibrosis. Luciferase reporter assay revealed a decrease in CaMKIIβ translation in the cells treated with miR-34c-5p mimic, indicating the direct effect of miR-34c-5p on mRNA and protein expression of CaMKIIβ (Park et al., 2018). Since Wnt signaling pathways are known to be associated with renal fibrosis, future studies are needed to examine the role of identified miRNAs and their target genes in the pathogenesis of aldosterone-mediated renal fibrosis.

4. Small RNA expression in common kidney diseases: Diabetic nephropathy, acute kidney injury, and chronic kidney disease

4.1 Diabetic nephropathy

Diabetic nephropathy (DN) is the leading cause of end-stage renal disease. Diabetes mellitus causes hyperglycemia, abnormal hemodynamics, and hypertension, resulting in the end-stage renal disease. Proteinuria caused by the damage of the glomerular basement membrane and reactive oxygen species (ROS) production caused by hyperglycemia are the main pathophysiological features of DN (Volpe, Villar-Delfino, Dos Anjos, & Nogueira-Machado, 2018; Williams, 2005). DN is also induced by hypoxia, which activates interstitial fibrosis (Miyata & de Strihou, 2010; Nayak et al., 2016). A previous study unraveled the miRNA profile of urinary exosomes from human patients (Delic et al., 2016). Microarray assay showed that expression of 14 miRNAs was increased and that of 2 miRNAs was decreased in the urinary exosomes from patients with DN compared to the miRNA levels in the healthy controls and patients with type II diabetes (Delic et al., 2016). In particular, results of quantitative real-time PCR demonstrated that levels of miR-320c and miR-6068 were the most strongly increased in patients with DN, which may affect TGF-β signaling by targeting thrombospondin 1 (Delic et al., 2016).

Krupa et al. (2010) demonstrated that miR-192 expression correlated inversely with both fibrosis score in renal biopsied tissues from patients with DN and estimated GFR. They also revealed that overexpression of miR-192 in HK-2 cells reduced the mRNA levels of ZEB1 and ZEB2 and increased mRNA and protein expression of E-cadherin (Krupa et al., 2010), suggesting that decreased expression of miR-192 is likely to play a role in the pathogenesis of renal fibrosis in DN associated with decreased estimated GFR. However, a later study revealed contrasting results showing that inhibition of miRNA-192 with a locked nucleic acid (LNA)-modified inhibitor reduced renal fibrosis in the mouse model of DN (Putta et al., 2012). When miR-192 expression was reduced in the kidney, increased expression of Zeb1 and Zeb2 and decreased expression of collagen, TGF-β, and fibronectin were observed in the kidneys of diabetic mice (Putta et al., 2012). There are some discrepancies regarding the effects of TGF-β on miR-192 expression in the kidney; some studies found that TGF-β decreased miR-192 expression in proximal tubular cells and other renal cell lines (Krupa et al., 2010; Wang et al., 2010), whereas other studies showed that TGF-β increased miR-192 expression in mesangial cells, glomeruli, endothelial cells, and podocytes (Kato et al., 2011, 2009, 2007; Putta et al., 2012). Future studies are needed to clarify these contrasting results.

In streptozotocin (STZ)-induced diabetic mice model, the miR-146a expression was significantly increased in both peritoneal and intrarenal macrophages (Bhatt et al., 2016). In contrast, in the diabetic miR-146a-depleted mice, expression of M1 activation markers was increased, whereas expression of M2 markers was decreased in the macrophages. Moreover, proinflammatory cytokines (IL-1β and IL-18) were concomitantly increased in the macrophages of diabetic miR-146a null mice (Bhatt et al., 2016), suggesting that miR-146a exerts an anti-inflammatory action in DN. Moreover, miR-25 levels were found to be significantly low in the peripheral blood from patients with diabetes and in the kidneys from streptozotocin-treated mice (type I diabetes) and *db/db* diabetic mice (type II diabetes) (Liu, Li, et al., 2017). In contrast, the systemic administration of miR-25 agomir in *db/db* mice improved proteinuria, glomerular fibrosis, and improved high blood pressure, suggesting a potential therapeutic application of miRNAs. RNA sequencing revealed that miR-25 agomir or miR-25 antagomir significantly changed the expression of their target genes, and among them, miR-25 directly interacted with 3′-UTR of CDC42, which is involved in the Ras signaling pathway (Liu, Li, et al., 2017).

4.2 Acute kidney injury

Acute kidney injury (AKI) has several causes, including hypovolemic shock, ischemia-reperfusion (I/R) injury, urinary tract obstruction, septic shock, and toxic nephropathy. The global miRNA transcriptome is significantly altered in AKI and this could affect cellular and molecular pathways. Conditional Dicer knock-out in the renal proximal tubules showed substantial resistance to ischemic AKI in mice, indicating that miRNAs play an important role in the pathophysiology of AKI (Godwin et al., 2010; Wei et al., 2010). Moreover, after I/R injury, miRNA profile was significantly changed (Godwin et al., 2010). In this study, nine miRNAs, miR-21, miR-20a, miR-146a, miR-199a-3p, miR214, miR-192, miR-187, miR-805, and miR-194, were differentially expressed, and among these nine miRNAs, miR-21 was significantly up-regulated in the renal tubular epithelial cells after I/R injury. This correlated with the decreased expression of tumor suppressor gene-programmed cell death 4 (PDCD4), a target of miR-21. Another study showed that miR-24 was significantly up-regulated in both mice kidneys subjected to I/R injury and human kidney transplant biopsy specimens (Lorenzen et al., 2014). Luciferase reporter assays demonstrated that miR-24 regulates the target genes H2A histone family, member X, and heme oxygenase 1 (Lorenzen et al., 2014). I/R injury in mice with miR-24 silencing was associated with a significant improvement in kidney function, reduced apoptosis, improved histologic tubular epithelial injury, and reduced infiltration of inflammatory cells (Lorenzen et al., 2014), indicating that miR-24 inhibition is likely to be a promising therapeutic option for the treatment of ischemic AKI. Moreover, miR-489 expression was induced in both ischemic AKI in mice and hypoxic condition in proximal tubule cells; its expression was primarily detected in the injured tubules localized at the outer stripe of the outer medulla (Wei et al., 2016). The expression of miR-489 was demonstrated to be regulated by HIF-1, as proximal tubule-specific HIF-1 knock-out reduced the induction of miR-489 after renal I/R injury (Wei et al., 2016).

4.3 Chronic kidney disease

The prominent features of chronic kidney disease (CKD) are tubulointerstitial fibrosis and glomerular fibrosis, leading to deposition of extracellular matrix and infiltration of inflammatory cells. Differential expression of miRNAs was demonstrated in progressive CKDs that showed glomerulosclerosis, interstitial fibrosis, and inflammation

(Rudnicki et al., 2016). Additional analysis also demonstrated differentially expressed miRNAs and their target genes analyzed by target prediction algorithms, which were associated with inflammation and apoptosis (Rudnicki et al., 2016). In particular, TGF-β is a pivotal cytokine, which mainly contributes to renal fibrosis and stimulates synthesis and deposition of extracellular matrix. miR-29, miR-21, miR-192, and miR-200 were showed to be involved in renal damages in different cell types under the TGF-β signaling pathway (Kato et al., 2007; Lan & Chung, 2012; Oba et al., 2010; Qin et al., 2011; Wang et al., 2011, 2012; Zhong et al., 2013). TGF-β2 down-regulated miR-200a expression in rat proximal tubular epithelial cells, whereas miR-200a down-regulated smad-3 activity and the expression of matrix proteins, preventing TGF-β-dependent epithelial-to-mesenchymal transition (Wang et al., 2011). This implies that miR-200a could be an important novel target for the inhibition of progressive CKD.

5. Conclusions and future directions

miRNAs, short non-coding RNAs interacting with 3′-UTR of mRNAs, regulate the target gene expression via degradation of target mRNAs or inhibition of their translation, leading to the down-regulation of protein expression (Krol, Loedige, & Filipowicz, 2010). Recent studies have revealed that miRNAs have multiple target mRNAs and also share target mRNAs with other miRNAs. Thus, miRNAs form complicated regulatory networks under physiological and pathophysiological conditions. miRNA-mediated regulation of protein expression in the kidney tubular epithelial cells could affect renal functions. For example, a significant decrease in AQP2 protein expression in kidney collecting duct was observed when miRNA mimics of miR-32 or miR-137 were overexpressed in mpkCCDc14 cells (Kim et al., 2015). Several studies have demonstrated that miRNAs are involved in various pathophysiological conditions in kidney, e.g., kidney development, renal cell carcinoma, acute kidney injury, chronic kidney disease, Alport nephropathy, and diabetic nephropathy (Gomez et al., 2015; Koga et al., 2015; Lai et al., 2015; Lee et al., 2017; Lu et al., 2015; Park et al., 2018). Therefore, miRNAs could be considered as biomarkers for progressive kidney diseases and novel therapeutic targets for the inhibition of progression of chronic kidney diseases. Further studies are warranted to identify novel miRNAs and their target genes for the better understanding of the pathogenesis and therapeutic targets of kidney diseases.

Acknowledgments

This study was supported by the National Research Foundation of Korea (NRF) funded by the Ministry of Science and ICT, Korea (2019R1A2C1007753 and 2017R1D1A3B03032262).

References

Agre, P., King, L. S., Yasui, M., Guggino, W. B., Ottersen, O. P., Fujiyoshi, Y., et al. (2002). Aquaporin water channels—From atomic structure to clinical medicine. *The Journal of Physiology*, *542*(Pt. 1), 3–16.

Ambros, V. (2001). microRNAs: Tiny regulators with great potential. *Cell*, *107*(7), 823–826.

Barad, O., Meiri, E., Avniel, A., Aharonov, R., Barzilai, A., Bentwich, I., et al. (2004). MicroRNA expression detected by oligonucleotide microarrays: System establishment and expression profiling in human tissues. *Genome Research*, *14*(12), 2486–2494. https://doi.org/10.1101/gr.2845604.

Bartel, D. P. (2004). MicroRNAs: Genomics, biogenesis, mechanism, and function. *Cell*, *116*(2), 281–297.

Bertero, T., Cottrill, K. A., Annis, S., Bhat, B., Gochuico, B. R., Osorio, J. C., et al. (2015). A YAP/TAZ-miR-130/301 molecular circuit exerts systems-level control of fibrosis in a network of human diseases and physiologic conditions. *Scientific Reports*, *5*, 18277. https://doi.org/10.1038/srep18277.

Bhatt, K., Lanting, L. L., Jia, Y., Yadav, S., Reddy, M. A., Magilnick, N., et al. (2016). Anti-inflammatory role of microRNA-146a in the pathogenesis of diabetic nephropathy. *Journal of the American Society of Nephrology*, *27*(8), 2277–2288. https://doi.org/10.1681/ASN.2015010111.

Bijkerk, R., Trimpert, C., van Solingen, C., de Bruin, R. G., Florijn, B. W., Kooijman, S., et al. (2018). MicroRNA-132 controls water homeostasis through regulating MECP2-mediated vasopressin synthesis. *American Journal of Physiology. Renal Physiology*, *315*(4), F1129–F1138. https://doi.org/10.1152/ajprenal.00087.2018.

Bonci, D., Coppola, V., Musumeci, M., Addario, A., Giuffrida, R., Memeo, L., et al. (2008). The miR-15a-miR-16-1 cluster controls prostate cancer by targeting multiple oncogenic activities. *Nature Medicine*, *14*(11), 1271–1277. https://doi.org/10.1038/nm.1880.

Borgnia, M., Nielsen, S., Engel, A., & Agre, P. (1999). Cellular and molecular biology of the aquaporin water channels. *Annual Review of Biochemistry*, *68*, 425–458. https://doi.org/10.1146/annurev.biochem.68.1.425.

Brem, A. S., Morris, D. J., Ge, Y., Dworkin, L., Tolbert, E., & Gong, R. (2010). Direct fibrogenic effects of aldosterone on normotensive kidney: An effect modified by 11beta-HSD activity. *American Journal of Physiology. Renal Physiology*, *298*(5), F1178–F1187. https://doi.org/10.1152/ajprenal.00532.2009.

Brown, N. J. (2013). Contribution of aldosterone to cardiovascular and renal inflammation and fibrosis. *Nature Reviews. Nephrology*, *9*(8), 459–469. https://doi.org/10.1038/nrneph.2013.110.

Castoldi, M., Schmidt, S., Benes, V., Noerholm, M., Kulozik, A. E., Hentze, M. W., et al. (2006). A sensitive array for microRNA expression profiling (miChip) based on locked nucleic acids (LNA). *RNA*, *12*(5), 913–920. https://doi.org/10.1261/rna.2332406.

Chandrasekaran, K., Karolina, D. S., Sepramaniam, S., Armugam, A., Wintour, E. M., Bertram, J. F., et al. (2012). Role of microRNAs in kidney homeostasis and disease. *Kidney International*, *81*(7), 617–627. https://doi.org/10.1038/ki.2011.448.

Chen, L., Lee, J. W., Chou, C. L., Nair, A. V., Battistone, M. A., Paunescu, T. G., et al. (2017). Transcriptomes of major renal collecting duct cell types in mouse identified by single-cell RNA-seq. *Proceedings of the National Academy of Sciences of the United States of America*, *114*(46), E9989–E9998. https://doi.org/10.1073/pnas.1710964114.

Choi, H. J., Jung, H. J., & Kwon, T. H. (2015). Extracellular pH affects phosphorylation and intracellular trafficking of AQP2 in inner medullary collecting duct cells. *American Journal of Physiology. Renal Physiology, 308*(7), F737–F748. https://doi.org/10.1152/ajprenal.00376.2014.

Choi, H. J., Yoon, Y. J., Kwon, Y. K., Lee, Y. J., Chae, S., Hwang, D., et al. (2012). Patterns of gene and metabolite define the effects of extracellular osmolality on kidney collecting duct. *Journal of Proteome Research, 11*(7), 3816–3828. https://doi.org/10.1021/pr300309d.

Chou, C. L., Christensen, B. M., Frische, S., Vorum, H., Desai, R. A., Hoffert, J. D., et al. (2004). Non-muscle myosin II and myosin light chain kinase are downstream targets for vasopressin signaling in the renal collecting duct. *The Journal of Biological Chemistry, 279*(47), 49026–49035. https://doi.org/10.1074/jbc.M408565200.

Coleman, R. A., Wu, D. C., Liu, J., & Wade, J. B. (2000). Expression of aquaporins in the renal connecting tubule. *American Journal of Physiology. Renal Physiology, 279*(5), F874–F883. https://doi.org/10.1152/ajprenal.2000.279.5.F874.

Cordes, K. R., & Srivastava, D. (2009). MicroRNA regulation of cardiovascular development. *Circulation Research, 104*(6), 724–732. https://doi.org/10.1161/CIRCRESAHA.108.192872.

Davison, T. S., Johnson, C. D., & Andruss, B. F. (2006). Analyzing micro-RNA expression using microarrays. *Methods in Enzymology, 411*, 14–34. https://doi.org/10.1016/S0076-6879(06)11002-2.

Delic, D., Eisele, C., Schmid, R., Baum, P., Wiech, F., Gerl, M., et al. (2016). Urinary exosomal miRNA signature in type II diabetic nephropathy patients. *PLoS One, 11*(3), e0150154. https://doi.org/10.1371/journal.pone.0150154.

Edinger, R. S., Coronnello, C., Bodnar, A. J., Labarca, M., Bhalla, V., LaFramboise, W. A., et al. (2014). Aldosterone regulates microRNAs in the cortical collecting duct to alter sodium transport. *Journal of the American Society of Nephrology, 25*(11), 2445–2457. https://doi.org/10.1681/ASN.2013090931.

Eichhorn, S. W., Guo, H., McGeary, S. E., Rodriguez-Mias, R. A., Shin, C., Baek, D., et al. (2014). mRNA destabilization is the dominant effect of mammalian microRNAs by the time substantial repression ensues. *Molecular Cell, 56*(1), 104–115. https://doi.org/10.1016/j.molcel.2014.08.028.

Godwin, J. G., Ge, X., Stephan, K., Jurisch, A., Tullius, S. G., & Iacomini, J. (2010). Identification of a microRNA signature of renal ischemia reperfusion injury. *Proceedings of the National Academy of Sciences of the United States of America, 107*(32), 14339–14344. https://doi.org/10.1073/pnas.0912701107.

Gomes, A., da Silva, I. V., Rodrigues, C. M. P., Castro, R. E., & Soveral, G. (2018). The emerging role of microRNAs in aquaporin regulation. *Frontiers in Chemistry, 6*, 238. https://doi.org/10.3389/fchem.2018.00238.

Gomez, I. G., MacKenna, D. A., Johnson, B. G., Kaimal, V., Roach, A. M., Ren, S., et al. (2015). Anti-microRNA-21 oligonucleotides prevent Alport nephropathy progression by stimulating metabolic pathways. *The Journal of Clinical Investigation, 125*(1), 141–156. https://doi.org/10.1172/JCI75852.

Grimson, A., Farh, K. K., Johnston, W. K., Garrett-Engele, P., Lim, L. P., & Bartel, D. P. (2007). MicroRNA targeting specificity in mammals: Determinants beyond seed pairing. *Molecular Cell, 27*(1), 91–105. https://doi.org/10.1016/j.molcel.2007.06.017.

Grivna, S. T., Beyret, E., Wang, Z., & Lin, H. (2006). A novel class of small RNAs in mouse spermatogenic cells. *Genes & Development, 20*(13), 1709–1714. https://doi.org/10.1101/gad.1434406.

Hajarnis, S., Yheskel, M., Williams, D., Brefort, T., Glaudemans, B., Debaix, H., et al. (2018). Suppression of microRNA activity in kidney collecting ducts induces partial loss of epithelial phenotype and renal fibrosis. *Journal of the American Society of Nephrology, 29*(2), 518–531. https://doi.org/10.1681/ASN.2017030334.

Halushka, M. K., Fromm, B., Peterson, K. J., & McCall, M. N. (2018). Big strides in cellular microRNA expression. *Trends in Genetics*, *34*(3), 165–167. https://doi.org/10.1016/j.tig.2017.12.015.

Harfe, B. D. (2005). MicroRNAs in vertebrate development. *Current Opinion in Genetics & Development*, *15*(4), 410–415. https://doi.org/10.1016/j.gde.2005.06.012.

Ho, J., & Kreidberg, J. A. (2013). MicroRNAs in renal development. *Pediatric Nephrology*, *28*(2), 219–225. https://doi.org/10.1007/s00467-012-2204-y.

Hoorn, E. J., Pisitkun, T., Yu, M. J., & Knepper, M. A. (2008). Proteomic approaches for the study of cell signaling in the renal collecting duct. *Contributions to Nephrology*, *160*, 172–185. https://doi.org/10.1159/000125981.

Huang, W., Liu, H., Wang, T., Zhang, T., Kuang, J., Luo, Y., et al. (2011). Tonicity-responsive microRNAs contribute to the maximal induction of osmoregulatory transcription factor OREBP in response to high-NaCl hypertonicity. *Nucleic Acids Research*, *39*(2), 475–485. https://doi.org/10.1093/nar/gkq818.

Hunt, E. A., Broyles, D., Head, T., & Deo, S. K. (2015). MicroRNA detection: Current technology and research strategies. *Annual Review of Analytical Chemistry (Palo Alto, California)*, *8*, 217–237. https://doi.org/10.1146/annurev-anchem-071114-040343.

Ichii, O., & Horino, T. (2018). MicroRNAs associated with the development of kidney diseases in humans and animals. *Journal of Toxicologic Pathology*, *31*(1), 23–34. https://doi.org/10.1293/tox.2017-0051.

Jang, K. J., Cho, H. S., Kang, D. H., Bae, W. G., Kwon, T. H., & Suh, K. Y. (2011). Fluid-shear-stress-induced translocation of aquaporin-2 and reorganization of actin cytoskeleton in renal tubular epithelial cells. *Integrative Biology*, *3*(2), 134–141. https://doi.org/10.1039/c0ib00018c.

Jung, H. J., & Kwon, T. H. (2016). Molecular mechanisms regulating aquaporin-2 in kidney collecting duct. *American Journal of Physiology. Renal Physiology*, *311*(6), F1318–F1328. https://doi.org/10.1152/ajprenal.00485.2016.

Jung, J. W., Lee, M. S., Choi, H. J., Jung, S., Lee, Y. J., Hwang, G. S., et al. (2016). Mass spectrometric imaging of metabolites in kidney tissues from rats treated with furosemide. *American Journal of Physiology. Renal Physiology*, *310*(11), F1317–F1327. https://doi.org/10.1152/ajprenal.00524.2015.

Kato, M., Arce, L., Wang, M., Putta, S., Lanting, L., & Natarajan, R. (2011). A microRNA circuit mediates transforming growth factor-beta1 autoregulation in renal glomerular mesangial cells. *Kidney International*, *80*(4), 358–368. https://doi.org/10.1038/ki.2011.43.

Kato, M., Putta, S., Wang, M., Yuan, H., Lanting, L., Nair, I., et al. (2009). TGF-beta activates Akt kinase through a microRNA-dependent amplifying circuit targeting PTEN. *Nature Cell Biology*, *11*(7), 881–889. https://doi.org/10.1038/ncb1897.

Kato, M., Zhang, J., Wang, M., Lanting, L., Yuan, H., Rossi, J. J., et al. (2007). MicroRNA-192 in diabetic kidney glomeruli and its function in TGF-beta-induced collagen expression via inhibition of E-box repressors. *Proceedings of the National Academy of Sciences of the United States of America*, *104*(9), 3432–3437. https://doi.org/10.1073/pnas.0611192104.

Kim, J. E., Jung, H. J., Lee, Y. J., & Kwon, T. H. (2015). Vasopressin-regulated miRNAs and AQP2-targeting miRNAs in kidney collecting duct cells. *American Journal of Physiology. Renal Physiology*, *308*(7), F749–F764. https://doi.org/10.1152/ajprenal.00334.2014.

Kim, J., Kim, Y. H., Cha, J. H., Tisher, C. C., & Madsen, K. M. (1999). Intercalated cell subtypes in connecting tubule and cortical collecting duct of rat and mouse. *Journal of the American Society of Nephrology*, *10*(1), 1–12.

Kim, Y. H., Kwon, T. H., Frische, S., Kim, J., Tisher, C. C., Madsen, K. M., et al. (2002). Immunocytochemical localization of pendrin in intercalated cell subtypes in rat and mouse kidney. *American Journal of Physiology. Renal Physiology*, *283*(4), F744–F754. https://doi.org/10.1152/ajprenal.00037.2002.

Kim, Y. K., Yu, J., Han, T. S., Park, S. Y., Namkoong, B., Kim, D. H., et al. (2009). Functional links between clustered microRNAs: Suppression of cell-cycle inhibitors by microRNA clusters in gastric cancer. *Nucleic Acids Research*, *37*(5), 1672–1681. https://doi.org/10.1093/nar/gkp002.

Knepper, M. A. (2012). Systems biology in physiology: The vasopressin signaling network in kidney. *American Journal of Physiology. Cell Physiology*, *303*(11), C1115–C1124. https://doi.org/10.1152/ajpcell.00270.2012.

Knepper, M. A., Kwon, T. H., & Nielsen, S. (2015). Molecular physiology of water balance. *The New England Journal of Medicine*, *372*(14), 1349–1358. https://doi.org/10.1056/NEJMra1404726.

Koga, K., Yokoi, H., Mori, K., Kasahara, M., Kuwabara, T., Imamaki, H., et al. (2015). MicroRNA-26a inhibits TGF-beta-induced extracellular matrix protein expression in podocytes by targeting CTGF and is downregulated in diabetic nephropathy. *Diabetologia*, *58*(9), 2169–2180. https://doi.org/10.1007/s00125-015-3642-4.

Kozomara, A., Birgaoanu, M., & Griffiths-Jones, S. (2019). miRBase: From microRNA sequences to function. *Nucleic Acids Research*, *47*(D1), D155–D162. https://doi.org/10.1093/nar/gky1141.

Krol, J., Loedige, I., & Filipowicz, W. (2010). The widespread regulation of microRNA biogenesis, function and decay. *Nature Reviews. Genetics*, *11*(9), 597–610. https://doi.org/10.1038/nrg2843.

Krupa, A., Jenkins, R., Luo, D. D., Lewis, A., Phillips, A., & Fraser, D. (2010). Loss of MicroRNA-192 promotes fibrogenesis in diabetic nephropathy. *Journal of the American Society of Nephrology*, *21*(3), 438–447. https://doi.org/10.1681/ASN.2009050530.

Kumar, M. S., Lu, J., Mercer, K. L., Golub, T. R., & Jacks, T. (2007). Impaired microRNA processing enhances cellular transformation and tumorigenesis. *Nature Genetics*, *39*(5), 673–677. https://doi.org/10.1038/ng2003.

Kwon, T. H., Frokiaer, J., Knepper, M. A., & Nielsen, S. (1998). Reduced AQP1, -2, and -3 levels in kidneys of rats with CRF induced by surgical reduction in renal mass. *The American Journal of Physiology*, *275*(5 Pt. 2), F724–F741.

Kwon, T. H., Frokiaer, J., & Nielsen, S. (2013). Regulation of aquaporin-2 in the kidney: A molecular mechanism of body-water homeostasis. *Kidney Research and Clinical Practice*, *32*(3), 96–102. https://doi.org/10.1016/j.krcp.2013.07.005.

Kwon, T. H., Hager, H., Nejsum, L. N., Andersen, M. L., Frokiaer, J., & Nielsen, S. (2001). Physiology and pathophysiology of renal aquaporins. *Seminars in Nephrology*, *21*(3), 231–238.

Kwon, T. H., Nielsen, J., Moller, H. B., Fenton, R. A., Nielsen, S., & Frokiaer, J. (2009). Aquaporins in the kidney. *Handbook of Experimental Pharmacology*, *190*, 95–132. https://doi.org/10.1007/978-3-540-79885-9_5.

Lagos-Quintana, M., Rauhut, R., Lendeckel, W., & Tuschl, T. (2001). Identification of novel genes coding for small expressed RNAs. *Science*, *294*(5543), 853–858. https://doi.org/10.1126/science.1064921.

Lagos-Quintana, M., Rauhut, R., Meyer, J., Borkhardt, A., & Tuschl, T. (2003). New microRNAs from mouse and human. *RNA*, *9*(2), 175–179.

Lai, J. Y., Luo, J., O'Connor, C., Jing, X., Nair, V., Ju, W., et al. (2015). MicroRNA-21 in glomerular injury. *Journal of the American Society of Nephrology*, *26*(4), 805–816. https://doi.org/10.1681/ASN.2013121274.

Lam, J. K., Chow, M. Y., Zhang, Y., & Leung, S. W. (2015). siRNA versus miRNA as therapeutics for gene silencing. *Molecular Therapy—Nucleic Acids*, *4*, e252, https://doi.org/10.1038/mtna.2015.23.

Lan, H. Y., & Chung, A. C. (2012). TGF-beta/Smad signaling in kidney disease. *Seminars in Nephrology*, *32*(3), 236–243. https://doi.org/10.1016/j.semnephrol.2012.04.002.

Lee, R. C., Feinbaum, R. L., & Ambros, V. (1993). The C. elegans heterochronic gene lin-4 encodes small RNAs with antisense complementarity to lin-14. *Cell*, *75*(5), 843–854.

Lee, H. W., Khan, S. Q., Khaliqdina, S., Altintas, M. M., Grahammer, F., Zhao, J. L., et al. (2017). Absence of miR-146a in podocytes increases risk of diabetic glomerulopathy via up-regulation of ErbB4 and notch-1. *The Journal of Biological Chemistry*, *292*(2), 732–747. https://doi.org/10.1074/jbc.M116.753822.

Lewis, B. P., Burge, C. B., & Bartel, D. P. (2005). Conserved seed pairing, often flanked by adenosines, indicates that thousands of human genes are microRNA targets. *Cell*, *120*(1), 15–20. https://doi.org/10.1016/j.cell.2004.12.035.

Liu, X., Edinger, R. S., Klemens, C. A., Phua, Y. L., Bodnar, A. J., LaFramboise, W. A., et al. (2017). A MicroRNA cluster miR-23-24-27 Is upregulated by aldosterone in the distal kidney nephron where it alters sodium transport. *Journal of Cellular Physiology*, *232*(6), 1306–1317. https://doi.org/10.1002/jcp.25599.

Liu, Y., Li, H., Liu, J., Han, P., Li, X., Bai, H., et al. (2017). Variations in MicroRNA-25 expression influence the severity of diabetic kidney disease. *Journal of the American Society of Nephrology*, *28*(12), 3627–3638. https://doi.org/10.1681/ASN.2015091017.

Lorenzen, J. M., Kaucsar, T., Schauerte, C., Schmitt, R., Rong, S., Hubner, A., et al. (2014). MicroRNA-24 antagonism prevents renal ischemia reperfusion injury. *Journal of the American Society of Nephrology*, *25*(12), 2717–2729. https://doi.org/10.1681/ASN.2013121329.

Lu, G. J., Dong, Y. Q., Zhang, Q. M., Di, W. Y., Jiao, L. Y., Gao, Q. Z., et al. (2015). miRNA-221 promotes proliferation, migration and invasion by targeting TIMP2 in renal cell carcinoma. *International Journal of Clinical and Experimental Pathology*, *8*(5), 5224–5229.

Lu, J., Getz, G., Miska, E. A., Alvarez-Saavedra, E., Lamb, J., Peck, D., et al. (2005). MicroRNA expression profiles classify human cancers. *Nature*, *435*(7043), 834–838. https://doi.org/10.1038/nature03702.

Lu, Z., Liu, M., Stribinskis, V., Klinge, C. M., Ramos, K. S., Colburn, N. H., et al. (2008). MicroRNA-21 promotes cell transformation by targeting the programmed cell death 4 gene. *Oncogene*, *27*(31), 4373–4379. https://doi.org/10.1038/onc.2008.72.

Lund, E., Guttinger, S., Calado, A., Dahlberg, J. E., & Kutay, U. (2004). Nuclear export of microRNA precursors. *Science*, *303*(5654), 95–98. https://doi.org/10.1126/science.1090599.

Luo, Y., Liu, Y., Liu, M., Wei, J., Zhang, Y., Hou, J., et al. (2014). Sfmbt2 10th intron-hosted miR-466(a/e)-3p are important epigenetic regulators of Nfat5 signaling, osmoregulation and urine concentration in mice. *Biochimica et Biophysica Acta*, *1839*(2), 97–106. https://doi.org/10.1016/j.bbagrm.2013.12.005.

McManus, M. T., & Sharp, P. A. (2002). Gene silencing in mammals by small interfering RNAs. *Nature Reviews. Genetics*, *3*(10), 737–747. https://doi.org/10.1038/nrg908.

Miska, E. A., Alvarez-Saavedra, E., Townsend, M., Yoshii, A., Sestan, N., Rakic, P., et al. (2004). Microarray analysis of microRNA expression in the developing mammalian brain. *Genome Biology*, *5*(9), R68. https://doi.org/10.1186/gb-2004-5-9-r68.

Miyakawa, H., Woo, S. K., Dahl, S. C., Handler, J. S., & Kwon, H. M. (1999). Tonicity-responsive enhancer binding protein, a rel-like protein that stimulates transcription in response to hypertonicity. *Proceedings of the National Academy of Sciences of the United States of America*, *96*(5), 2538–2542.

Miyata, T., & de Strihou, C. (2010). Diabetic nephropathy: A disorder of oxygen metabolism? *Nature Reviews. Nephrology*, *6*(2), 83–95. https://doi.org/10.1038/nrneph.2009.211.

Morizane, R., Fujii, S., Monkawa, T., Hiratsuka, K., Yamaguchi, S., Homma, K., et al. (2014). miR-34c attenuates epithelial-mesenchymal transition and kidney fibrosis with ureteral obstruction. *Scientific Reports*, *4*, 4578. https://doi.org/10.1038/srep04578.

Nagalakshmi, V. K., Ren, Q., Pugh, M. M., Valerius, M. T., McMahon, A. P., & Yu, J. (2011). Dicer regulates the development of nephrogenic and ureteric compartments in the mammalian kidney. *Kidney International*, *79*(3), 317–330. https://doi.org/10.1038/ki.2010.385.

Nayak, B. K., Shanmugasundaram, K., Friedrichs, W. E., Cavaglierii, R. C., Patel, M., Barnes, J., et al. (2016). HIF-1 mediates renal fibrosis in OVE26 type 1 diabetic mice. *Diabetes, 65*(5), 1387–1397. https://doi.org/10.2337/db15-0519.

Nelson, P. T., Baldwin, D. A., Scearce, L. M., Oberholtzer, J. C., Tobias, J. W., & Mourelatos, Z. (2004). Microarray-based, high-throughput gene expression profiling of microRNAs. *Nature Methods, 1*(2), 155–161. https://doi.org/10.1038/nmeth717.

Nielsen, S., Frokiaer, J., Marples, D., Kwon, T. H., Agre, P., & Knepper, M. A. (2002). Aquaporins in the kidney: From molecules to medicine. *Physiological Reviews, 82*(1), 205–244. https://doi.org/10.1152/physrev.00024.2001.

Nielsen, J., Kwon, T. H., Christensen, B. M., Frokiaer, J., & Nielsen, S. (2008). Dysregulation of renal aquaporins and epithelial sodium channel in lithium-induced nephrogenic diabetes insipidus. *Seminars in Nephrology, 28*(3), 227–244. https://doi.org/10.1016/j.semnephrol.2008.03.002.

Nielsen, J., Kwon, T. H., Masilamani, S., Beutler, K., Hager, H., Nielsen, S., et al. (2002). Sodium transporter abundance profiling in kidney: Effect of spironolactone. *American Journal of Physiology. Renal Physiology, 283*(5), F923–F933. https://doi.org/10.1152/ajprenal.00015.2002.

Nishihara, T., Zekri, L., Braun, J. E., & Izaurralde, E. (2013). miRISC recruits decapping factors to miRNA targets to enhance their degradation. *Nucleic Acids Research, 41*(18), 8692–8705. https://doi.org/10.1093/nar/gkt619.

Oba, S., Kumano, S., Suzuki, E., Nishimatsu, H., Takahashi, M., Takamori, H., et al. (2010). miR-200b precursor can ameliorate renal tubulointerstitial fibrosis. *PLoS One, 5*(10), e13614. https://doi.org/10.1371/journal.pone.0013614.

Park, E. J., Jung, H. J., Choi, H. J., Cho, J. I., Park, H. J., & Kwon, T. H. (2018). miR-34c-5p and CaMKII are involved in aldosterone-induced fibrosis in kidney collecting duct cells. *American Journal of Physiology. Renal Physiology, 314*(3), F329–F342. https://doi.org/10.1152/ajprenal.00358.2017.

Peter, M. E. (2010). Targeting of mRNAs by multiple miRNAs: The next step. *Oncogene, 29*(15), 2161–2164. https://doi.org/10.1038/onc.2010.59.

Piatek, M. J., & Werner, A. (2014). Endogenous siRNAs: Regulators of internal affairs. *Biochemical Society Transactions, 42*(4), 1174–1179. https://doi.org/10.1042/BST20140068.

Procino, G., Mastrofrancesco, L., Tamma, G., Lasorsa, D. R., Ranieri, M., Stringini, G., et al. (2012). Calcium-sensing receptor and aquaporin 2 interplay in hypercalciuria-associated renal concentrating defect in humans. An in vivo and in vitro study. *PLoS One, 7*(3), e33145. https://doi.org/10.1371/journal.pone.0033145.

Putta, S., Lanting, L., Sun, G., Lawson, G., Kato, M., & Natarajan, R. (2012). Inhibiting microRNA-192 ameliorates renal fibrosis in diabetic nephropathy. *Journal of the American Society of Nephrology, 23*(3), 458–469. https://doi.org/10.1681/ASN.2011050485.

Qin, W., Chung, A. C., Huang, X. R., Meng, X. M., Hui, D. S., Yu, C. M., et al. (2011). TGF-beta/Smad3 signaling promotes renal fibrosis by inhibiting miR-29. *Journal of the American Society of Nephrology, 22*(8), 1462–1474. https://doi.org/10.1681/ASN.2010121308.

Ranieri, M., Zahedi, K., Tamma, G., Centrone, M., Di Mise, A., Soleimani, M., et al. (2018). CaSR signaling down-regulates AQP2 expression via a novel microRNA pathway in pendrin and NaCl cotransporter knockout mice. *The FASEB Journal, 32*(4), 2148–2159. https://doi.org/10.1096/fj.201700412RR.

Regev, A., Teichmann, S. A., Lander, E. S., Amit, I., Benoist, C., Birney, E., et al. (2017). The human cell atlas. *eLife, 6*, e27041. https://doi.org/10.7554/eLife.27041.

Remuzzi, G., Cattaneo, D., & Perico, N. (2008). The aggravating mechanisms of aldosterone on kidney fibrosis. *Journal of the American Society of Nephrology, 19*(8), 1459–1462. https://doi.org/10.1681/ASN.2007101079.

Riccardi, D., Park, J., Lee, W. S., Gamba, G., Brown, E. M., & Hebert, S. C. (1995). Cloning and functional expression of a rat kidney extracellular calcium/polyvalent cation-sensing receptor. *Proceedings of the National Academy of Sciences of the United States of America, 92*(1), 131–135.

Rojek, A., Fuchtbauer, E. M., Kwon, T. H., Frokiaer, J., & Nielsen, S. (2006). Severe urinary concentrating defect in renal collecting duct-selective AQP2 conditional-knockout mice. *Proceedings of the National Academy of Sciences of the United States of America, 103*(15), 6037–6042. https://doi.org/10.1073/pnas.0511324103.

Rossier, B. C., Bochud, M., & Devuyst, O. (2017). The hypertension pandemic: An evolutionary perspective. *Physiology (Bethesda), 32*(2), 112–125. https://doi.org/10.1152/physiol.00026.2016.

Rudnicki, M., Perco, P., Haene, B. D., Leierer, J., Heinzel, A., Muhlberger, I., et al. (2016). Renal microRNA- and RNA-profiles in progressive chronic kidney disease. *European Journal of Clinical Investigation, 46*(3), 213–226. https://doi.org/10.1111/eci.12585.

Saini, H. K., Griffiths-Jones, S., & Enright, A. J. (2007). Genomic analysis of human microRNA transcripts. *Proceedings of the National Academy of Sciences of the United States of America, 104*(45), 17719–17724. https://doi.org/10.1073/pnas.0703890104.

Sandoval, P. C., Claxton, J. S., Lee, J. W., Saeed, F., Hoffert, J. D., & Knepper, M. A. (2016). Systems-level analysis reveals selective regulation of Aqp2 gene expression by vasopressin. *Scientific Reports, 6*, 34863. https://doi.org/10.1038/srep34863.

Schrier, R. W., Berl, T., & Anderson, R. J. (1979). Osmotic and nonosmotic control of vasopressin release. *The American Journal of Physiology, 236*(4), F321–F332. https://doi.org/10.1152/ajprenal.1979.236.4.F321.

Sequeira-Lopez, M. L., Weatherford, E. T., Borges, G. R., Monteagudo, M. C., Pentz, E. S., Harfe, B. D., et al. (2010). The microRNA-processing enzyme dicer maintains juxtaglomerular cells. *Journal of the American Society of Nephrology, 21*(3), 460–467. https://doi.org/10.1681/ASN.2009090964.

Tahiri, A., Aure, M. R., & Kristensen, V. N. (2018). MicroRNA networks in breast cancer cells. *Methods in Molecular Biology, 1711*, 55–81. https://doi.org/10.1007/978-1-4939-7493-1_4.

Thomson, J. M., Parker, J., Perou, C. M., & Hammond, S. M. (2004). A custom microarray platform for analysis of microRNA gene expression. *Nature Methods, 1*(1), 47–53. https://doi.org/10.1038/nmeth704.

Treiber, T., Treiber, N., & Meister, G. (2019). Regulation of microRNA biogenesis and its crosstalk with other cellular pathways. *Nature Reviews. Molecular Cell Biology, 20*(1), 5–20. https://doi.org/10.1038/s41580-018-0059-1.

Volpe, C. M. O., Villar-Delfino, P. H., Dos Anjos, P. M. F., & Nogueira-Machado, J. A. (2018). Cellular death, reactive oxygen species (ROS) and diabetic complications. *Cell Death & Disease, 9*(2), 119. https://doi.org/10.1038/s41419-017-0135-z.

Wang, B., Herman-Edelstein, M., Koh, P., Burns, W., Jandeleit-Dahm, K., Watson, A., et al. (2010). E-cadherin expression is regulated by miR-192/215 by a mechanism that is independent of the profibrotic effects of transforming growth factor-beta. *Diabetes, 59*(7), 1794–1802. https://doi.org/10.2337/db09-1736.

Wang, B., Koh, P., Winbanks, C., Coughlan, M. T., McClelland, A., Watson, A., et al. (2011). miR-200a prevents renal fibrogenesis through repression of TGF-beta2 expression. *Diabetes, 60*(1), 280–287. https://doi.org/10.2337/db10-0892.

Wang, B., Komers, R., Carew, R., Winbanks, C. E., Xu, B., Herman-Edelstein, M., et al. (2012). Suppression of microRNA-29 expression by TGF-beta1 promotes collagen expression and renal fibrosis. *Journal of the American Society of Nephrology, 23*(2), 252–265. https://doi.org/10.1681/ASN.2011010055.

Wang, Y., Luo, J., Zhang, H., & Lu, J. (2016). microRNAs in the same clusters evolve to coordinately regulate functionally related genes. *Molecular Biology and Evolution*, *33*(9), 2232–2247. https://doi.org/10.1093/molbev/msw089.

Wang, X. Y., Masilamani, S., Nielsen, J., Kwon, T. H., Brooks, H. L., Nielsen, S., et al. (2001). The renal thiazide-sensitive Na-Cl cotransporter as mediator of the aldosterone-escape phenomenon. *The Journal of Clinical Investigation*, *108*(2), 215–222. https://doi.org/10.1172/JCI10366.

Wei, Q., Bhatt, K., He, H. Z., Mi, Q. S., Haase, V. H., & Dong, Z. (2010). Targeted deletion of dicer from proximal tubules protects against renal ischemia-reperfusion injury. *Journal of the American Society of Nephrology*, *21*(5), 756–761. https://doi.org/10.1681/ASN.2009070718.

Wei, Q., Liu, Y., Liu, P., Hao, J., Liang, M., Mi, Q. S., et al. (2016). MicroRNA-489 induction by hypoxia-inducible factor-1 protects against ischemic kidney injury. *Journal of the American Society of Nephrology*, *27*(9), 2784–2796. https://doi.org/10.1681/ASN.2015080870.

Williams, M. E. (2005). Diabetic nephropathy: The proteinuria hypothesis. *American Journal of Nephrology*, *25*(2), 77–94. https://doi.org/10.1159/000084286.

Wilson, J. L., Miranda, C. A., & Knepper, M. A. (2013). Vasopressin and the regulation of aquaporin-2. *Clinical and Experimental Nephrology*, *17*(6), 751–764. https://doi.org/10.1007/s10157-013-0789-5.

Witwer, K. W., & Halushka, M. K. (2016). Toward the promise of microRNAs—Enhancing reproducibility and rigor in microRNA research. *RNA Biology*, *13*(11), 1103–1116. https://doi.org/10.1080/15476286.2016.1236172.

Yu, J. (2014). miRNAs in mammalian ureteric bud development. *Pediatric Nephrology*, *29*(4), 745–749. https://doi.org/10.1007/s00467-013-2734-y.

Zhong, X., Chung, A. C., Chen, H. Y., Dong, Y., Meng, X. M., Li, R., et al. (2013). miR-21 is a key therapeutic target for renal injury in a mouse model of type 2 diabetes. *Diabetologia*, *56*(3), 663–674. https://doi.org/10.1007/s00125-012-2804-x.

CHAPTER SEVEN

Aquaporin regulation: Lessons from secretory vesicles

Bhanu P. Jena*
Department of Physiology, Wayne State University School of Medicine, Detroit, MI, United States
*Corresponding author: e-mail address: bjena@med.wayne.edu

Contents

Abstract

Secretory vesicle swelling has been demonstrated to be a requirement in cell secretion. In the past 25 years, the quest to elucidate the molecular mechanism of secretory vesicle swelling, serendipitously revealed the presence of water channels or aquaporins at the secretory vesicle membrane and their involvement in rapid gaiting of water into secretory vesicles and their swelling during secretion. These studies further provided an understanding of aquaporin regulation at the secretory vesicle membrane. Secretory vesicles within live cells, isolated secretory vesicles, single vesicle electrophysiological patch clamp studies and the swelling complex reconstituted into liposomes, have all been utilized to elucidate the mechanism and regulation of secretory vesicle swelling. Results from these studies collectively demonstrate the involvement of β-adrenergic receptor, heterotrimeric GTP-binding G-proteins such as Gαl; PLA2 and potassium and chloride channels, in the regulation of aquaporins at the secretory vesicle membrane.

Vesicle volume increases following stimulation of cell secretion had been proposed from earlier electrophysiological measurements in mast cells and studies on bovine adrenal chromaffin cells (Alvarez de Toledo, Fernández-Chaócn, & Fernández, 1993; Curran & Brodwick, 1991; Fernandez, Villalon, & Verdugo, 1991; Monck, Oberhauser, Alverez de Toledo, & Fernandez, 1991). However, a direct observation of secretory vesicle swelling in live cells following a secretory stimulus at nm resolution and in real time, was first demonstrated in acinar cells of the exocrine pancreas (Fig. 1; Kelly, Cho, Jeremic, Abu-Hamdah, & Jena, 2004), using atomic force microscopy (AFM). Live exocrine pancreatic acinar cells placed

ISSN 0083-6729
https://doi.org/10.1016/bs.vh.2019.08.007

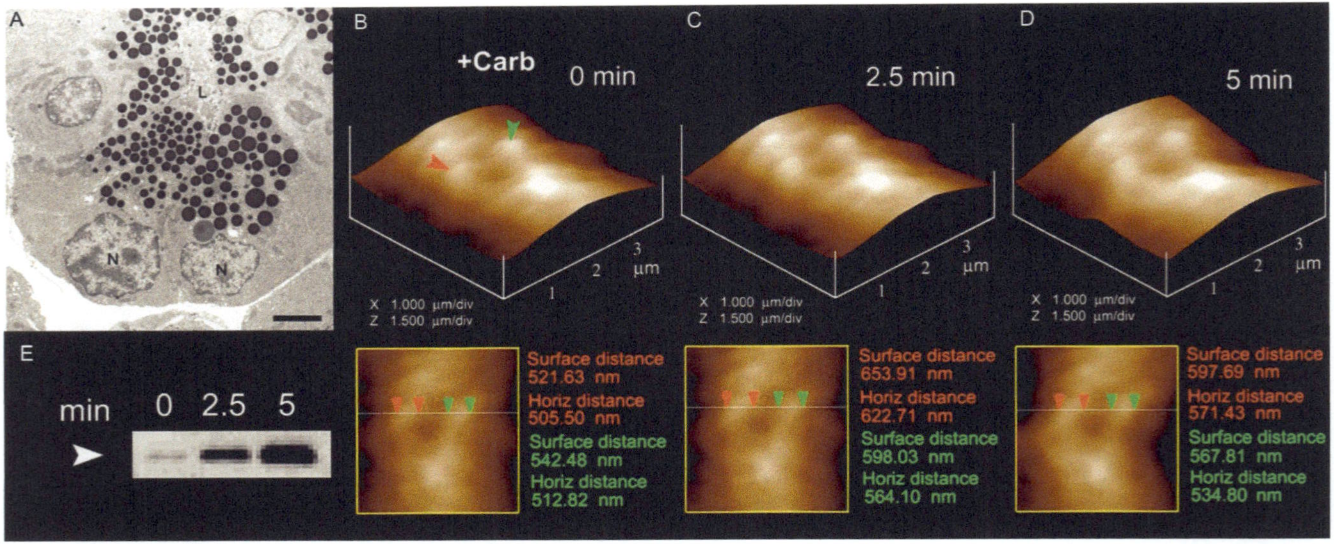

Fig. 1 The swelling dynamics of ZGs in live pancreatic acinar cells. (A) Electron micrograph of pancreatic acinar cells showing the basolaterally located nucleus (N) and the apically located ZGs. The apical end of the cell faces the acinar lumen (L). Bar = 2.5 µm. (B–D) The apical ends of live pancreatic acinar cells were imaged by AFM, showing ZGs (red and green arrowheads) lying just below the apical plasma membrane. Exposure of the cell to a secretory stimulus using 1 µM carbamylcholine, resulted in ZG swelling within 2.5 min, followed by a decrease in ZG size after 5 min. The decrease in size of ZGs after 5 min is due to the release of secretory products such as a-amylase, as demonstrated by the immunoblot assay (E). *Kelly, M., Cho, W. J., Jeremic, A., Abu-Hamdah, R., & Jena, B. P. (2004). Vesicle swelling regulates content expulsion during secretion.* Cell Biology International, 28, *709–716. ©Bhanu Jena.*

in physiological buffered solution when imaged at the nano scale using the atomic force microscope (AFM), provide dynamic changes in the size of secretory vesicles in the exocrine pancreas called zymogen granules (ZGs) lying immediately beneath the apical plasma membrane of the cell (Fig. 1). Following a 2.5 min exposure to a secretory stimulus, majority of the ZGs within acinar cells swell and grow larger, followed by secretion and a concomitant decrease in ZG size (Fig. 1). Interestingly, no loss of secretory vesicles is observed following secretion, demonstrating transient or "kiss-and-run" mechanism of secretion. Measurements of intracellular ZG size reveal that different vesicles swell differently following a secretory stimulus, explaining why following stimulation of cell secretion, secretory vesicles demonstrate the presence of varying vesicular content. Similarly, freshly isolated secretory vesicles from the exocrine pancreas (Cho et al., 2002; Jena et al., 1997) and neurons (Chen, Lee, Shin, Cho, & Jena, 2010; Jeremic, Cho, & Jena, 2005; Lee, Cho, Shin, & Jena, 2010; Shin et al., 2010), single vesicle patch (Kelly, Abu-Hamdah, Cho, Ilie, & Jena, 2005), and reconstituted swelling-competent proteoliposomes (Abu-Hamdah et al., 2004) have been utilized to determine the mechanism and regulation of secretory vesicle swelling. Results from these studies collectively demonstrate the presence and involvement of aquaporins at the secretory vesicle membrane in vesicle volume increase, and its regulation involving β-adrenergic receptor, heterotrimeric GTP-binding G-proteins, PLA2, potassium and chloride channels.

Exposure of isolated ZGs (Fig. 2) to GTP and to the Gi-protein stimulating peptide Mastoparan (Mas7), results in rapid gating of water via the water channel AQP1 at the ZG membrane, resulting in ZG swelling (Cho et al., 2002; Jena et al., 1997; Kelly et al., 2004). This swelling of isolated ZG is similar to observations in live acinar cells (Cho et al., 2002; Fig. 1), where each isolated ZG respond differently to the same stimulus. Mastoparan is an amphiphilic tetradecapeptide from wasp venom and is known to potentiate the GTPase activity of Gi/Go proteins (Higashijima, Uzu, Nakajima, & Ross, 1988; Vitale et al., 1993). Stimulation of G proteins is believed to occur by the peptide inserting into the phospholipid membrane and forming a highly structured α-helix that resembles the intracellular loops of G protein-coupled receptors. Analogous to receptor activation, mastoparan is thought to interact with the COOH-terminal domain of the G protein α subunit (Weingarten, Ransnas, Mueller, Sklar, & Bokoch, 1990). Active mastoparan (Mas7) and inactive mastoparan (Mas17) analogs in the presence of [γ-32P] GTP, demonstrate a significant

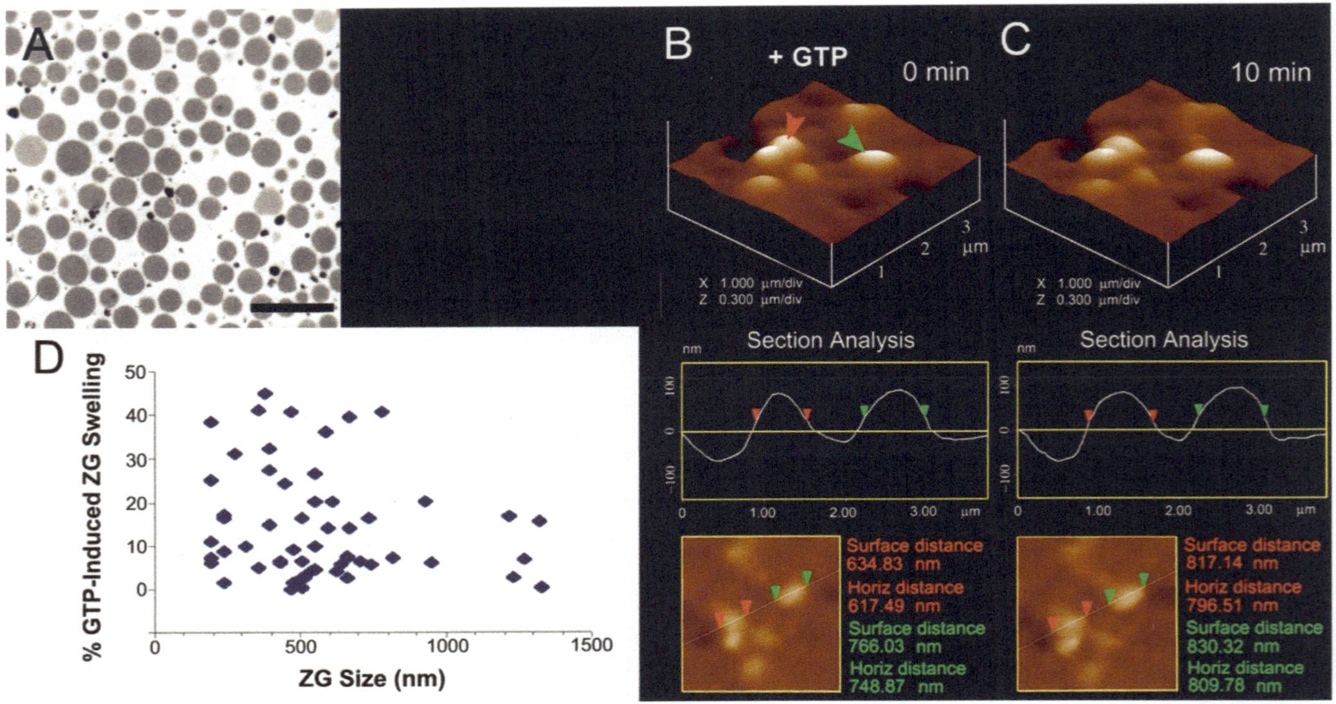

Fig. 2 Swelling of isolated ZGs. (A) Electron micrograph of isolated ZGs demonstrating a homogeneous preparation. Bar = 2.5 µm. (B, C) Isolated ZGs, on exposure to 20 µM GTP, swell rapidly. Note the enlargement of ZGs as determined by AFM section analysis of two vesicles (red and green arrowheads). (D) Percent ZG volume increase in response to 20 µM GTP. Note how different ZGs respond to the GTP-induced swelling differently. *Kelly, M., Cho, W. J., Jeremic, A., Abu-Hamdah, R., & Jena, B. P. (2004). Vesicle swelling regulates content expulsion during secretion.* Cell Biology International, 28, *709–716. ©Bhanu Jena.*

increase in swelling and GTPase activity in ZG, in the presence of Mas7 over control (Mas17), confirming the presence of Gi proteins with ZGs (Cho et al., 2002; Jena et al., 1997). The GTP and Mas7 induced ZG volume increase is abrogated in the presence of AQP inhibitors such as mercuric chloride, silver nitrate and AQP1-specific antibody (Cho et al., 2002), suggesting that Gi-protein at the ZG membrane functions as the AQP1 "on" switch. Furthermore, since Mas7 induces ZG swelling, suggests the presence of a β-adrenergic receptor at the ZG membrane via which the Gi is activated. PLA2, K and Cl channel have also been implicated (Figs. 3 and 4). Similarly, Mas7 induces synaptic vesicle swelling via a Go protein and AQP6 (Chen et al., 2010; Jeremic et al., 2005) and the presence of a β-adrenergic receptor has been demonstrated (Fig. 5–8).

Detergent-solubilized ZGs immunoisolated using monoclonal AQP-1 antibody, co-isolates AQP1, PLA2, $G_{\alpha i3}$, potassium channel IRK-8, and the chloride channel ClC-2 (Figs. 9 and 10; Abu-Hamdah et al., 2004). Exposure of ZGs to either the potassium channel blocker glyburide, or the PLA2 inhibitor ONO-RS-082, blocks GTP-induced ZG swelling. Red blood cells known to possess AQP1 at the plasma membrane also swell on exposure to the $G_{\alpha I}$ agonist Mas7 and responds similarly to ONO-RS-082 and glyburide, as do ZGs (Abu-Hamdah et al., 2004). Artificial liposomes reconstituted with the AQP1 immunoisolated complex from solubilized ZG preparation exhibit swell on exposure to GTP. Glyburide or ONO-RS-082 are found to abrogate the GTP effect on the reconstituted liposomes. AQP1 immunoisolates, reconstituted to planar lipid membrane, demonstrate conductance, which is sensitive to glyburide and an AQP1-specific antibody. These results demonstrate a $G_{\alpha i3}$-PLA2 mediated pathway and potassium channel involvement in AQP1 regulation at the ZG membrane (Abu-Hamdah et al., 2004), contributing to ZG swelling. Similarly, AQP-6 involvement has been demonstrated in GTP-induced and G_o-mediated synaptic vesicle swelling in neurons (Jeremic et al., 2005).

Single ZG patch clamp studies have been performed to further characterize the ion channel present at the ZG membrane (Kelly et al., 2005). These studies confirm earlier findings of the presence of both potassium and chloride ion channels at the ZG membrane. In single ZG patch studies, the electrical activity at the ZG membrane displays a range of sensitivity both to chloride and potassium channel blockers. Whole vesicle conductance decreased with the addition of the chloride channel blocker DIDS and the ATP $K+$ channel blocker glyburide, in both vesicle patches and indirect analysis, supporting the hypothesis for the presence of more than one

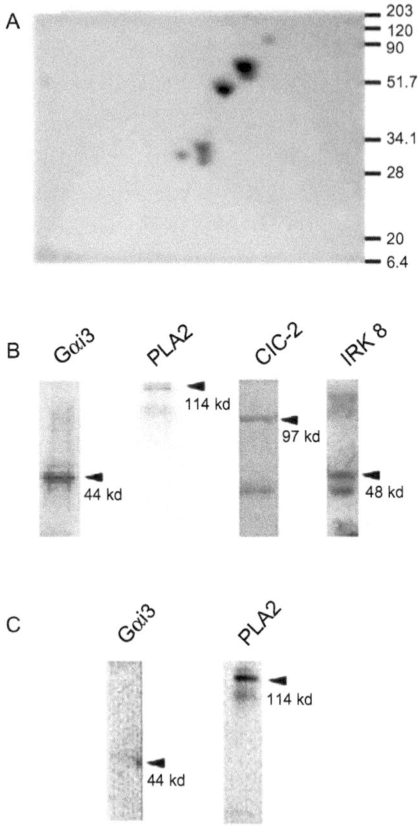

Fig. 3 One (1D) and two dimensionally (2D) resolved, AQP1-immunoisolated proteins from solubilized pancreatic ZG preparations (A, B) and red blood cells (C), using an AQP1-specific antibody. The 2D-resolved proteins were Coomassie stained and the 1D resolved proteins transferred to nitrocellulose membranes for immunoblot analysis. Note the identification in the immunoisolates of seven spots in the Coomassie-stained 2D-resolved gel (A). Immunoblot analysis of the 1D-resolved immunoisolated proteins demonstrate the presence of PLA_2, $G_{\alpha i3}$, the potassium channel IRK-8, and chloride channel ClC-2 (B). Similarly, in red blood cells, PLA_2 and $G_{\alpha i3}$ immunoreactive bands are detected (C). Lower molecular weight bands may represent proteolytic cleavage products. *Abu-Hamdah, R., Cho, W.-J., Cho, S.-J., Jeremic, A., Kelly, M., Ilie, A.E., et al. (2004). Regulation of the water channel aquaporin-1: Isolation and reconstitution of the regulatory complex.* Cell Biology International, 28, *7–17. ©Bhanu Jena.*

channel type (Kelly et al., 2005). These findings were further confirmed by Western Blot analysis and as speculated, the presence of two chloride channels, CLC-2 and CLC-3 was demonstrated (Kelly et al., 2005). Consistent with pharmacological evidence is the presence of ATP sensitive potassium channel, Kir6.1 in Western Blot analysis of ZGs.

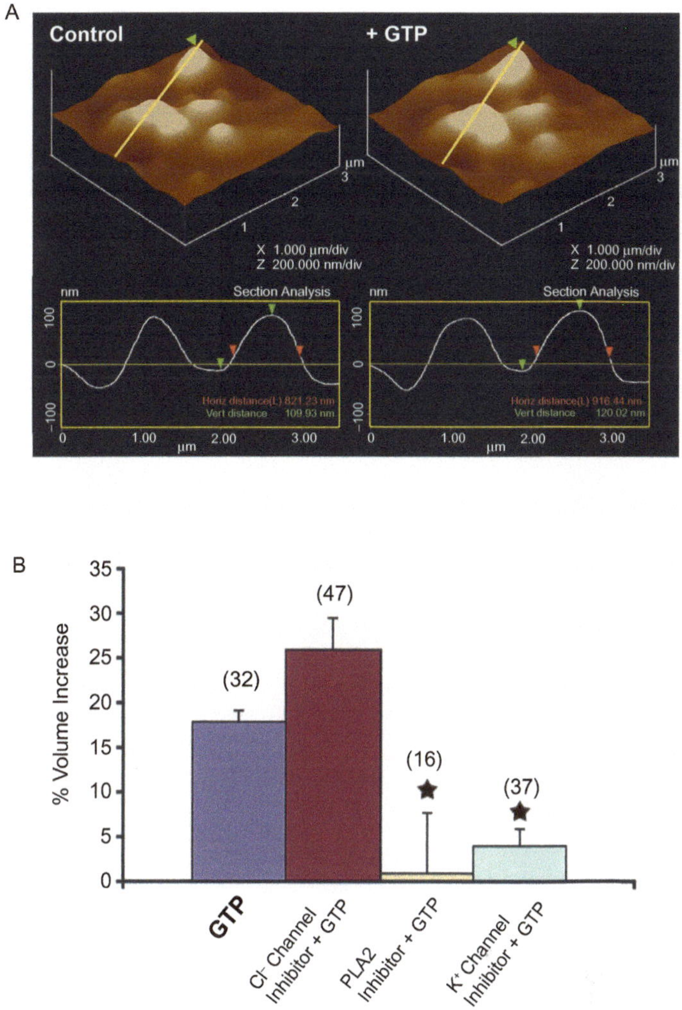

Fig. 4 ZG volume changes measured using AFM. AFM micrographs demonstrating swelling of isolated ZGs in response to 40 µM GTP (A). Histogram showing changes in ZG volume following exposure of isolated ZGs to GTP, and to the inhibitor of the chloride channel DIDS, the inhibitor of the potassium channel, glyburide, or to the inhibitor of PLA_2, ONO-RS-082 (B). Note the significant inhibition of GTP-induced ZG swelling, in the presence of glyburide, or ONO-RS-082. DIDS has little effect on GTP-induced ZG swelling. Values represent mean ± SE of mean of the number of ZGs (in parenthesis), which were randomly selected. A significant ($P < 0.01$) inhibition (*) in the presence of glyburide, or ONO-RS-082 was demonstrated. *Abu-Hamdah, R., Cho, W.-J., Cho, S.-J., Jeremic, A., Kelly, M., Ilie, A. E., et al. (2004). Regulation of the water channel aquaporin-1: Isolation and reconstitution of the regulatory complex.* Cell Biology International, *28, 7–17. ©Bhanu Jena.*

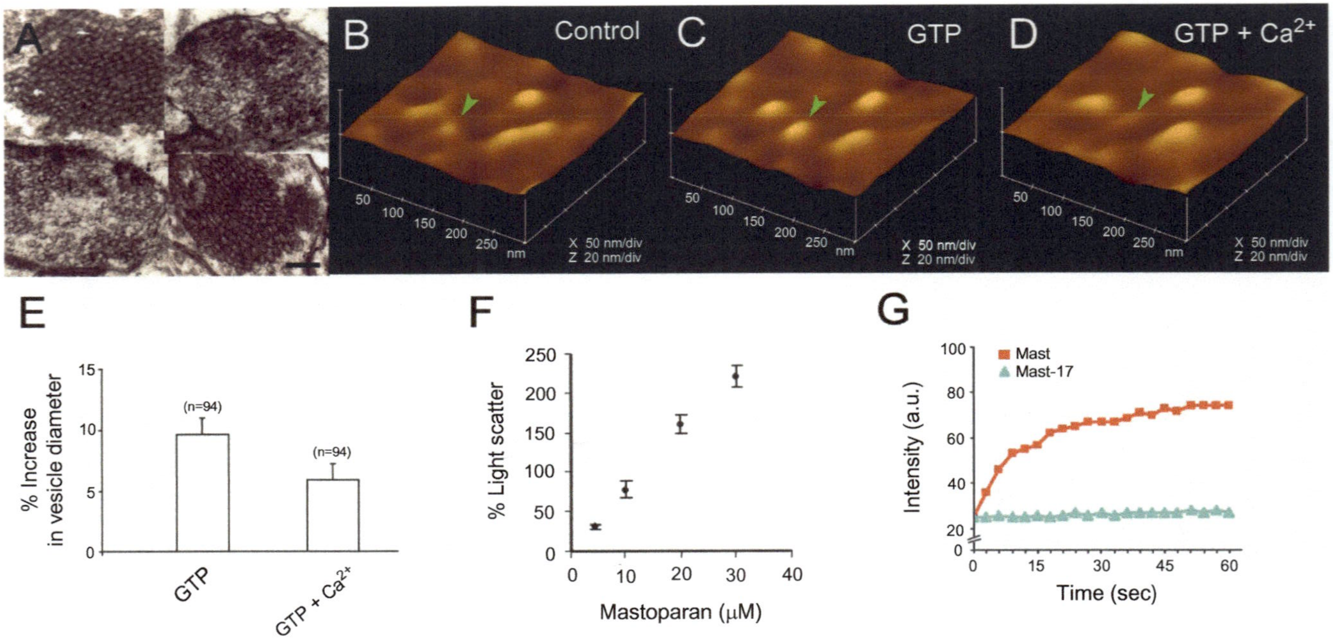

Fig. 5 Synaptic vesicles swell in response to GTP and mastoparan (Mas7), and vesicle swelling is required for neurotransmitter release. (A) Electron micrographs of brain synaptosomes demonstrating the presence of 40–50 nm synaptic vesicles within. Bar = 200 nm. (B) AFM micrographs of synaptosomal membrane, demonstrating the presence of 40–50 nm synaptic vesicles docked to the cytosolic face of the presynaptic membrane. (C) Exposure of the synaptic vesicles to 20 μM GTP results in vesicle swelling (blue arrowhead). (D, E) Addition of calcium, results in the transient fusion of the synaptic vesicles at porosomes in the presynaptic membrane of the nerve terminal, and expulsion of intra-vesicular contents. Note the decrease in size of the synaptic vesicle following content expulsion. (F) Light scattering assays on isolated synaptic vesicles demonstrates the Mas7 dose-dependent increase in vesicle swelling ($n = 5$), and further confirms the AFM results. (G) Exposure of isolated synaptic vesicles to 20 μM Mas7, demonstrates a time-dependent (in seconds) increase in their swelling. Note the control peptide mast-17 has little or no effect on synaptic vesicle swelling. *Kelly, M., Cho, W. J., Jeremic, A., Abu-Hamdah, R., & Jena, B. P. (2004). Vesicle swelling regulates content expulsion during secretion.* Cell Biology International, 28, *709–716. ©Bhanu Jena.*

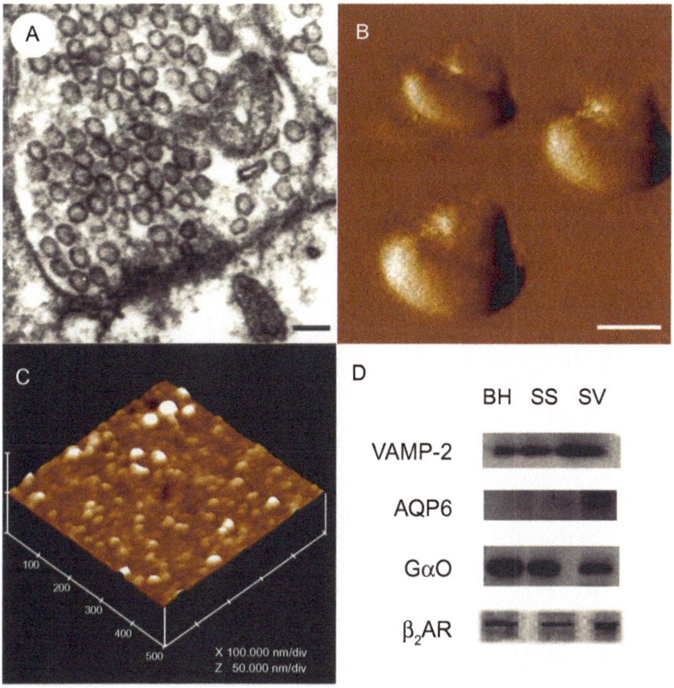

Fig. 6 Association of β_2 adrenergic receptor with synaptic vesicles (SV). Purity of the synaptosome (A, B) and SV (C) was determined using transmission electron microscopy (A, Bar = 100 nm), atomic force microscopy (B, Bar = 1 μm, and C), and immunoblot analysis (D) on total rat brain homogenate (BH), synaptosome (SS) and SV (SV) preparation. Note the clean 1–2 μm mushroom-shaped synatosomes (B) and the 35–50 nm SV (C) preparations in the atomic force micrographs. Immunoblot analysis of 10 μg protein each, of BH, SS and SV, demonstrate the presence of $G_{\alpha o}$ protein, and the enriched presence of SV proteins VAMP-2 and the water channel AQP6. Note the enriched presence of β_2AR in the SV fraction. *Chen, Z.-H., Lee, J.-S., Shin, L., Cho, W.-J., & Jena, B. P. (2010). Involvement of β-adrenergic receptor in synaptic vesicle swelling and implication in neurotransmitter release.* Journal of Cellular and Molecular Medicine, 15, *572–576. ©Bhanu Jena.*

Similar to in the ZG of the exocrine pancreas, Mas7 the amphiphilic tetradecapeptide from wasp venom activates G_o protein GTPase and stimulates synaptic vesicle swelling, hence the presence of β-adrenergic receptor at the synaptic vesicle membrane was hypothesized. Stimulation of G proteins is believed to occur via insertion of Mas7 into the phospholipid membrane to form a highly structured α-helix that resembles the intracellular loops of G protein-coupled adrenergic receptors. Consequently, the presence of adrenoceptors and the presence of an

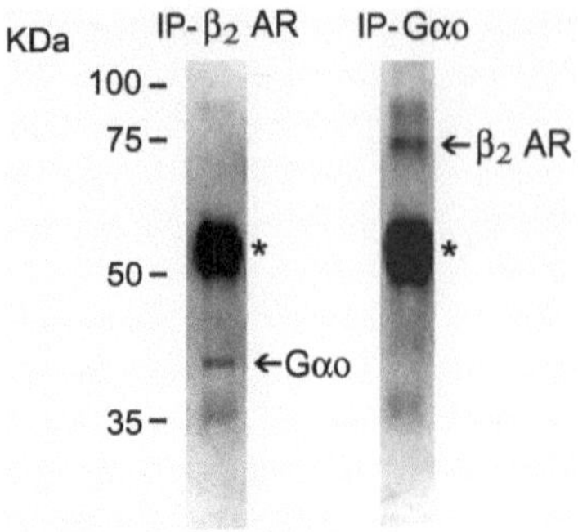

Fig. 7 Interaction of β_2 adrenergic receptor with Gαo protein in synaptic vesicles (SV). Detergent-solubilized SV preparation immunoprecipitated using β_2AR-specific antibody results in co-isolation of the 45 kD $G_{\alpha o}$ protein, suggesting their physical association in the SV. Similarly, when detergent-solubilized SV preparation was immunoprecipitated using $G_{\alpha o}$-specific antibody and the resultant isolate probed with the β_2AR-specific antibody, the 68 kD β_2AR protein is found co-isolated, supporting the physical association between the proteins in the native state at the SV membrane. The dark band represented by (*) is the heavy chain of the antibody used in immunoprecipitation reaction. *Chen, Z.-H., Lee, J.-S., Shin, L., Cho, W.-J., & Jena, B. P. (2010). Involvement of β-adrenergic receptor in synaptic vesicle swelling and implication in neurotransmitter release.* Journal of Cellular and Molecular Medicine, 15, *572–576. ©Bhanu Jena.*

endogenous b-adrenergic agonist at the synaptic vesicle membrane has been investigated. Immunoblot analysis of synaptic vesicles using β-adrenergic receptor specific antibody (Figs. 6 and 7), and vesicle swelling experiments using β-adrenergic agonists and antagonists, demonstrate the presence of functional β-adrenergic receptors at the synaptic vesicle membrane (Chen et al., 2010). Electron microscopy (EM) (Fig. 6A), AFM (Fig. 6B and C), and immunoblot analysis (Fig. 6D), demonstrated a highly enriched synaptosome (SS) and synaptic vesicle (SV) preparation. Immunoblot analysis, demonstrate the SV preparation to be enriched in VAMP-2 and AQP-6 (Fig. 6D), both of which are SV-specific marker proteins (Jeremic et al., 2005). Additionally, in conformation with previous findings (Jeremic et al., 2005), the GTP-binding Gαo protein was found to associate with the SV-enriched preparation. Collectively, these studies demonstrated the isolation of a highly enriched SV preparation from rat brain tissue for SV swelling and β_2AR immunolocalization assays. Western blot analysis confirmed that β_2AR is enriched in the SV fraction (Fig. 6D).

Upon binding to endogenous activators, adrenergic receptors are known to undergo a conformational change that leads to the activation of hetrotrimeric GTP-binding G-proteins. Different groups of adrenergic

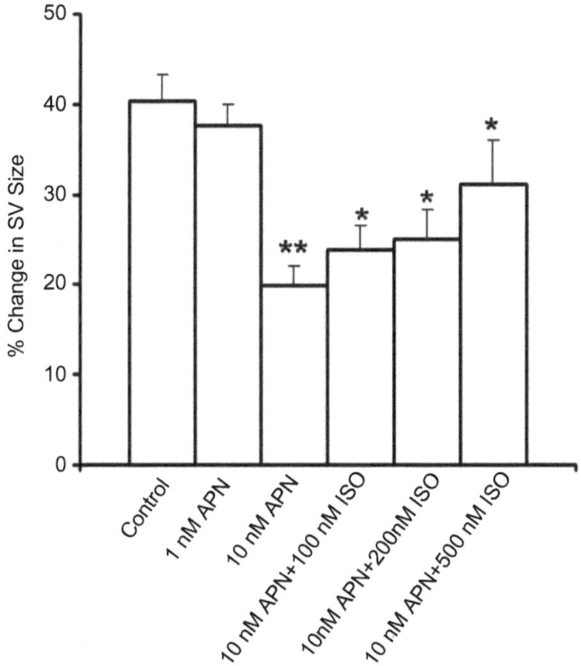

Fig. 8 β_2 adrenergic receptors in synaptic vesicles (SV) are functional. Photon correlation spectroscopy (PCS) used in the determination of SV size. Exposure of SV to 40 μM GTP and 40 μM mastoparan results a significant increase in SV size (control). However, exposure of SV to the β_2AR antagonist alprenolol (APN), demonstrates inhibition of the GTP-mastoparan mediated SV swelling. Note the significant ($n=6$, $^{**}P<0.001$) inhibition of GTP-mastoparan mediated SV swelling in the presence of 10 nM APN. Exposure of the 10 nM APN-treated SV to the β_2AR agonist isoproterenol (ISO), demonstrated a dose-dependent increase ($n=6$, $^{*}P<0.05$) in SV size. *Chen, Z.-H., Lee, J.-S., Shin, L., Cho, W.-J., & Jena, B. P. (2010). Involvement of β-adrenergic receptor in synaptic vesicle swelling and implication in neurotransmitter release.* Journal of Cellular and Molecular Medicine, 15, *572–576. ©Bhanu Jena.*

receptors couple to and activate only certain G protein types, thus leading to specific intracellular signals (Wettschureck & Offermanns, 2005). Our immunoblot assays demonstrate the enriched presence of β2 adrenergic receptors in the SV fraction therefore, the physical interaction of SV-associated $G_{\alpha o}$ and β2 adrenergic receptors was investigated. To determine the co-association of $G_{\alpha o}$ and β_2AR at the SV, immunoprecipitation, and immunoblot analysis were carried out. Detergent-solubilized SV preparation was immunoprecipitated using β_2AR-specific antibody, and the resultant isolate probed with a $G_{\alpha o}$-specific primary antibody. Results from

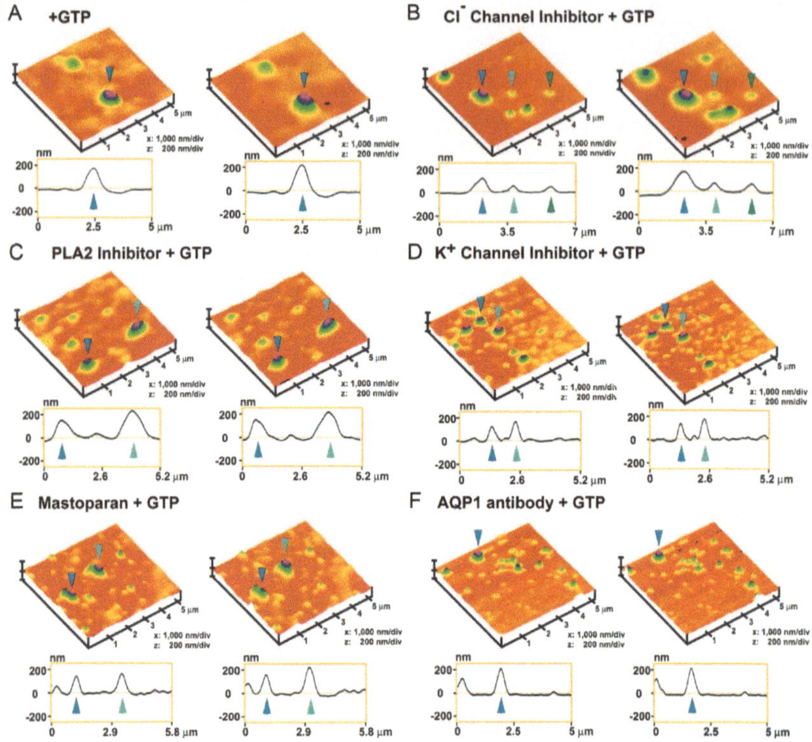

Fig. 9 AFM micrographs of liposomes demonstrating functional reconstitution of the immunoisolated AQP1-complex. Note swelling of the reconstituted liposomes in response to GTP (A) panel left is control and panel right is 5 min following GTP exposure. Exposure of the chloride channel inhibitor DIDS, did not inhibit GTP-induced swelling (B); however, the PLA_2 inhibitor ONO-RS-082 (C), the potassium channel inhibitor glyburide (D), and the inhibitory AQP1-specific antibody inhibited GTP-induced swelling (F). The $G_{\alpha i}$-stimulable peptide mastoparan, demonstrates no additional effect on swelling over GTP (E), hence the GTP effect is mediated through the $G_{\alpha i3}$ protein. *Abu-Hamdah, R., Cho, W.-J., Cho, S.-J., Jeremic, A., Kelly, M., Ilie, A. E., et al. (2004). Regulation of the water channel aquaporin-1: Isolation and reconstitution of the regulatory complex.* Cell Biology International, 28, *7–17. ©Bhanu Jena.*

the study demonstrate the co-isolation of a 45 kD $G_{\alpha o}$-specific protein, suggesting the physical association of $G_{\alpha o}$ and the β_2AR in SV (Fig. 7). Further confirmation of this interaction was demonstrated when detergent-solubilized SV preparation was once again immunoprecipitated, this time using $G_{\alpha o}$-specific antibody followed by probing with a β_2AR-specific antibody. In agreement, the co-isolation of the 68 kD β_2AR and $G_{\alpha o}$, suggested the physical association of $G_{\alpha o}$ and β_2AR in SV (Fig. 7).

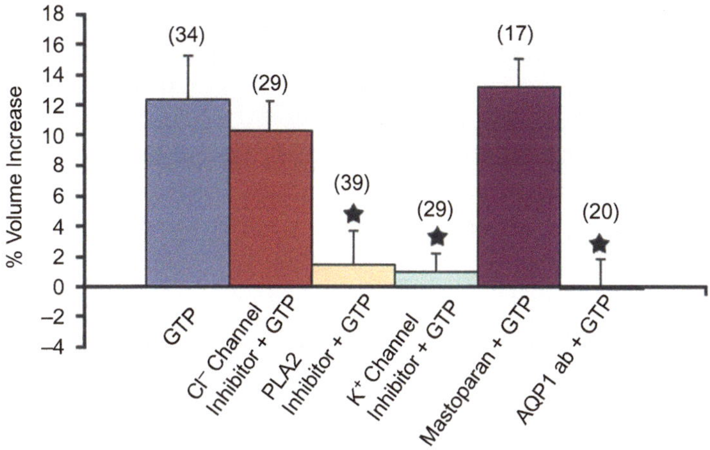

Fig. 10 Percent volume changes in reconstituted liposomes measured by AFM. Estimation of the changes in liposome volume demonstrates a 12–15% increase, 5 min following exposure to 40 μM GTP. No significant change in GTP-induced liposome swelling is observed following exposure to the chloride channel inhibitor DIDS or to the $G_{\alpha i}$-stimulable peptide mastoparan. Similar to the responses in ZGs, exposure of the reconstituted liposomes to the PLA_2 inhibitor ONO-RS-082, the potassium channel inhibitor glyburide, and the AQP1-specific antibody, significantly inhibits the GTP effect. Values represent mean ± SE of mean of randomly selected liposomes (in parenthesis). A significant ($P < 0.01$) inhibition (*) in the presence of glyburide, ONO-RS-082, and the AQP1-specific antibody, is demonstrated. *Abu-Hamdah, R., Cho, W.-J., Cho, S.-J., Jeremic, A., Kelly, M., Ilie, A. E., et al. (2004). Regulation of the water channel aquaporin-1: Isolation and reconstitution of the regulatory complex.* Cell Biology International, 28, *7–17. ©Bhanu Jena.*

The functional state of the β_2AR in SV and its role in regulating SV swelling, was determined using photon correlation spectroscopy (PCS), in the presence and absence of both β_2AR agonist (isoproterenol) and antagonist (alprenolol). Exposure of SV to 40 mM GTP and 40 mM Mas7 results in a significant increase in SV size. When SV preparations are exposed to various concentrations of the β_2 adrenergic receptor antagonist alprenolol, GTP- and Mas7-induced swelling was attenuated in a dose-dependent manner. Ten nano molar alprenolol significantly decrease the GTP/Mas7-induced SV swelling (Fig. 8; Chen et al., 2010). To determine the effect of β_2 adrenergic receptor agonist, isoproterenol, increasing concentrations (100, 200 and 500 nM) of the ligand was used following alprenolol inhibition. Results this study demonstrate that all three concentrations of the agonist significantly stimulate GTP/Mas7-induced SV swelling, with 500 nM isoproterenol having the greatest effect (Fig. 8). Pre-exposure of

SV to alprenolol, results in isoproterenol stimulation of the GTP/Mas7-induced SV swelling (Fig. 8), suggesting that perhaps β_2 adrenergic receptors are already preoccupied by some endogenous ligand, and the receptor-ligand complex is present in a desensitized state at the SV membrane. However, following exposure to alprenolol, the antagonist is able to dislodge this endogenous ligand from the receptor, enabling the agonist isoproterenol to bind and stimulate the SV-associated β_2 adrenergic receptor, causing SV swelling.

In summary, studies on secretory vesicle swelling in vivo in live cells and in vitro on isolated vesicles and in swelling-competent proteoliposomes (Figs. 9 and 10), and single vesicle patch clamp studies, demonstrate that functional β_2 adrenergic receptors and endogenous ligands are associated with secretory vesicles and regulate water transport via Aquaporins to induce

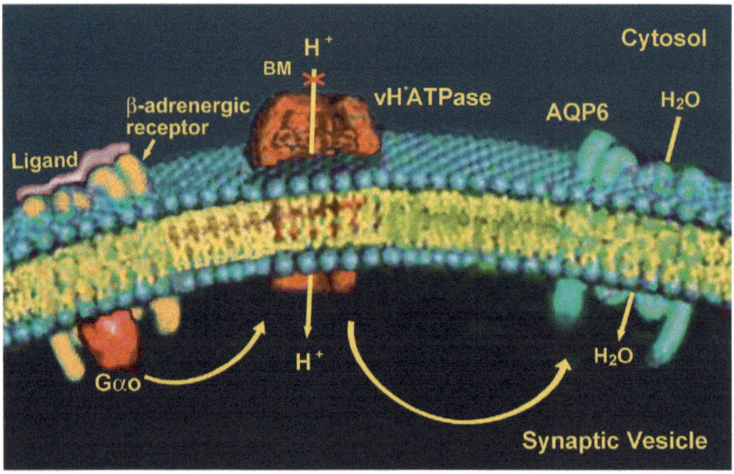

Fig. 11 Schematic diagram of synaptic vesicle membrane, depicting the presence and participation of β_2AR-$G_{\alpha o}$, vH$^+$-ATPase, and the water channel AQP6, is vesicle swelling. The current study shows that GTP-binding $G_{\alpha o}$ protein is stimulated by the activated β_2AR via its endogenous ligand. Earlier studies demonstrate the involvement of vH$^+$-ATPase in GTP-$G_{\alpha o}$-mediated synaptic vesicle swelling. Bafilomycin (BM)-sensitive (red X) vesicle acidification following the GTP-$G_{\alpha o}$ stimulus has also been demonstrated, and since water channels are bidirectional and the vH$^+$-ATPase inhibitor BM decreases both the volume of isolated SV and GTP-mastoparan stimulated swelling, suggests vH$^+$-ATPase to be upstream of AQP-6. Therefore, the SV swelling pathway involves occupancy of the β_2AR and as a consequence, the activation of the GTP-binding $G_{\alpha o}$ protein followed by activation of vH$^+$-ATPase, and transport of water via AQP6 into SV. *Chen, Z.-H., Lee, J.-S., Shin, L., Cho, W.-J., & Jena, B. P. (2010). Involvement of β-adrenergic receptor in synaptic vesicle swelling and implication in neurotransmitter release.* Journal of Cellular and Molecular Medicine, 15, *572–576. ©Bhanu Jena.*

vesicle swelling. Additionally, this vesicle-associated Aquaporin regulation involves GTP-binding G-proteins, PLA2, potassium and chloride channels (Fig. 11). In SV, since vH^+-ATPase activity is upstream of AQP-6 in the pathway leading from $G_{\alpha o}$-stimulated swelling of SV, binding of the endogenous β-adrenergic agonist to its receptors at the SV membrane is hypothesized to initiate the SV swelling cascade.

Acknowledgments

The author thanks the many students and collaborators who have participated in the various studies discussed in this article. Support from the National Institutes of Health (USA), the National Science Foundation (USA), and Wayne State University, is greatly appreciated.

References

Abu-Hamdah, R., Cho, W.-J., Cho, S.-J., Jeremic, A., Kelly, M., Ilie, A. E., et al. (2004). Regulation of the water channel aquaporin-1: Isolation and reconstitution of the regulatory complex. *Cell Biology International, 28*, 7–17.

Alvarez de Toledo, G., Fernández-Chaócn, R., & Fernández, J. M. (1993). Release of secretory products during transient vesicle fusion. *Nature, 363*, 554–558.

Chen, Z.-H., Lee, J.-S., Shin, L., Cho, W.-J., & Jena, B. P. (2010). Involvement of β-adrenergic receptor in synaptic vesicle swelling and implication in neurotransmitter release. *Journal of Cellular and Molecular Medicine, 15*, 572–576.

Cho, S. –. J., Sattar, A. K., Jeong, E. H., Satchi, M., Cho, J., Dash, S., et al. (2002). Aquaporin 1 regulates GTP-induced rapid gating of water in secretory vesicles. *Proceedings of the National Academy of Sciences of the United States of America, 99*, 4720–4724.

Curran, M. J., & Brodwick, M. S. (1991). Ionic control of the size of the vesicle matrix of beige mouse mast cells. *The Journal of General Physiology, 98*, 771–790.

Fernandez, J. M., Villalon, M., & Verdugo, P. (1991). Reversible conductance of mast cell secretory products in vitro. *Biophysical Journal, 59*, 1022–1027.

Higashijima, T., Uzu, S., Nakajima, T., & Ross, E. (1988). Mastoparan, a peptide toxin from wasp venom, mimics receptors by activating GTP-binding regulatory proteins (G proteins). *The Journal of Biological Chemistry, 263*, 6491–6494.

Jena, B. P., Schneider, S. W., Geibel, J. P., Webster, P., Oberleithner, H., & Sritharan, K. C. (1997). G_i regulation of secretory vesicle swelling examined by atomic force microscopy. *Proceedings of the National Academy of Sciences of the United States of America, 94*, 13317–13322.

Jeremic, A., Cho, W.-J., & Jena, B. P. (2005). Involvement of water channels in synaptic vesicle swelling. *Experimental Biology and Medicine, 230*, 674–680.

Kelly, M., Abu-Hamdah, R., Cho, S.-J., Ilie, A. L., & Jena, B. P. (2005). Patch clamped single pancreatic zymogen granules: Direct measurement of ion channel activities at the granule membrane. *Pancreatology, 5*, 443–449.

Kelly, M., Cho, W. J., Jeremic, A., Abu-Hamdah, R., & Jena, B. P. (2004). Vesicle swelling regulates content expulsion during secretion. *Cell Biology International, 28*, 709–716.

Lee, J.-S., Cho, W.-J., Shin, L., & Jena, B. P. (2010). Involvement of cholesterol in synaptic vesicle swelling. *Experimental Biology and Medicine, 235*, 470–477.

Monck, J. R., Oberhauser, A. F., Alverez de Toledo, G., & Fernandez, J. M. (1991). Is swelling of the secretory granule matrix the force that dilates the exocytotic fusion pore? *Biophysical Journal, 59*, 39–47.

Shin, L., Basi, N., Lee, J.-S., Cho, W.-J., Chen, Z., Abu-Hamdah, R., et al. (2010). Involvement of vH^+-ATPase in synaptic vesicle swelling. *Journal of Neuroscience Research*, *88*, 95–101.
Vitale, N., Mukai, H., Rouot, B., Thiersé, D., Aunis, D., & Bader, M. F. (1993). Exocytosis in chromaffin cells. Possible involvement of the heterotrimeric GTP-binding protein G(o). *The Journal of Biological Chemistry*, *268*, 14715–14723.
Weingarten, R., Ransnas, L., Mueller, H., Sklar, L., & Bokoch, G. (1990). Mastoparan interacts with the carboxyl terminus of the alpha subunit of G_i. *The Journal of Biological Chemistry*, *265*, 11044–11049.
Wettschureck, N., & Offermanns, S. (2005). Mammalian G proteins and their cell type specific functions. *Physiological Reviews*, *85*, 1159–1204.

CHAPTER EIGHT

CFTR regulation of aquaporin-mediated water transport

David F. Carrageta[a], Raquel L. Bernardino[a], Marco G. Alves[a], Pedro F. Oliveira[a,b,c,*]

[a]Department of Microscopy, Laboratory of Cell Biology, Unit for Multidisciplinary Research in Biomedicine (UMIB), Institute of Biomedical Sciences Abel Salazar (ICBAS), University of Porto, Porto, Portugal
[b]Department of Genetics, Faculty of Medicine, University of Porto, Porto, Portugal
[c]i3S-Instituto de Investigação e Inovação em Saúde, University of Porto, Porto, Portugal
*Corresponding author: e-mail address: pfobox@gmail.com

Contents

Abstract

The cystic fibrosis transmembrane conductance regulator (CFTR) is an anion channel responsible for the direct transport of bicarbonate and chloride. CFTR-dependent ionic transport is crucial for pH regulation and fluid homeodynamics among epithelial surfaces. Particularly, CFTR performs an essential role in the male reproductive tract, which requires a tight regulation of water and electrolytes in order to produce healthy spermatozoa. The absence or malfunction of CFTR results in cystic fibrosis, the most common lethal disease among Caucasians, that is characterized by an impaired fluid and ionic homeostasis in the whole organism. Due to the wide expression and importance of CFTR, the male reproductive tract is highly affected by cystic fibrosis, resulting in male infertility. Although CFTR is not permeable to water, this protein acts as a regulator of other protein channels, such as aquaporins. In fact, CFTR acts as a molecular partner of aquaporins in epithelial cells, regulating fluid homeodynamics. Herein, up-to-date data concerning the regulation of aquaporin-mediated water transport by CFTR will be discussed, highlighting the role of both channels in the male reproductive tract.

Vitamins and Hormones, Volume 112
ISSN 0083-6729
https://doi.org/10.1016/bs.vh.2019.08.004

Abbreviations

ABC ATP binding cassette
AQP aquaporin
CFTR cystic fibrosis transmembrane conductance regulator
NBC Na^+/HCO_3^- cotransporter
NKCC1 Na^+ K^+ $2Cl^-$ cotransporter
PKA cAMP-dependent protein kinase A

1. Cystic fibrosis transmembrane conductance regulator (CFTR)

1.1 Structure and (mal)function

The cystic fibrosis transmembrane conductance regulator (CFTR) is a protein that functions as a cAMP-activated anion channel. CFTR is part of the ATP binding cassette (ABC) transporter superfamily and is responsible for the direct transport of HCO_3^- and Cl^-, with subsequent regulation of other Cl^-/HCO_3^- exchangers (Anderson et al., 1991; Poulsen, Fischer, Illek, & Machen, 1994; Shcheynikov et al., 2008). CFTR was firstly identified in 1989, as the protein encoded by the *CFTR* gene, located on chromosome 7, whose mutations are responsible for cystic fibrosis (Riordan et al., 1989). Interestingly, CFTR was firstly described as a protein specifically involved in the regulation of other anion transporters and only later it was discovered that CFTR was, in fact, also an anion channel (Bear et al., 1992). Cystic fibrosis is one of the most prevalent monogenic disorders among Caucasians, affecting nearly 1 in 2500 newborns in Europe and 1 in 3500 newborns in the United States (Csanady, Vergani, & Gadsby, 2019). Cystic fibrosis is characterized by the malfunction or absence of CFTR, resulting in impaired transport of Cl^- and HCO_3^- and, thus, impaired fluid and ionic regulation. Due to the wide expression of CFTR, cystic fibrosis leads to an impaired fluid an ionic homeostasis in the whole organism, which results in airway blockage by thickened mucus and subsequent chronic lung inflammation, obstruction of the bowel, distal intestinal obstruction syndrome, high salt sweat, obstruction of the *vas deferens* and male infertility (Csanady et al., 2019; Saint-Criq & Gray, 2017).

CFTR-dependent ionic transport is crucial for pH regulation and fluid dynamics among epithelial surfaces (Saint-Criq & Gray, 2017). Structurally, CFTR is described as an asymmetric hourglass-shaped pore constituted by two transmembrane domains which are, in turn, regulated by three cytosolic domains: two nucleotide binding domains and a regulatory domain

(Callebaut, Chong, & Forman-Kay, 2018; Liu, Zhang, Csanady, Gadsby, & Chen, 2017). Curiously, among the entire ABC superfamily, CFTR is the only protein that is able to act as a transmembrane ion pore, whereas the remaining family proteins act as active transporters. Indeed, CFTR's anion permeation follows a simple ohmic behavior, accordingly to the ionic gradient (Berger et al., 1991). In addition, CFTR is reported to yield a typical bursting behavior, characterized by either long periods of channel opening with brief closures or long closures interrupted by brief openings (Zhang, Yu, Yeh, & Hwang, 2018).

1.2 Regulation of ion and water transport

The mechanism of ion and fluid transport is dependent on the expression and distribution of ion channels, exchangers, cotransporters, and pumps in the plasma membrane of epithelial cells. Epithelial tissues are constituted by one or more layers of closely assembled and polarized epithelial cells, where the basolateral membrane is faced to the interstitium and the apical or luminal membrane composes the physical and chemical barrier to the external environment. Both membranes display a distinct protein profile, accordingly to their specific roles. In order to maintain fluid ionic homeostasis, both water and ions are able to move either between cells (paracellular pathway) or through cells (transcellular pathway) (Saint-Criq & Gray, 2017). The transcellular pathway is dependent on active (or ATP-dependent) or passive ionic transport. Generally, ionic absorption depends on the active transepithelial absorption of Na^+ ions, creating an electrochemical gradient that leads to the passive Cl^- transport. This ionic movement is accompanied by the absorption of water by simple diffusion or by diffusion mediated by aquaporins (Bernardino et al., 2016). On the other hand, transcellular ionic and water transport is mediated by Cl^- efflux through the apical membrane, which is mainly conducted by CFTR (Park & Lee, 2012). While CFTR is highly conductive for Cl^-, which is the main responsible for fluid secretion in most tissues, this channel also mediates the transport of HCO_3^-, an important ion for pH regulation. Despite CFTR providing maximal conductance for Cl^- and HCO_3^-, this channel is also able to conduct cations (Tabcharani, Linsdell, & Hanrahan, 1997). Notwithstanding, CFTR displays high anion selectivity to the detriment of cations and is able to transport several other anions, such as NO_3^- or SCN^- (Linsdell, 2001). CFTR mediates the efflux of Cl^- and HCO_3^-, which stimulates the transport of Na^+ and increases its concentration on the luminal surface. Ultimately, Na^+

accumulation generates an osmotic driving force for water movement, usually either through the paracellular pathway or mediated by aquaporins (Fischbarg, 2010; Shachar-Hill & Hill, 2002; Verkman, Matthay, & Song, 2000). Thus, CFTR regulates water secretion by mediating transcellular ionic transport, with Cl^- as the main protagonist. Indeed, the rate of Cl^- transport is thought to be the rate-limiting step in fluid movement. Cl^- transport is described as a two-steps mechanism that begins with the active accumulation of Cl^- across the basolateral membrane. Due to the activity of the Na^+ K^+ $2Cl^-$ cotransporter (NKCC1), a protein regulated by the Na^+ and K^+ gradient created by the Na^+/K^+-ATPase, a Cl^- accumulation above the electrochemical equilibrium is generated. Additionally, the basolateral Cl^-/HCO_3^- anion exchanger plus the Na^+/HCO_3^- cotransporter (NBC) are reported to further mediate the accumulation of Cl^- (Shan et al., 2012). Then, CFTR mediates the transport of the accumulated Cl^-, which is dependent of three distinct factors: the abundance of CFTR channels, membrane potential and Cl^- concentration, and CFTR gating and activity. The CFTR activation process is regulated by two distinct processes. At first, CFTR becomes active due to phosphorylation of the regulatory domain by the cAMP-dependent protein kinase A (PKA). Then, the opening or closing of previously phosphorylated channels is regulated by ATP binding or hydrolysis at nucleotide binding domains (Anderson et al., 1991; Berger, Ikuma, & Welsh, 2005). Accompanying the CFTR-mediated transport of Cl^-, Na^+ is paracellularly transported, which stimulates the osmotic movement of water. It is also important to establish that the transport of Cl^- will lead to the depolarization of the apical membrane, which will inhibit Cl^- secretion. Simultaneously, intracellular Cl^- may also decrease, resulting in the activation of signaling pathways that regulate CFTR's activity and permeability (Saint-Criq & Gray, 2017). Thus, CFTR is a major regulator of ion and, indirectly, water transport, highly contributing to fluid homeodynamics.

2. CFTR regulation of aquaporin-mediated water transport

2.1 Aquaporins (AQPs)

Aquaporins (AQPs) are a family of transmembrane channel proteins mainly responsible for the transport of water, but also may transport small uncharged molecules such as glycerol (Borgnia & Agre, 2001), urea (Litman, Sogaard, & Zeuthen, 2009), ammonia (Saparov, Liu, Agre, & Pohl, 2007), hydrogen peroxide (Bienert et al., 2007), arsenite (Liu et al., 2002), and certain gases across biological membranes (Bernardino et al., 2016). Since the discover of AQPs in 1992, 13 distinct isoforms have been identified (AQP0–12) and classified into three main groups (orthodox, aquaglyceroporins, and superaquaporins) based on their homology and biophysical properties (Agre, 2004). Seven AQPs belong in the orthodox group: AQP0, AQP1, AQP2, AQP4, AQP5, AQP6, and AQP8. Orthodox AQPs are described as selective to water molecules, although there are few exceptions. For instance, AQP6 was reported to be involved in the transport of Cl^- at acidic conditions (Yasui, 2009) and AQP8 has a role in the transport of ammonia (Soria et al., 2010) and hydrogen peroxide, which justify the designation of ammoniaporin or peroxiporin (Chauvigne, Boj, Finn, & Cerda, 2015). The second group concerns aquaglyceroporins and includes four homologs: AQP3, AQP7, AQP9, and AQP10. Aquaglyceroporins have a bigger pore size (Sales, Lobo, Carvalho, Moura, & Rodrigues, 2013) and are permeable not only to water but also to small uncharged solutes, such as glycerol, urea, or hydrogen peroxide (Agre et al., 2002; Watanabe, Moniaga, Nielsen, & Hara-Chikuma, 2016). The last group is constituted by the unorthodox AQPs, AQP11 and AQP12, known as superaquaporins. Superaquaporins present low homology in comparison to orthodox AQPs and aquaglyceroporins (Gorelick, Praetorius, Tsunenari, Nielsen, & Agre, 2006). Moreover, superaquaporins are specifically localized in the membrane of intracellular organelles. Thus, it has been suggested that the role of AQP11 and AQP12 is linked with intracellular water and glycerol transport, regulating organelles volume and homeostasis (Badaut, Fukuda, Jullienne, & Petry, 2014; Madeira et al., 2014; Nozaki, Ishii, & Ishibashi, 2008).

AQPs are matter of intense investigation for the past couple of decades, which permitted to further elucidate ionic and fluid movement processes and identify new molecular markers for several diseases. However, the investigation concerning the regulation of AQPs just reached the surface, mostly due to its high complexity. For instance, although several studies addressed the interaction of AQPs with CFTR, the mechanism is largely unknown. On the following topics, up-to-date data concerning AQPs regulation by CFTR will be further discussed.

2.2 CFTR and AQPs as molecular partners

CFTR is known as one of the major regulators of fluids dynamics and ion homeostasis in several tissues. It is clear that CFTR modulates the function of several membrane transport proteins due to its permeation to Cl^- and HCO_3^-. As a result of its dysfunction, there is a disrupted ion transport in epithelial tissues and subsequent impaired water transport. Thus, cystic fibrosis is characterized by abnormal fluid dynamics. Although the ionic transport of CFTR is well defined, its participation in water transport still needs further investigation. The main hypothesis is that CFTR is able to mediate water transport indirectly, through promotion of ion exchanges (Schreiber, Greger, Nitschke, & Kunzelmann, 1997). As aforementioned, CFTR mediates the transport of Cl^- and HCO_3^-, which leads to the transport and increased concentration of Na^+ on the luminal surface. Na^+ accumulation generates and osmotic driving force for water movement, which is transported either by paracellular pathways or transcellularly through AQPs (Fig. 1).

On the other hand, compelling evidence suggests that CFTR directly regulates AQPs function. One of the earliest evidences toward the interaction of CFTR and AQPs was observed in human lungs, specifically in the airway epithelial cells. In this study, it was reported that CFTR promotes the transcellular water movement through AQP3 (Schreiber, Nitschke, Greger, & Kunzelmann, 1999). In fact, CFTR is reported to be a major regulator of the airway surface liquid composition (Knowles & Boucher, 2002; Tarran et al., 2002). In the absence or in case of a defective CFTR, the

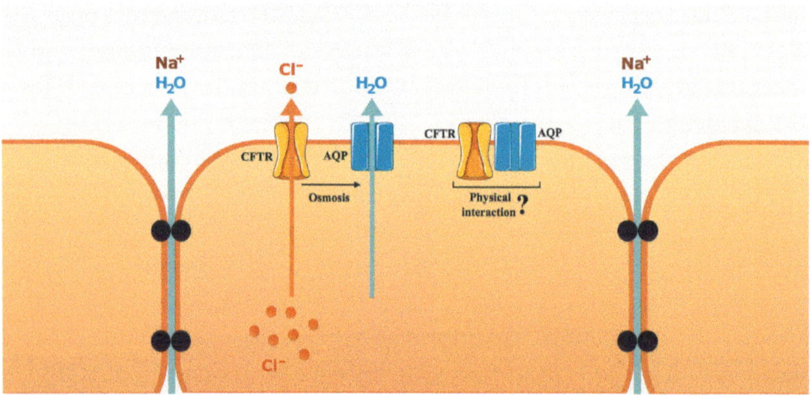

Fig. 1 Schematic illustration of the interaction between CFTR and AQPs. CFTR mediates the efflux of Cl^-, which stimulates the transport of Na^+ and increases its concentration on the luminal surface. Ultimately, Na^+ accumulation generates an osmotic driving force for water movement, usually either through the paracellular pathway or mediated by AQPs. However, it is also hypothesized that CFTR is able to regulate the AQP-mediated water transport directly, through a possible physical interaction between both channels. *Abbreviations*: *AQP*, aquaporin; *CFTR*, cystic fibrosis transmembrane conductance regulator.

airway surface liquid constitution and function are disrupted. Indeed, a malfunction of CFTR affects mucin secretion, leading to a severe mucus plugging within the glands. Due to the lack of Cl^- secretion and consequent lack of an osmotic gradient, fluid secretion is impaired (Inglis, Corboz, & Ballard, 1998). Besides, CFTR also has a role in the fluid regulation in the alveolar epithelium, where this channel is reported to have a role in fluid resorption. For instance, it has been reported that the inhibition of Cl^- channels hindered fluid clearance by 50% (Fang et al., 2002). Also in this study, it was shown that increasing intracellular cAMP levels resulted in higher fluid clearance, which were disrupted upon the action of glibenclamide, an CFTR inhibitor. In a posterior study, these results were confirmed in primary human alveolar type II cells. When these cells were treated with CFTR-Inh172, a specific CFTR inhibitor, an inhibition of 35% of cAMP-induced fluid absorption was observed (Fang et al., 2006). These findings are supported by another study in the human lung cell line NCI-H441, reporting that the inhibition of CFTR also reduced basal water resorption (Korbmacher et al., 2014). In sum, CFTR is a crucial player in fluid and ionic homeodynamics in the lungs and its action in fluid movement may be partially due to the expression of AQPs. However, data concerning the interaction between CFTR and AQPs in the lungs is still scarce and further studies are needed.

Conversely, compelling evidence from the male reproductive tract further supports the interaction between CFTR and AQPs. In these tissues, both CFTR and AQPs are responsible for a tightly regulated environment, which is essential for a proper spermatozoa production. On the following topic, the physical interaction between CFTR and AQPs in the male reproductive tract will be further discussed.

2.2.1 CFTR and AQPs as molecular partners: Evidence from the male reproductive tract

The main function of the male reproductive tract is to confer the reproductive potential to men and perpetuate the human species. For this purpose, the testis produces spermatozoa, the male gamete, during a complex and finely regulated process named spermatogenesis. However, spermatozoa maturation only occurs in the excurrent ducts and epididymis. During the whole process, the establishment of proper environment is crucial. The regulation of ion and water movement throughout the male reproductive tract is vital for the production of healthy spermatozoa and for its transport and maturation (Rato, Socorro, Cavaco, & Oliveira, 2010). In this sense, CFTR activity is crucial for the male reproductive tract. CFTR is highly expressed in the male reproductive tract, namely in the apical region of non-ciliated cells in the efferent ducts, and in the epididymal and *vas deferens* epithelium (Hihnala et al., 2006; Tizzano, Silver, Chitayat, Benichou, & Buchwald, 1994). On the other hand, CFTR presents a lower expression in the testis (Tizzano et al., 1994). Herein, CFTR was identified in Sertoli cells (Boockfor, Morris, DeSimone, Hunt, & Walsh, 1998; Jesus et al., 2014a, 2014b). Sertoli cells play an important role in the testis, being responsible for the homeodynamics of the seminiferous tubules, and physical and nutritional support of spermatogenesis (Oliveira, Sousa, Barros, Moura, & da Costa, 2009; Rato et al., 2010). However, there is evidence that CFTR mainly function in Sertoli cells as a transporter of HCO_3^-, highlighting its function in pH control (Xu et al., 2011). Regulation of Sertoli cells pH is essential on seminiferous tubular fluid homeostasis. Furthermore, CFTR is reported to be expressed in germ cells of various developmental stages, primary spermatocytes, round spermatids, and elongating spermatids (Gong et al., 2001; Teixeira et al., 2013; Trezise & Buchwald, 1991). As CFTR is widely expressed in the male reproductive tract, it is no surprise that male infertility is considered a hallmark in cystic fibrosis (Chen, Ruan, Xu, Chen, & Chan, 2012). Mutations in *CFTR* can lead to anatomical abnormalities of the reproductive tract, mainly a congenital bilateral absence of the *vas deferens* (de Souza, Faucz, Pereira-Ferrari, Sotomaior, & Raskin, 2018). Additionally, a defective CFTR leads to an abnormal production of germ cells, decreasing their quality and number (Boockfor et al., 1998).

Compelling evidence associates AQPs as a molecular partner for CFTR in the regulation of fluid homeostasis in male reproductive tract (Alves, Sá, Jesus, Sousa, & Oliveira, 2015). Similarly to CFTR, AQPs are widely expressed in the male reproductive tract (Yeste, Morato, Rodriguez-Gil, Bonet, & Prieto-Martinez, 2017). For instance, our group described the expression of CFTR, AQP4 and AQP9 in cultured rat primary Sertoli cells (Jesus et al., 2014a, 2014b). Moreover, in this study we investigated the physical interaction between CFTR and AQP4 and AQP9 in Sertoli cells. Taking advantage of co-immunoprecipitation techniques, we observed a molecular interaction between CFTR and both AQP4 and AQP9. Since AQP4 and AQP9 are reported as crucial for water and ion homeostasis in several tissues (Pastor-Soler et al., 2010; Yeung, Callies, Rojek, Nielsen, & Cooper, 2009; Yeung, Callies, Tuttelmann, Kliesch, & Cooper, 2010), it is also expectable that both these AQPs regulate water homeostasis in Sertoli cells and consequently inside seminiferous tubules. Taken together, these results support the hypothesis that CFTR is, at least in Sertoli cells, a regulator of AQPs activity in water homeodynamics. Additionally, there is further evidence that CFTR interacts with AQP9. Both CFTR and AQP9 are co-localized in the luminal membrane of the principal cells in the epididymis of rat and humans (Cheung, Leung, Leung, & Wong, 2003; Ruz, Andonian, & Hermo, 2004). In fact, *in vivo* and *in vitro* inhibition of both AQP9 and CFTR reduced water permeability (Cheung et al., 2003). Thus, it has been suggested that CFTR may potentiate water permeability of AQP9 in the epididymal epithelium, which is essential for proper sperm maturation and movement. On the other hand, the distinct expression pattern of CFTR found in germ cells during spermatid differentiation into spermatozoa points out to a possible role of CFTR in spermiogenesis. Spermiogenesis is the final process in spermatogenesis, where round spermatids change their spherical morphology into an elongated and, ultimately, into a flagellar shape. Along this process, spermatids suffer a markedly reduction of volume. Thus, it has been suggested that CFTR may be essential for volume reduction and water efflux during spermiogenesis (Gong et al., 2001). At the same time, CFTR is reported as essential for human sperm capacitation and the impairment of CFTR expression in spermatozoa is correlated with a reduction of sperm quality (Li et al., 2010). In addition, AQPs are expressed in human spermatozoa and its absence is correlated with infertility (Yeste et al., 2017). Hence, it is possible for CFTR to interact with AQPs in spermatozoa, although the role of both channels in these cells needs further investigation. On a side note, it was also described that AQP7 and AQP8 are expressed with in a highly similar pattern with CFTR in the testis, highlighting a possible interaction (Huang et al., 2006; Suzuki-Toyota, Ishibashi, & Yuasa, 1999). In support of this hypothesis, CFTR was co-localized with several other AQPs in different tissues, such as AQP1 and AQP5 in the pancreas (Burghardt et al., 2003). Notwithstanding, it is still unclear how CFTR physically regulates AQPs function.

3. Conclusions and future perspectives

For the past couple of decades, both CFTR and AQPs are under intense debate. CFTR participates mainly in the regulation of ion transport, due to its high permeability to Cl^- and HCO_3^-. However, compelling evidence associates CFTR with AQPs, highlighting a possible mechanism of aquaporin-mediated water transport regulated by CFTR. Recent studies, namely in the male reproductive tract, have shown that CFTR and AQPs physically interact, although the exact molecular mechanism remains to be elucidated. Even more recently, our group identified for the first time the expression of AQP3 and AQP9 in mouse Sertoli cells (Bernardino et al., 2018). In this study, we have shown that these aquaglyceroporins are the main responsible for glycerol transport in Sertoli cells. Since CFTR regulates AQP-mediated water transport, it is possible that CFTR would also regulate the transport of other AQP-mediated solutes, although further evidence is necessary to supports this hypothesis.

In sum, compelling evidence shows that CFTR mediates water transport through AQPs, either indirectly through Cl^- and HCO_3^- transport or, possibly, directly through physical interaction with CFTR. In this sense, functional studies are needed to further unveil the regulation of aquaporin-mediated water transport regulated by CFTR.

Acknowledgments

This work was supported by "Fundação para a Ciência e a Tecnologia"—FCT to David F. Carrageta (SFRH/BD/136779/2018). The work was co-funded by FEDER through the COMPETE/QREN, FSE/POPH to Marco G. Alves (IFCT2015, PTDC/BIM-MET/4712/2014, and PTDC/MEC-AND/28691/2017); Pedro F. Oliveira (IFCT2015 and PTDC/BBB-BQB/1368/2014); UMIB (PEst-OE/SAU/UI0215/2019); co-funded by the EU Framework Programme for Research and Innovation H2020 (POCI/COMPETE2020).

Conflict of interest

None to declare.

References

Agre, P. (2004). Aquaporin water channels (Nobel Lecture). *Angewandte Chemie International Edition in English*, *43*(33), 4278–4290. https://doi.org/10.1002/anie.200460804.

Agre, P., King, L. S., Yasui, M., Guggino, W. B., Ottersen, O. P., Fujiyoshi, Y., et al. (2002). Aquaporin water channels—From atomic structure to clinical medicine. *Journal of Physiology*, *542*(Pt. 1), 3–16. https://doi.org/10.1113/jphysiol.2002.020818.

Alves, M. G., Sá, R., Jesus, T. T., Sousa, M., & Oliveira, P. F. (2015). CFTR regulation of aquaporin-mediated water transport: A target in male fertility. *Current Drug Targets*, *16*(9), 993–1006. https://doi.org/10.2174/1573399811666150615144108.

Anderson, M. P., Berger, H. A., Rich, D. P., Gregory, R. J., Smith, A. E., & Welsh, M. J. (1991). Nucleoside triphosphates are required to open the CFTR chloride channel. *Cell*, *67*(4), 775–784. https://doi.org/10.1016/0092-8674(91)90072-7.

Anderson, M. P., Gregory, R. J., Thompson, S., Souza, D. W., Paul, S., Mulligan, R. C., et al. (1991). Demonstration that CFTR is a chloride channel by alteration of its anion selectivity. *Science*, *253*(5016), 202–205. https://doi.org/10.1126/science.1712984.

Badaut, J., Fukuda, A. M., Jullienne, A., & Petry, K. G. (2014). Aquaporin and brain diseases. *Biochimica et Biophysica Acta, 1840*(5), 1554–1565. https://doi.org/10.1016/j.bbagen.2013.10.032.

Bear, C. E., Li, C. H., Kartner, N., Bridges, R. J., Jensen, T. J., Ramjeesingh, M., et al. (1992). Purification and functional reconstitution of the cystic fibrosis transmembrane conductance regulator (CFTR). *Cell, 68*(4), 809–818. https://doi.org/10.1016/0092-8674(92)90155-6.

Berger, H. A., Anderson, M. P., Gregory, R. J., Thompson, S., Howard, P. W., Maurer, R. A., et al. (1991). Identification and regulation of the cystic fibrosis transmembrane conductance regulator-generated chloride channel. *Journal of Clinical Investigation, 88*(4), 1422–1431. https://doi.org/10.1172/JCI115450.

Berger, A. L., Ikuma, M., & Welsh, M. J. (2005). Normal gating of CFTR requires ATP binding to both nucleotide-binding domains and hydrolysis at the second nucleotide-binding domain. *Proceedings of the National Academy of Sciences of the United States of America, 102*(2), 455–460. https://doi.org/10.1073/pnas.0408575102.

Bernardino, R. L., Carrageta, D. E., Silva, A. M., Calamita, G., Alves, M. G., Soveral, G., et al. (2018). Estrogen modulates glycerol permeability in sertoli cells through down-regulation of aquaporin-9. *Cell,* 7(10), 153. https://doi.org/10.3390/cells7100153.

Bernardino, R. L., Marinelli, R. A., Maggio, A., Gena, P., Cataldo, I., Alves, M. G., et al. (2016). Hepatocyte and sertoli cell aquaporins, recent advances and research trends. *International Journal of Molecular Sciences, 17*(7), 1096. https://doi.org/10.3390/ijms17071096.

Bienert, G. P., Moller, A. L., Kristiansen, K. A., Schulz, A., Moller, I. M., Schjoerring, J. K., et al. (2007). Specific aquaporins facilitate the diffusion of hydrogen peroxide across membranes. *Journal of Biological Chemistry, 282*(2), 1183–1192. https://doi.org/10.1074/jbc.M603761200.

Boockfor, F. R., Morris, R. A., DeSimone, D. C., Hunt, D. M., & Walsh, K. B. (1998). Sertoli cell expression of the cystic fibrosis transmembrane conductance regulator. *American Journal of Physiology, 274*(4 Pt. 1), C922–C930.

Borgnia, M. J., & Agre, P. (2001). Reconstitution and functional comparison of purified GlpF and AqpZ, the glycerol and water channels from *Escherichia coli. Proceedings of the National Academy of Sciences of the United States of America, 98*(5), 2888–2893. https://doi.org/10.1073/pnas.051628098.

Burghardt, B., Elkaer, M. L., Kwon, T. H., Racz, G. Z., Varga, G., Steward, M. C., et al. (2003). Distribution of aquaporin water channels AQP1 and AQP5 in the ductal system of the human pancreas. *Gut, 52*(7), 1008–1016. https://doi.org/10.1136/gut.52.7.1008.

Callebaut, I., Chong, P. A., & Forman-Kay, J. D. (2018). CFTR structure. *Journal of Cystic Fibrosis, 17*(2S), S5–S8. https://doi.org/10.1016/j.jcf.2017.08.008.

Chauvigne, F., Boj, M., Finn, R. N., & Cerda, J. (2015). Mitochondrial aquaporin-8-mediated hydrogen peroxide transport is essential for teleost spermatozoon motility. *Scientific Reports, 5*, 7789. https://doi.org/10.1038/srep07789.

Chen, H., Ruan, Y. C., Xu, W. M., Chen, J., & Chan, H. C. (2012). Regulation of male fertility by CFTR and implications in male infertility. *Human Reproduction Update, 18*(6), 703–713. https://doi.org/10.1093/humupd/dms027.

Cheung, K. H., Leung, C. T., Leung, G. P., & Wong, P. Y. (2003). Synergistic effects of cystic fibrosis transmembrane conductance regulator and aquaporin-9 in the rat epididymis. *Biology of Reproduction, 68*(5), 1505–1510. https://doi.org/10.1095/biolreprod.102.010017.

Csanady, L., Vergani, P., & Gadsby, D. C. (2019). Structure, gating, and regulation of the CFTR anion channel. *Physiological Reviews, 99*(1), 707–738. https://doi.org/10.1152/physrev.00007.2018.

de Souza, D. A. S., Faucz, F. R., Pereira-Ferrari, L., Sotomaior, V. S., & Raskin, S. (2018). Congenital bilateral absence of the vas deferens as an atypical form of cystic fibrosis: Reproductive implications and genetic counseling. *Andrology, 6*(1), 127–135. https://doi.org/10.1111/andr.12450.

Fang, X., Fukuda, N., Barbry, P., Sartori, C., Verkman, A. S., & Matthay, M. A. (2002). Novel role for CFTR in fluid absorption from the distal airspaces of the lung. *Journal of General Physiology*, *119*(2), 199–207. https://doi.org/10.1085/jgp.119.2.199.

Fang, X., Song, Y., Hirsch, J., Galietta, L. J., Pedemonte, N., Zemans, R. L., et al. (2006). Contribution of CFTR to apical-basolateral fluid transport in cultured human alveolar epithelial type II cells. *American Journal of Physiology—Lung Cellular and Molecular Physiology*, *290*(2), L242–L249. https://doi.org/10.1152/ajplung.00178.2005.

Fischbarg, J. (2010). Fluid transport across leaky epithelia: Central role of the tight junction and supporting role of aquaporins. *Physiological Reviews*, *90*(4), 1271–1290. https://doi.org/10.1152/physrev.00025.2009.

Gong, X. D., Li, J. C., Cheung, K. H., Leung, G. P., Chew, S. B., & Wong, P. Y. (2001). Expression of the cystic fibrosis transmembrane conductance regulator in rat spermatids: Implication for the site of action of antispermatogenic agents. *Molecular Human Reproduction*, 7(8), 705–713.

Gorelick, D. A., Praetorius, J., Tsunenari, T., Nielsen, S., & Agre, P. (2006). Aquaporin-11: A channel protein lacking apparent transport function expressed in brain. *BMC Biochemistry*, 7, 14. https://doi.org/10.1186/1471-2091-7-14.

Hihnala, S., Kujala, M., Toppari, J., Kere, J., Holmberg, C., & Hoglund, P. (2006). Expression of SLC26A3, CFTR and NHE3 in the human male reproductive tract: Role in male subfertility caused by congenital chloride diarrhoea. *Molecular Human Reproduction*, *12*(2), 107–111. https://doi.org/10.1093/molehr/gal009.

Huang, H. F., He, R. H., Sun, C. C., Zhang, Y., Meng, Q. X., & Ma, Y. Y. (2006). Function of aquaporins in female and male reproductive systems. *Human Reproduction Update*, *12*(6), 785–795. https://doi.org/10.1093/humupd/dml035.

Inglis, S. K., Corboz, M. R., & Ballard, S. T. (1998). Effect of anion secretion inhibitors on mucin content of airway submucosal gland ducts. *American Journal of Physiology*, *274* (5 Pt. 1), L762–L766.

Jesus, T. T., Bernardino, R. L., Martins, A. D., Sa, R., Sousa, M., Alves, M. G., et al. (2014a). Aquaporin-4 as a molecular partner of cystic fibrosis transmembrane conductance regulator in rat sertoli cells. *Biochemical and Biophysical Research Communications*, *446*(4), 1017–1021. https://doi.org/10.1016/j.bbrc.2014.03.046.

Jesus, T. T., Bernardino, R. L., Martins, A. D., Sa, R., Sousa, M., Alves, M. G., et al. (2014b). Aquaporin-9 is expressed in rat sertoli cells and interacts with the cystic fibrosis transmembrane conductance regulator. *IUBMB Life*, *66*(9), 639–644. https://doi.org/10.1002/iub.1312.

Knowles, M. R., & Boucher, R. C. (2002). Mucus clearance as a primary innate defense mechanism for mammalian airways. *Journal of Clinical Investigation*, *109*(5), 571–577. https://doi.org/10.1172/Jci200215217.

Korbmacher, J. P., Michel, C., Neubauer, D., Thompson, K., Mizaikoff, B., Frick, M., et al. (2014). Amiloride-sensitive fluid resorption in NCI-H441 lung epithelia depends on an apical Cl(-) conductance. *Physiological Reports*, *2*(1), e00201. https://doi.org/10.1002/phy2.201.

Li, C. Y., Jiang, L. Y., Chen, W. Y., Li, K., Sheng, H. Q., Ni, Y., et al. (2010). CFTR is essential for sperm fertilizing capacity and is correlated with sperm quality in humans. *Human Reproduction*, *25*(2), 317–327. https://doi.org/10.1093/humrep/dep406.

Linsdell, P. (2001). Relationship between anion binding and anion permeability revealed by mutagenesis within the cystic fibrosis transmembrane conductance regulator chloride channel pore. *Journal of Physiology (London)*, *531*(1), 51–66. https://doi.org/10.1111/j.1469-7793.2001.0051j.x.

Litman, T., Sogaard, R., & Zeuthen, T. (2009). Ammonia and urea permeability of mammalian aquaporins. *Handbook of Experimental Pharmacology*, *190*, 327–358. https://doi.org/10.1007/978-3-540-79885-9_17.

Liu, Z., Shen, J., Carbrey, J. M., Mukhopadhyay, R., Agre, P., & Rosen, B. P. (2002). Arsenite transport by mammalian aquaglyceroporins AQP7 and AQP9. *Proceedings of the National Academy of Sciences of the United States of America*, *99*(9), 6053–6058. https://doi.org/10.1073/pnas.092131899.

Liu, F., Zhang, Z., Csanady, L., Gadsby, D. C., & Chen, J. (2017). Molecular structure of the human CFTR ion channel. *Cell*, *169*(1), 85–95.e88. https://doi.org/10.1016/j.cell.2017.02.024.

Madeira, A., Fernandez-Veledo, S., Camps, M., Zorzano, A., Moura, T. F., Ceperuelo-Mallafre, V., et al. (2014). Human aquaporin-11 is a water and glycerol channel and localizes in the vicinity of lipid droplets in human adipocytes. *Obesity (Silver Spring)*, *22*(9), 2010–2017. https://doi.org/10.1002/oby.20792.

Nozaki, K., Ishii, D., & Ishibashi, K. (2008). Intracellular aquaporins: Clues for intracellular water transport? *Pflügers Archiv—European Journal of Physiology*, *456*(4), 701–707. https://doi.org/10.1007/s00424-007-0373-5.

Oliveira, P. F., Sousa, M., Barros, A., Moura, T., & da Costa, A. R. (2009). Intracellular pH regulation in human sertoli cells: Role of membrane transporters. *Reproduction*, *137*(2), 353–359.

Park, H. W., & Lee, M. G. (2012). Transepithelial bicarbonate secretion: Lessons from the pancreas. *Cold Spring Harbor Perspectives in Medicine*, *2*(10), a009571. https://doi.org/10.1101/cshperspect.a009571.

Pastor-Soler, N. M., Fisher, J. S., Sharpe, R., Hill, E., Van Hoek, A., Brown, D., et al. (2010). Aquaporin 9 expression in the developing rat epididymis is modulated by steroid hormones. *Reproduction*, *139*(3), 613–621. https://doi.org/10.1530/Rep-09-0284.

Poulsen, J. H., Fischer, H., Illek, B., & Machen, T. E. (1994). Bicarbonate conductance and pH regulatory capability of cystic fibrosis transmembrane conductance regulator. *Proceedings of the National Academy of Sciences of the United States of America*, *91*(12), 5340–5344. https://doi.org/10.1073/pnas.91.12.5340.

Rato, L., Socorro, S., Cavaco, J. E., & Oliveira, P. F. (2010). Tubular fluid secretion in the seminiferous epithelium: Ion transporters and aquaporins in sertoli cells. *Journal of Membrane Biology*, *236*(2), 215–224. https://doi.org/10.1007/s00232-010-9294-x.

Riordan, J. R., Rommens, J. M., Kerem, B., Alon, N., Rozmahel, R., Grzelczak, Z., et al. (1989). Identification of the cystic fibrosis gene: Cloning and characterization of complementary DNA. *Science*, *245*(4922), 1066–1073. https://doi.org/10.1126/science.2475911.

Ruz, R., Andonian, S., & Hermo, L. (2004). Immunolocalization and regulation of cystic fibrosis transmembrane conductance regulator in the adult rat epididymis. *Journal of Andrology*, *25*(2), 265–273. https://doi.org/10.1002/j.1939-4640.2004.tb02787.x.

Saint-Criq, V., & Gray, M. A. (2017). Role of CFTR in epithelial physiology. *Cellular and Molecular Life Sciences*, *74*(1), 93–115. https://doi.org/10.1007/s00018-016-2391-y.

Sales, A. D., Lobo, C. H., Carvalho, A. A., Moura, A. A., & Rodrigues, A. P. (2013). Structure, function, and localization of aquaporins: Their possible implications on gamete cryopreservation. *Genetics and Molecular Research*, *12*(4), 6718–6732. https://doi.org/10.4238/2013.December.13.5.

Saparov, S. M., Liu, K., Agre, P., & Pohl, P. (2007). Fast and selective ammonia transport by aquaporin-8. *Journal of Biological Chemistry*, *282*(8), 5296–5301. https://doi.org/10.1074/jbc.M609343200.

Schreiber, R., Greger, R., Nitschke, R., & Kunzelmann, K. (1997). Cystic fibrosis transmembrane conductance regulator activates water conductance in *Xenopus* oocytes. *Pflügers Archiv—European Journal of Physiology*, *434*(6), 841–847. https://doi.org/10.1007/s004240050473.

Schreiber, R., Nitschke, R., Greger, R., & Kunzelmann, K. (1999). The cystic fibrosis transmembrane conductance regulator activates aquaporin 3 in airway epithelial cells. *Journal of Biological Chemistry*, *274*(17), 11811–11816. https://doi.org/10.1074/jbc.274.17.11811.

Shachar-Hill, B., & Hill, A. E. (2002). Paracellular fluid transport by epithelia. In: *International review of cytology: Vol. 215* (pp. 319–350). Elsevier.

Shan, J., Liao, J., Huang, J., Robert, R., Palmer, M. L., Fahrenkrug, S. C., et al. (2012). Bicarbonate-dependent chloride transport drives fluid secretion by the human airway epithelial cell line Calu-3. *Journal of Physiology, 590*(21), 5273–5297. https://doi.org/10.1113/jphysiol.2012.236893.

Shcheynikov, N., Yang, D., Wang, Y., Zeng, W., Karniski, L. P., So, I., et al. (2008). The Slc26a4 transporter functions as an electroneutral Cl-/I-/HCO3- exchanger: Role of Slc26a4 and Slc26a6 in I- and HCO3- secretion and in regulation of CFTR in the parotid duct. *Journal of Physiology, 586*(16), 3813–3824. https://doi.org/10.1113/jphysiol.2008.154468.

Soria, L. R., Fanelli, E., Altamura, N., Svelto, M., Marinelli, R. A., & Calamita, G. (2010). Aquaporin-8-facilitated mitochondrial ammonia transport. *Biochemical and Biophysical Research Communications, 393*(2), 217–221. https://doi.org/10.1016/j.bbrc.2010.01.104.

Suzuki-Toyota, F., Ishibashi, K., & Yuasa, S. (1999). Immunohistochemical localization of a water channel, aquaporin 7 (AQP7), in the rat testis. *Cell and Tissue Research, 295*(2), 279–285.

Tabcharani, J. A., Linsdell, P., & Hanrahan, J. W. (1997). Halide permeation in wild-type and mutant cystic fibrosis transmembrane conductance regulator chloride channels. *Journal of General Physiology, 110*(4), 341–354.

Tarran, R., Loewen, M. E., Paradiso, A. M., Olsen, J. C., Gray, M. A., Argent, B. E., et al. (2002). Regulation of murine airway surface liquid volume by CFTR and Ca2+-activated Cl- conductances. *Journal of General Physiology, 120*(3), 407–418.

Teixeira, S., Sa, R., Grangeia, A., Silva, J., Oliveira, C., Ferraz, L., et al. (2013). Immunohystochemical analysis of CFTR in normal and disrupted spermatogenesis. *Systems Biology in Reproductive Medicine, 59*(1), 53–59. https://doi.org/10.3109/19396368.2012.718851.

Tizzano, E. F., Silver, M. M., Chitayat, D., Benichou, J. C., & Buchwald, M. (1994). Differential cellular expression of cystic fibrosis transmembrane regulator in human reproductive tissues. Clues for the infertility in patients with cystic fibrosis. *American Journal of Pathology, 144*(5), 906–914.

Trezise, A. E., & Buchwald, M. (1991). In vivo cell-specific expression of the cystic fibrosis transmembrane conductance regulator. *Nature, 353*(6343), 434–437. https://doi.org/10.1038/353434a0.

Verkman, A. S., Matthay, M. A., & Song, Y. (2000). Aquaporin water channels and lung physiology. *American Journal of Physiology—Lung Cellular and Molecular Physiology, 278*(5), L867–L879. https://doi.org/10.1152/ajplung.2000.278.5.L867.

Watanabe, S., Moniaga, C. S., Nielsen, S., & Hara-Chikuma, M. (2016). Aquaporin-9 facilitates membrane transport of hydrogen peroxide in mammalian cells. *Biochemical and Biophysical Research Communications, 471*(1), 191–197. https://doi.org/10.1016/j.bbrc.2016.01.153.

Xu, W. M., Chen, J., Chen, H., Diao, R. Y., Fok, K. L., Dong, J. D., et al. (2011). Defective CFTR-dependent CREB activation results in impaired spermatogenesis and azoospermia. *PLoS One, 6*(5), e19120. https://doi.org/10.1371/journal.pone.0019120.

Yasui, M. (2009). pH regulated anion permeability of aquaporin-6. *Handbook of Experimental Pharmacology, 190*, 299–308. https://doi.org/10.1007/978-3-540-79885-9_15.

Yeste, M., Morato, R., Rodriguez-Gil, J. E., Bonet, S., & Prieto-Martinez, N. (2017). Aquaporins in the male reproductive tract and sperm: Functional implications and cryobiology. *Reproduction in Domestic Animals, 52*(Suppl. 4), 12–27. https://doi.org/10.1111/rda.13082.

Yeung, C. H., Callies, C., Rojek, A., Nielsen, S., & Cooper, T. G. (2009). Aquaporin isoforms involved in physiological volume regulation of murine spermatozoa. *Biology of Reproduction*, *80*(2), 350–357. https://doi.org/10.1095/biolreprod.108.071928.

Yeung, C. H., Callies, C., Tuttelmann, F., Kliesch, S., & Cooper, T. G. (2010). Aquaporins in the human testis and spermatozoa—Identification, involvement in sperm volume regulation and clinical relevance. *International Journal of Andrology*, *33*(4), 629–641. https://doi.org/10.1111/j.1365-2605.2009.00998.x.

Zhang, J. Y., Yu, Y. C., Yeh, J. T., & Hwang, T. C. (2018). Functional characterization reveals that zebrafish CFTR prefers to occupy closed channel conformations. *PLoS One*, *13*(12), e0209862. https://doi.org/10.1371/journal.pone.0209862.

CHAPTER NINE

Glucocorticoid gene regulation of aquaporin-7

Laura González-Dávalos[a], Mariana Álvarez-Pérez[a], Tania Quesada-López[b,c], Rubén Cereijo[b,c], Laura Campderrós[b,c], Enrique Piña[d], Armando Shimada[a], Francesc Villarroya[b,c], Alfredo Varela-Echavarria[e], Ofelia Mora[a,*]

[a]Laboratorio de Rumiología y Metabolismo Nutricional (RuMeN), Facultad de Estudios Superiores Cuautitlán, UNAM, Cuautitlán, Mexico
[b]Department of Biochemistry and Molecular Biomedicine, Institute of Biomedicine of the University of Barcelona, Barcelona, Spain
[c]CIBER Fisiopatología de la Obesidad y Nutrición, Barcelona, Spain
[d]Departamento de Bioquímica, Facultad de Medicina, UNAM, Ciudad de México, Mexico
[e]Laboratorio de Diferenciación Neural y Axogénesis, Instituto de Neurobiología, UNAM, Querétaro, Mexico
*Corresponding author: e-mail address: ofemora66@unam.mx

Contents

Abstract

AQP7 is the primary glycerol transporter in white (WAT) and brown (BAT) adipose tissues. There are immediate and quantitatively important actions of cortisone over the expression of *AQP7* in murine and human adipocytes. Short-term response (minutes) of cortisone treatment result in an mRNA overexpression in white and brown differentiated adipocytes (between 1.5 and 6 folds). Conversely, long-term response

Vitamins and Hormones, Volume 112
ISSN 0083-6729
https://doi.org/10.1016/bs.vh.2019.08.005

(hours or days) result in decreased mRNA expression. The effects observed on *AQP7* mRNA expression upon cortisone treatment in brown and white differentiated adipocytes are concordant with those observed for *GK* and *HSD1B11*.

1. Aquaporin family

Aquaporins (AQPs) are a class of membrane water channels whose primary function is to facilitate the passive transport of water across the plasma membrane of the cell in response to osmotic gradients that are created by the active transport of solutes (Verkman, Anderson, & Papadopoulos, 2014). They transport water and small molecules, such as urea and glycerol, for the maintenance of fluid homeostasis in various environments (Ishibashi, Morishita, & Tanaka, 2017). Moreover, some AQPs transport nutrients, metabolic precursors, waste products, toxins and gases. AQPs are ubiquitous in almost all living systems including virus, but excluding some prokaryotes and protozoa (Gazzarrini et al., 2006; Ishibashi et al., 2017).

1.1 Classification

Since the discovery of mammalian AQPs in 1992, much effort has been made to understand their structure and function (Pérez Di Giorgio et al., 2014). The idea of AQPs as a protein family was formally introduced with a phylogenetic analysis of putative proteins that can transport water and solutes in bacteria and eukaryotes, hypothetically evolutionally related to previously characterized AQPs (Zardoya & Villalba, 2001). Bioinformatic tools have allowed to identify putative functional domains which helped to almost automatically identify new potential AQPs. This superfamily of integral membrane channel proteins was called major intrinsic protein (MIP). The AQP family could be constituted by non-homologous proteins with similar structure and motifs due to evolutionary convergence (Pérez Di Giorgio et al., 2014).

However, structural analysis of several AQPs has established that these protein channels share common structural features. The functional aquaporin unit is a homotetramer, which comprises six α-helix transmembrane domains with two conserved asparagine–proline–alanine (NPA) motifs embedding into the plasma membrane, a signature sequence of water channels, five loops (A–E) and intracellular N- and C-termini. Human AQPs have been reported to share approximately 30–50% homology in their amino acid sequences. And interestingly, conformational changes of AQP protein allow the traffic through the plasma membrane of other molecules (i.e., urea, glycerol, H_2O_2, NH_3, CO_2) (Li & Wang, 2017).

Based on their structure and functional similarities, AQPs are initially subdivided into two subfamilies, classical AQPs (water-selective) and aquaglyceroporins (glycerol channel) aquaporins. However, this viewpoint was challenged by recent evidence revealing that both subfamilies overlap functionally, for example, some classical AQPs transport water and other small solutes, e.g., glycerol. In addition, a new group of AQPs discovered recently showed that their structure is highly deviated from the previous AQPs especially around the AQP NPA box (Ishibashi, Tanaka, & Morishita, 2014; Li & Wang, 2017). This subfamily was later named super-aquaporin (also called unorthodox aquaporin) as it has very low homology with the previous two subfamilies, the three families of human aquaporins are shown in Fig. 1 (Li & Wang, 2017).

Aquaglyceroporins, which form a subset of the 13 mammalian AQPs, also facilitate the passive transport of glycerol and possibly other small solutes such as urea and carbon dioxide. While water-selective AQPs are involved in many biological functions, including transepithelial fluid transport, cell migration, brain edema and neuroexcitation, aquaglyceroporins are involved in cell proliferation, adipocyte metabolism and epidermal water retention (Verkman et al., 2014).

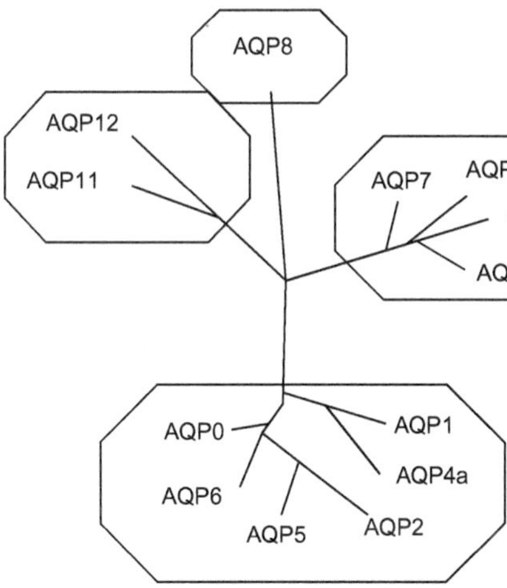

Fig. 1 The phylogenetic tree of human AQPs. *Modified by Li, C., & Wang, W. (2017). Molecular biology of aquaporins. In: Yang, B. (Eds.), Aquaporins. Advances in experimental medicine and biology. Springer, Dordrecht, 1–34. 10.1007/978-94-024-1057-0_1.*

There is also evidence (some of which is conflicting) suggesting that, in addition to the transport of water and glycerol, aquaglyceroporins could potentially transport gases such as carbon dioxide, ammonia and nitric oxide; as well as larger polar solutes such as sugars, hydrogen peroxide and even some ions. However, it is not known whether the transport of molecules other than water and glycerol is biologically important (Verkman et al., 2014).

1.2 Tissue expression and its relationship with glands secretion

AQPs are expressed in a wide range of tissues (Day et al., 2014), often spatially located within a certain region of the cell. Their expression is particularly important in cell types that are involved in fluid transport, such as epithelial cells; however in some other organs that do not have an obvious role in fluid transport, such as adipose tissue, their expression has proved to have a critical role in homeostatic maintenance (Verkman et al., 2014). This enables them to play a central role in the flow of water through those tissues, which typically triggers cell volume regulation mechanisms. While transport through all AQPs utilizes a common passive mechanism; their regulation and cellular distribution vary significantly between systems, specifically in constituent tissues and cells (Day et al., 2014).

Regarding its relationship with glands, AQPs are expressed in most exocrine and endocrine secretory glands. By participating to transcellular water flow, AQPs are involved in the mechanisms leading to exocrine glandular fluid secretion. A first step involves the creation of a transepithelial osmotic gradient, formed by the accumulation of ions, mostly NaCl. A second step involves the movement of water from epithelial cells to the gland lumen though AQPs. Water flows from secretory epithelial cells according to a transepithelial osmotic gradient. In addition, AQPs could somehow participate in endocrine secretion, however more studies will needed to clarify the role played by them in glandular secretory processes (Delporte, 2017).

2. Aquaporin-7

AQP7 is the primary glycerol transporter in white (WAT) and brown (BAT) adipose tissues. This aquaglyceroporin resides predominantly in the cytoplasm and is abundantly expressed in the plasma membrane of adipocytes. The differentiation of 3T3-L1 rodent cell line to adipocytes, has proven that *AQP7* mRNA expression was increased in parallel with differentiation, and that this induction was correlated with glycerol release from these cells suggesting that plasma membrane glycerol permeability may mediate the accumulation of fat in adipocytes (Iena & Lebeck, 2018). Also, fasting conditions caused an increase *AQP7* mRNA expression in rodent's adipocytes (Day et al., 2014). Regarding the other members of the subfamily, *AQP3* and *AQP9* are also expressed in human adipose tissue. *A*QP3 is predominantly localized in the cytoplasm and *AQP9* constitutively expressed in the plasma membrane. Interestingly, a positive correlation between adipose transcript levels of AQPs 3, 7 and 9 and body weight (BMI) has been suggested (Day et al., 2014; Rodríguez et al., 2011).

Regarding *AQP7* gene expression and metabolic diseases, it has been shown to be downregulated in the WAT of obese human subjects compared with normal controls, but unchanged in type 2 diabetes (T2DM) (Day et al., 2014). Rodríguez et al. (2011) showed that obese patients with T2DM showed higher expression of *AQP7* in visceral adipose tissue and lower expression in subcutaneous adipose tissue. Conversely, hepatic *AQP9* showed a decreased expression. However, the link between human AQPs and obesity is contradictory; further information can be consulted in a review by Maeda (2012). On the other hand, *AQP10* is predominantly expressed in the intestine, it has recently been proposed as an alternative pathway for glycerol efflux in human adipocytes (Day et al., 2014).

The aquaglyceroporins *AQP7* (expressed primarily in adipocytes) and *AQP9* (expressed in hepatocytes) are involved in adipose metabolism. *AQP7*-null mice have a progressive age-related increase in adipose mass and adipocyte hypertrophy, with the accumulation of glycerol and triglycerides in adipocytes. *AQP7* deficiency leads to reduced permeability of the plasma membrane to glycerol, which causes cellular accumulation of glycerol and triglyceride as well as upregulation of glycerol kinase expression. In humans, *AQP7* expression in adipocytes is downregulated in obesity (Rodríguez et al., 2011). These results suggest that increasing adipocyte *AQP7* expression and/or function may be a novel approach for treating obesity; however, there is a lack of critical proof-of-concept data, such as the effects of *AQP7* overexpression (induced by gene transfer) on adipocyte hypertrophy in obese mice. There is evidence that AQP9 is involved in the hepatic uptake of glycerol and perhaps urea, and that AQP7 and AQP9 act as metabolic regulators in diabetes and obesity (Verkman et al., 2014).

3. Hormone regulation of aquaporin-7 and obesity

Long-term regulation by AQPs is modulated (increased/decreased) by systemic hormones (vasopressin, insulin, ANP, angiotensin II), local molecules (purines, prostaglandins, bradykinin, dopamine), and other common microenvironment signals (pH, divalent cation concentrations and osmolality) (Boone, Kortenoeven, Robben, Tamma, & Deen, 2011; Choi, Jung, & Kwon, 2015; Gunnarson et al., 2005; Lanaspa et al., 2010; Li & Wang, 2017; Olesen, Rutzler, Moeller, Praetorius, & Fenton, 2011; Tamma, Carmosino, Svelto, & Valenti, 2005; Zhang et al., 2015). These regulations of AQPs are often associated with certain physiological or pathophysiological conditions.

Adipose tissue AQPs are selectively regulated by hormones implicated in the pathogenesis of insulin resistance and dyslipidemia, as shown in different reports (da Silva & Soveral, 2017; Miranda et al., 2010; Rodríguez et al., 2011). Recent studies evidenced that AQPs are key players in adipose tissue biology and involved in obesity onset. Among the various mammalian aquaglyceroporins, AQP7 is the most representative glycerol channel and was the first to be detected in human and mice adipose tissue and adipocytes. AQP7 was localized in the vascular tissue surrounding fat rather that in adipose tissue membranes, a double localization in adipocytes and endothelial cells was also described and *AQP7* expression in both human subcutaneous and visceral adipose tissue was reported (da Silva & Soveral, 2017; Miranda et al., 2010; Rodríguez et al., 2011; Rojek, Praetorius, Frokiaer, Nielsen, & Fenton, 2008).

However, a correlation of adipose *AQP7* expression with glycerol metabolism and related metabolic complications was not so obvious in human studies (Ceperuelo-Mallafre et al., 2007). Albeit a link between adipocyte *AQP7* expression and insulin resistance was found by a genome-wide study, where *AQP7* decreased gene expression was linked to T2DM and metabolic syndrome as well as with obesity but only for the female. Gender differences in the role of *AQP7* in adipose tissue metabolism were supported by higher fasting circulating levels of glycerol in women than in men, probably due to higher percentage of subcutaneous fat in females, higher lipolytic rates and higher *AQP7* expression levels (da Silva & Soveral, 2017; Lindgren et al., 2002; Loos et al., 2003; Prudente et al., 2007; Rodríguez, Marinelli, Tesse, Fruhbeck, & Calamita, 2015).

In humans and mice, *AQP7* gene expression is upregulated by fasting or exercise, leading to glycerol production from endogenous triacylglycerols, whereas during the feeding state it is downregulated. So, AQP7 abundance showed to be related to plasma insulin levels (Kishida et al., 2000) and *AQP7* gene transcription was inhibited by the increase of insulin levels through a negative insulin response element (IRE) identified in the promoter region of *AQP7* gene in mice and humans and by blockage of the phosphatidylinositol-3 kinase (PI3K) pathway (da Silva & Soveral, 2017). On the contrary, an upregulation of *AQP7* by peroxisome proliferator-activated receptor gamma (*PPARγ*) was demonstrated in mice and human adipocytes. Thiazolidinediones (synthetic *PPARγ* ligands) and insulin sensitizers were reported to upregulate *AQP7*, whereas leptin, tumor necrosis factor alpha (TNF-α), adrenergic agonists and steroids, being insulin resistance inducers, downregulate *AQP7* expression. Concordant with these findings, it has also been shown that under fasting conditions, concomitant with lipolysis, low plasma insulin levels and catecholamine stimulation, there is an induction of *AQP7* gene transcription and intracellular AQP7 translocation to the plasma membrane, thus potentiating glycerol release from adipocytes.

Rodríguez et al., (2011) showed that isoproterenol treatment of 3T3-L1 differentiated adipocytes caused intracellular AQP7 translocation to the plasma membrane. A similar translocation was observed for AQP3 upon isoproterenol treatment, whereas the AQP9 immunoreactivity in the plasma membrane was maintained. However, the isoproterenol (10 μmol/L) stimulation for 4h showed no significant differences in AQPs expression in human omental adipocytes.

4. Cortisone short-term regulation of *AQP7* is correlated to *GK* and *HSD1B11* gene expression in white and brown adipocytes

We have previously described (Quesada-López, González-Dávalos, Piña, & Mora, 2016) that there is an immediate and quantitatively important action of cortisone over the expression of *AQP7* and 11β-hydroxysteroid dehydrogenase type 1 (*HSD1B11*) in 3T3-L1 adipocytes. The treatment with cortisone (0.1, 1, and 10 μM) during different times (0, 5, 10, 15, and 20 min, and 48 h) to 3T3-L1 differentiated cells resulted in a time-dependent response of *AQP7*, a dose-dependent response of *HSD1B11*, and an increased expression of both genes during the first minute of treatment (five- and sixfold, respectively), followed by a decrease during the following 5–10 min ($P < 0.05$).

These studies suggest an important hormone-dependent response, which would be worthwhile to explore in greater depth, and the study also questions the physiological significance of real, fast regulation mediated by corticosteroids that favors a dramatic increase in expression during short time periods. We have now identified an important effect on genomic responses mediated by glucocorticoids (GCs), and its impact on metabolic complications should be explored. The response during short intervals should be explored in metabolically compromised cells in order to determine the role of cortisone in metabolism and the pathogenesis of obesity in adipose tissue.

With the aim to compare the action of cortisone over the expression of the two genes analyzed previously: *AQP7* and *HSD1B11*, and to test if there were differences between white adipocytes (WA) and brown adipocytes (BA), two different primary cultures from CD-1 mice were performed (following the Mexican government's technical specifications for production, care, and use of laboratory animals, NOM-062-ZOO-1999, 1999). Brown adipose tissue from 1-day-old mice was dissected and digested for precursors' extraction and 21-day-old mice subcutaneous WAT was dissected to obtain white precursors. All cell culture assays were differentiated into white or brown adipocytes (when corresponding, as previously reported by Carrière et al., 2014) and, once differentiated, cortisone response was characterized. In addition, the human SGBS cell line was employed, these precursors were differentiated into white adipocytes (as previously described by Moure et al., 2016) in order to characterize potential differences between mice and human cells. Our working hypothesis was that cortisone favored *HSD1B11* expression and reduced *AQP7* expression; effects which, by means of two complementary mechanisms, may impact metabolism of pro-diabetic adipocytes. Our previous results (Quesada-López, González-Dávalos, et al., 2016) showed, contrary to our hypothesis, the ability of cortisone to induce immediate and nearly parallel messenger RNA (mRNA) expression for both genes under study.

After cell culture treatment, total RNA was isolated employing the TRIzol reagent and processed to obtain the complementary DNA (cDNA) strand in order to performed the quantitative PCR (qPCR) analyses as previously described (Quesada-López, González-Dávalos, et al., 2016). Quantitative RNA expression for *AQP7* and *HSD1B11* was determined employing a StepOne Real-Time PCR System (Applied Biosystems, Foster City, CA, USA) and the LightCycler FastStart DNA Master Sybr Green I kit (Roche Applied Science, Mannheim, Germany). The program was performed as follows: a denaturation at 96 °C for 10 min followed by 40 cycles of 10 s at 95 °C for denaturation; 10 s at 65 °C for primer alignment (for *AQP7*, *HSD1B11*, *TBP*, *MnSOD* and *PPIA* (*Table 1*)), and 10 s at 63 °C for *GK*, and 10 s at 72 °C for elongation; the melting curve started at 60 °C for 1 min and ended with 15 s at 95 °C. Subsequently, the relative

Table 1 PCR primers used in mice cells.

Gene	Primer sequence	Accession number	Fragment size (bp)
HSD1B11	Forward 5′ TCAGACCAGAAATGCTCCAG Reverse 5′ GCCAGCAATGTAGTGAGCAG	NM_008288.2	174
AQP7	Forward 5′ TATGGTGCGAGAGTTTCTGG Reverse 5′ GCCTAGTGCACAATTGGTGA	NM_007473.4	200
GK	Forward 5′ ATCCGCTGGCTAAGAGACAA Reverse 5′ AGGCGCATATAACCCTGAAA	NM_008194.3	126
PPIA	Forward 5′ AGCACTGGGGAGAAAGGATT Reverse 5′ AGCCACTCAGTCTTGGCAGT	NM_017101.1	248
TBP	Forward 5′ AACAGCCTTCCACCTTATGC Reverse 5′ GCTGCTGTCTTTGTTGCTCT	NM_013684.3	166
MnSOD	Forward 5′ GGACCCATTGCAAGGAACAA Reverse 5′ GTAGTAAGCGTGCTCCCACAC	NM_013671.3	69
18 S	Forward 5′ GGAGCGATTTGTCTGGGTTA Reverse 5′ GTAGGGTAGGCACACGCTGA	NR_036642.1	196

abundance of the mRNA was calculated by normalization to *MnSOD* and *PPIA* for BA, and *TBP* and *PPIA* for WA [best housekeeping gene candidates suggested by NormFinder (MDL, Aarhus, Denmark)] according to the $2^{-\Delta\Delta CT}$ method.

In the case of SGBS cells 1 μL of the previously synthesized cDNA, 1 μL of the TaqMan probes solution (Table 2, Applied Biosystems) and 10 μL Master mix reagents (Platinum Quantitative PCR SuperMix-UDG with ROX reagent; Thermo Fisher Scientific) were used.

Table 2 TaqMan qRT-PCR probes used for SGSB cells.

Gene	**Reference number (Applied Biosystems)**
SOD2	Hs00167309_m1
HSD11B1	Hs00194153_m1
AQP7	Hs00927261_g1

Quantitative RNA expression for the previously mentioned genes was determined using 7500 Fast Real-time PCR systems and 7500 Software (Applied Biosystems). The program was adjusted as follows: a pre-holding stage of 2 min at 50 °C and a denaturation stage at 96 °C for 10 min; followed by 40 cycles of 15 s at 95 °C for denaturation and 1 min at 60 °C for alignment and elongation. Subsequently, the relative abundance of mRNA was calculated by normalization to *RPLP0* using the $2^{-\Delta\Delta CT}$ method.

Glycerol quantification was performed following the protocol suggested by the free glycerol determination kit (Sigma-Aldrich).

All results were analyzed with the SAS general linear model, and a least square means comparison testing was performed.

4.1 Cortisone effects over adipose *AQP7* and *GK* mRNA expression and metabolites

The acute and long-term effects of GCs on adipose tissue remain unclear (Peckett, Wright, & Riddell, 2011; Quesada-López, González-Dávalos, et al., 2016). It is known that AQP7 facilitates the secretion of glycerol from adipocytes. Glycerol constitutes a key metabolite in the control of fat accumulation (as the carbon backbone of triglycerides, TG) and glucose metabolism (as the major substrate for hepatic gluconeogenesis during fasting), for these reasons it was proposed that the glycerol channel, AQP7, plays a pivotal role in adipose tissue enlargement and function as well as in glucose homeostasis (Hibuse et al., 2005; Rodríguez, Catalán, Gómez-Ambrosi, & Frühbeck, 2006).

In a previous report, using 3T3-L1 adipocytes, we observed that *AQP7* mRNA expression was increased sixfold after 1 min treatment with 1 μM cortisone and threefold after 5 min with 0.1 μM cortisone, but similar expressions of mRNA after 5 min until 48 h, were observed with both treatments ($P < 0.05$). Quantification of mRNA relative expression by qPCR was not possible with the 10 μM cortisone treatment.

Our present results with murine WA showed an increase near to twofold induction after 10 min, but only for 10 μM cortisone, a situation that is maintained up to 20 min for doses of 0.1–10 μM cortisone (Fig. 2). Fasshauer et al. (2003) using 3T3-L1 adipocytes, observed that with 10 μM isoproterenol, 10 ng/mL TNF-α, and 100 nM dexamethasone for 16 h inhibited *AQP7* gene expression by 62%, 60%, and 39%, respectively. The inhibitory effects were dose dependent with significant suppression detectable at concentrations as low as 1 nM isoproterenol, 1 ng/mL TNF-α, and 10 nM dexamethasone.

In the case of SGBS adipocytes, the induction of AQP7 mRNA by cortisone was lower but immediate, similar to those reported by Quesada-López, González-Dávalos, et al. (2016) and also the effect was observed at 10 min ($P=0.09$). This diminished observed response, could be due to fact that they were affected by a previous GCs stimulus, SGBS cells precise the presence of 25 nM dexamethasone and 100 nM cortisol in cell culture medium in order to achieve a proper differentiation. This previous long-term exposure may affect these adipocytes' response to cortisone reflected in a lower *AQP7* mRNA induction compared to murine differentiated white adipocytes. Regardless of this detail, *AQP7* mRNA induction by cortisone in WA (murine, SGBS and 3T3-L1) showed the same tendency, an increase in the short-time (1–20 min) with doses among 0.1–1 μM, over expression was higher in 3T3-L1 (sixfold) and faster (1 min, Quesada-López, González-Dávalos, et al., 2016). In the case of BA, the induction observed was greater than that observed for WA, during the first 10 min of treatment at doses of 0.1- and 1-μM cortisone, but not for 15 and 20 min where the effect was only observed with the highest dose (10 μM).

In all three protocols for the distinct adipocytes used there was an important decrease in the expression of *AQP7* after 48 h of treatment (Fig. 2), 0.5-fold induction compared with basal (−5 min), those results were similar with the ones observed in 3T3-L1 where after 48 h (Quesada-López, González-Dávalos, et al., 2016). We could highlight from all these in vitro experiments that *AQP7* expression is increased in the short-term (minutes) by exposure to cortisone but decreases when exposure is medium or long-term (hours or days).

To complete the scheme, we measured the glycerol concentration in the medium of the cells exposed to cortisone and observed (Fig. 3) in WA (A) a slight time-dependent decrease in glycerol concentration at the beginning of the treatment, but an increase at 48 h with 1 μM cortisone. In the case of SGBS adipocytes (B), glycerol levels in the media were practically non-detected. On the other side, for BA (C), cortisone treatment caused a sevenfold increase in glycerol concentration in the media with 0.1 μM

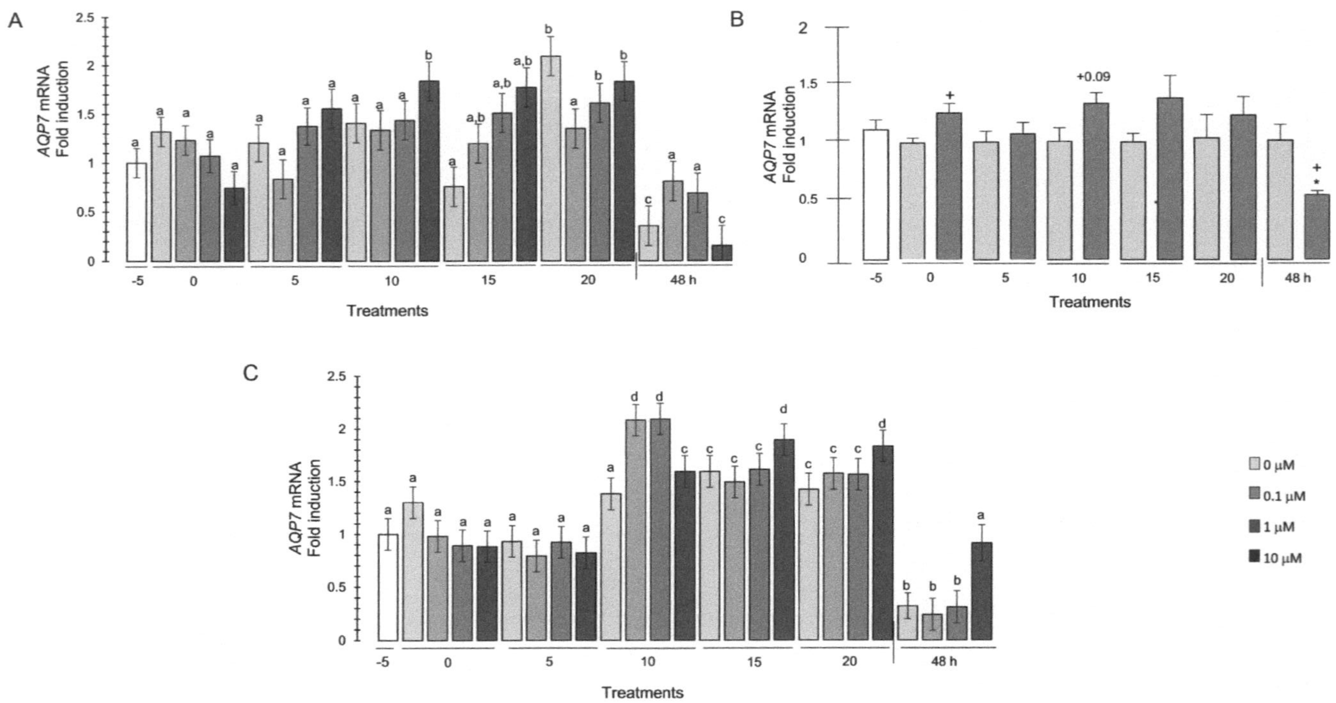

Fig. 2 Effect of cortisone on *AQP7* mRNA in adipocytes of mice and human. (A) *AQP7* in mice WA. Bars show the relative expression of mRNA from −5 min up to 48 h, as indicated in the figure, for treatments with 0, 0.1, 1 and 10 µM cortisone. (B) *AQP7* in SGBS adipocytes. Bars show the relative expression of mRNA from −5 min up to 48 h, as indicated in the figure, for treatments with 0 and 1 µM cortisone. (C) *AQP7* in mice BA. Bars show the relative expression of mRNA from −5 min up to 48 h, as indicated in the figure, for treatments with 0, 0.1, 1 and 10 µM cortisone. The intensity of the bars corresponds to the increase in cortisone concentration. All error bars represent mean ± SEM, $P < 0.05$ ($n = 4$, SGBS adipocytes $n = 3$).

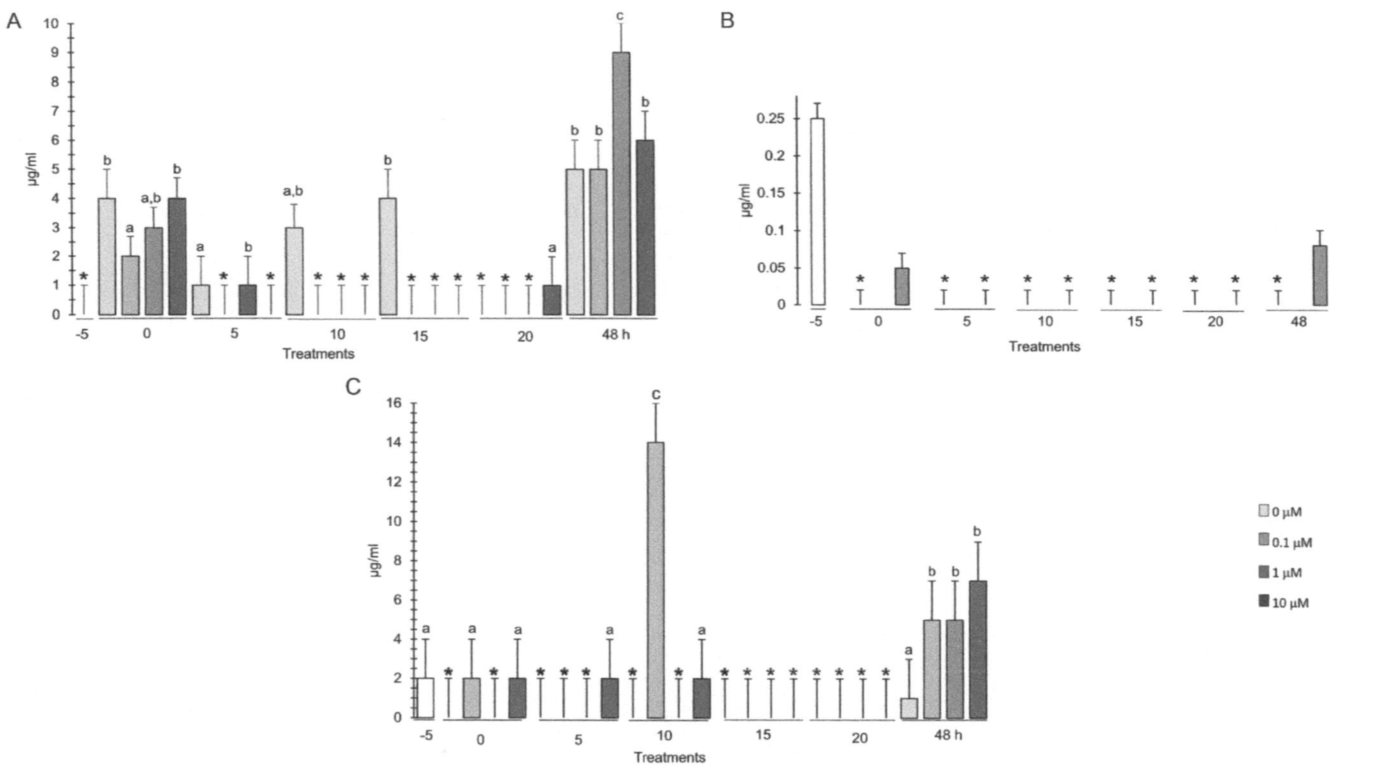

Fig. 3 Effect of cortisone on glycerol quantification in the medium of mice and human adipose cells. (A) Glycerol concentration in the medium of mice WA. Bars show the quantification from −5 min up to 48 h for treatments with 0, 0.1, 1 and 10 μM cortisone. (B) Glycerol concentration in the medium of SGBS adipocytes. Bars show the quantification from −5 min up to 48 h for treatments with 0 and 1 μM cortisone. (C) Glycerol concentration in the medium of mice BA. Bars show the quantification from −5 min up to 48 h for treatments with 0, 0.1, 1 and 10 μM cortisone. The intensity of the bars corresponds to the increase in cortisone concentration. All error bars represent mean ± SEM, $P < 0.05$ ($n = 4$, SGBS adipocytes $n = 3$). *Non-detectable.

concentration of the hormone. Yet, after 48 h of treatment, glycerol concentration in the culture media increased twofold under all cortisone treatments ($P < 0.05$). Taking into consideration these results, we considered appropriate to monitor glycerol kinase (*GK*) mRNA in our models.

GK is a key enzyme that catalyzes the phosphorylation of glycerol to glycerol-3-phosphate. Particularly in fasted liver, GK enhances the conversion of glycerol to glycerol-3-P for gluconeogenesis in parallel with the upregulation of phosphoenolpyruvate carboxykinase (Hibuse et al., 2005). Yeh et al. (2004) analyzed the crystal structure of GK in Gram-positive *Enterococcus casseliflavus* and determined the phosphorylation site which, when phosphorylated, resulted in 10–15-fold increase of GK enzymatic activity. They also demonstrated the catalytic cleft of GK, in which glycerol and ATP bind. Their results indicate that GK enzymatic activity is induced by glycerol itself. Hibuse et al. (2005) showed that intracellular accumulation of glycerol in *AQP7*-KO and -knockdown adipocytes increased GK enzymatic activity, resulting in enhancement of fatty acid uptake and triglyceride accumulation. *AQP7* KO mice exhibited adult-onset and diet induced obesity associated with severe insulin resistance. Disruption of adipose AQP7 is associated with elevated GK enzymatic activity through increased intracellular glycerol, increased uptake of fatty acid, and acceleration of triglycerides synthesis.

Our results show (Fig. 4) that, upon cortisone treatment, rodent WA (A) show a slight increase of *GK* mRNA (1.5-fold at 10 μM cortisone) at 10 min and a decrease (0.5-fold) in all doses of cortisone at 48 h. In the case of SGSB differentiated adipocytes (B), *GK* mRNA expression decrease was observed only at 48 h. This was similar to what was observed for murine BA (C) which decreased AQP7 expression after 48 h of treatment.

The consistency between *GK* and *AQP7* mRNA expression observed for WA results of special interest, suggesting a correlation in the regulation of both genes in parallel to cortisone response. On the contrary, for BA, the drop in *GK* mRNA expression at times 0 and 10 min could be the trigger for the induction of *AQP7* at 10 min. This evidence may suggest that for WA cortisone promotes *AQP7* mRNA induction firstly and as a consequence of this induction *GK* is modified. Conversely, in BA the effect over mRNA expression is observed early for *GK* and followed by *AQP7*. These contrary effects observed for both types of adipocytes are also observed for proliferation. Barclay et al. (2015) demonstrated that GCs appear to exert opposite effects on the proliferation of brown and white preadipocytes, stimulating growth of former and inhibiting that of the latter.

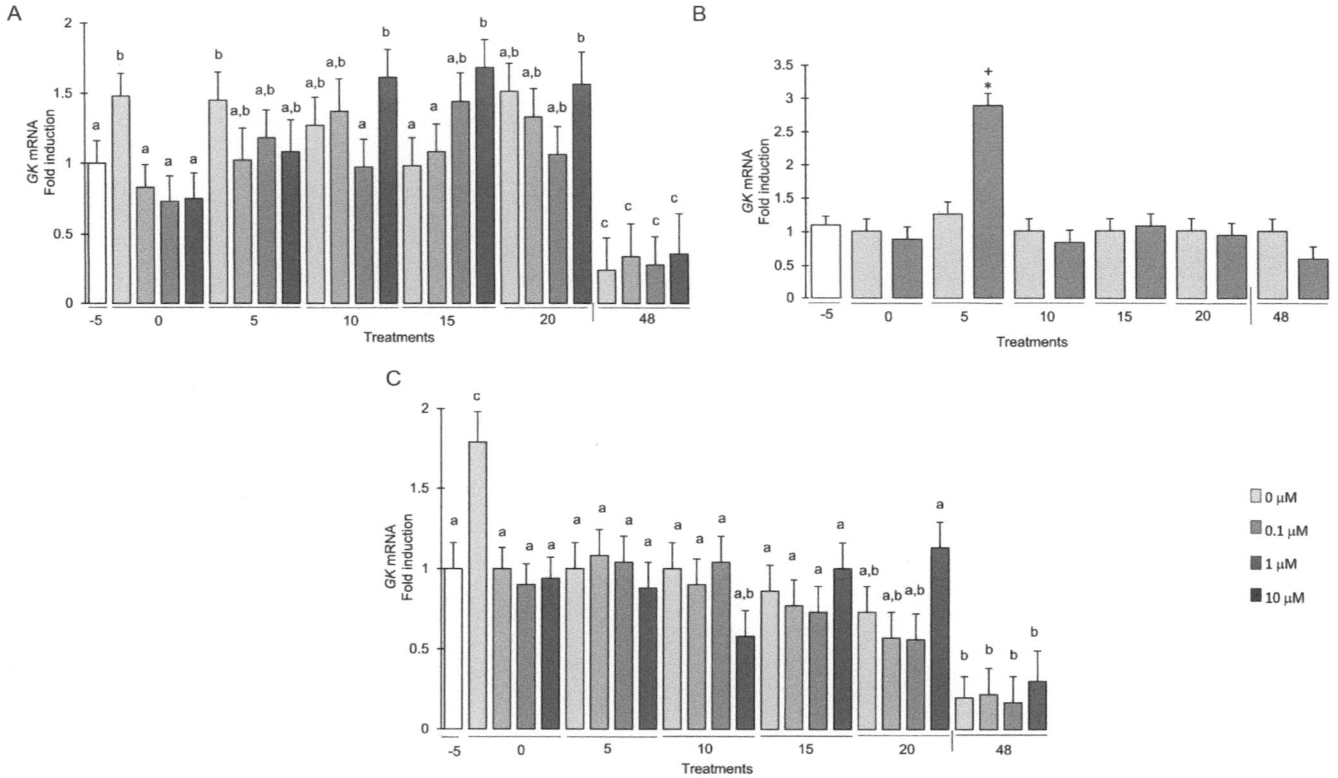

Fig. 4 Effect of cortisone on *GK* mRNA in adipocytes of mice and human. (A) *GK* in mice WA. Bars show the relative expression of mRNA from −5 min up to 48 h, as indicated in the figure, for treatments with 0, 0.1, 1 and 10 μM cortisone. (B) *GK* in SGBS adipocytes. Bars show the relative expression of mRNA from −5 min up to 48 h, as indicated in the figure, for treatments with 0 and 1 μM cortisone. (C) *GK* in mice BA. Bars show the relative expression of mRNA from −5 min up to 48 h, as indicated in the figure, for treatments with 0, 0.1, 1 and 10 μM cortisone. The intensity of the bars corresponds to the increase in cortisone concentration. All error bars represent mean ± SEM, $P<0.05$ ($n=4$). $^{+,*}P<0.01$ ($n=3$).

Finally, we wanted to know the differential levels of expression of *AQP7* between murine WA and BA, so we compared the $2^{-\Delta Ct}$ at time −5 min (Fig. 5A). Results suggest that BA express 100% more *AQP7* than WA ($P<0.01$). Furthermore, a comparison between WA and BA among all treatments (Fig. 5B) was performed showing a decreased expression in *AQP7* upon cortisone treatment at 5 min (for the doses 1 and 10 μM). And at 10 min an overexpression promoted by cortisone treatment at 0.1 and 1 μM; which was detected at 15 min at dose 0.1 μM. Lastly, at 48 h of cortisone treatment (0.1 and 1 μM) BA expressed less than 50% of *AQP7* when compared with WA ($P<0.05$).

Kroon et al. (2018) observed activity of murine BA was inhibited by corticosterone, that is, decreased NE-induced *UCP-1* expression upon 10 nM and 1 μM corticosterone exposure. GC receptors inhibitor CORT125281 and RU486 were able to (partially) prevent corticosterone-induced inhibition of BAT activity, as coincubation with the GR antagonists results in enhanced *UCP-1* expression. Although we did not quantify *UCP-1* mRNA expression nor adrenergic stimulation response in BA, GK is an important marker of BA thermogenic activation, thus the result after 48 h may suggest a decrease thermogenic machinery expression (Festuccia et al., 2003).

Finally, glucocorticoid response elements (GREs) were identified in the lastly updated available gene sequences for *AQP7* in *Mus musculus* and *Homo sapiens*. Using MatInspector computer software, we entered both sequences. This software was able to identify GREs in both genes. In the case of the murine gen for *AQP7*, two potential response elements at positions 98 and 359 were identified. On the other hand, in the reported human gene for *AQP7*, three potential response elements were found, at 830, 1017, and 1087. The presence of these GREs strongly suggests that cortisone can certainly regulate the transcription of the *AQP7* gene which supports the results obtained in this work.

4.2 Cortisone effect on *HSD1B11* mRNA expression in white and brown adipocytes

Exogenous administration or endogenous overproduction of GCs has shown to have lipogenic effects, this has been linked to increased adiposity. The link between GCs and excess adiposity is well established clinically and is clearly demonstrated in individuals with Cushing Syndrome or those on exogenous corticosteroid treatment. These individuals show increased weight gain, hypertension, and visceral adiposity, and are at an increased risk for developing T2DM (Peckett et al., 2011; Stewart & Petersenn, 2009).

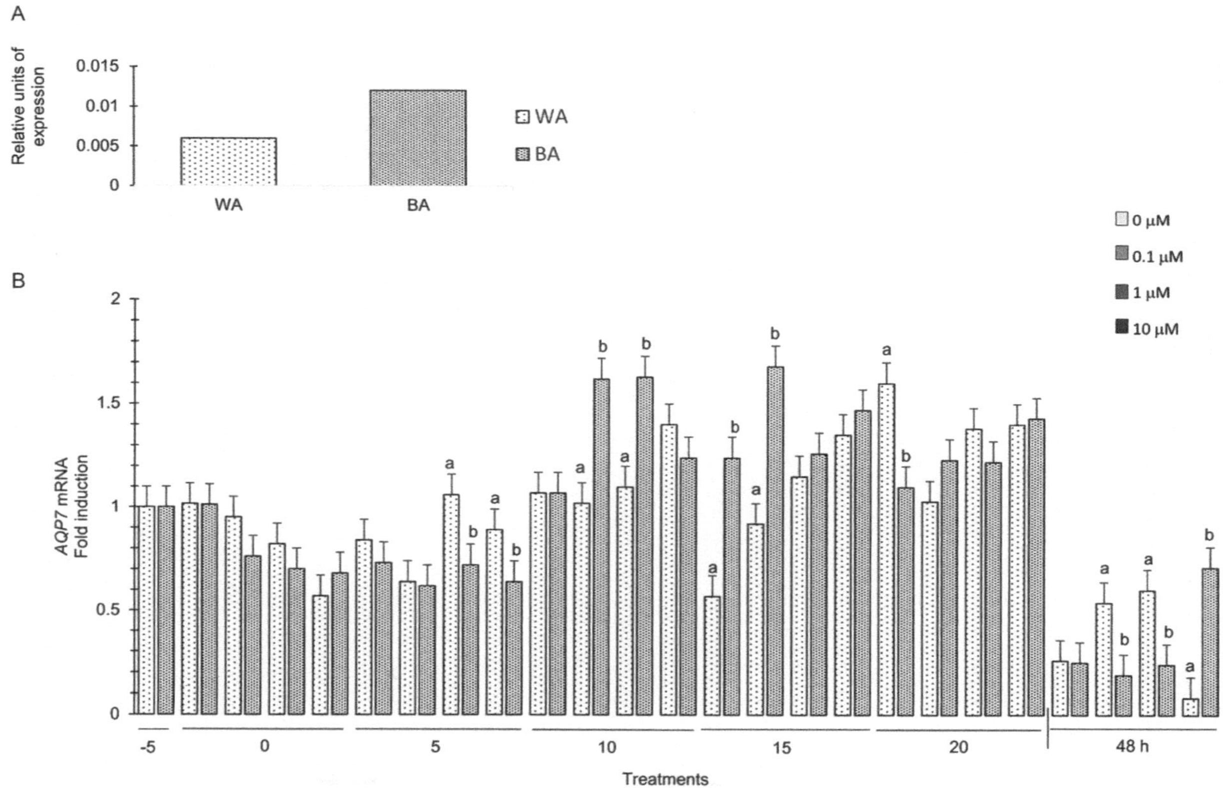

Fig. 5 Relation between *AQP7* mRNA expression on WA and BA. (A) ΔCt mRNA at −5 min. (B) Bars represents *AQP7* mRNA fold induction in WA (first column) and BA (second column), compared between treatments. All error bars represent mean ± SEM, $P < 0.05$ ($n = 4$).

Adipose tissue-specific increases in GCs action, via the upregulation of the enzyme HSD1B11 has also shown to promote and increase in visceral fat accumulation, thereby further suggesting that GCs may have a primary role in the development of central obesity (Desbriere et al., 2006). When GCs are released, both active and inactive forms of the hormone are sent into circulation. Inactive GCs, cortisone in humans and rodents, travel freely in plasma in high concentrations. When these hormones reach the target tissues, these tissues possess varying levels of activating enzymes that are able to convert between the active and inactive forms. Two microsomal enzymes, collectively referred to as the 11β-hydroxysteroid dehydrogenase (11β-HSD) system, interconvert receptor-active cortisol and inert cortisone as shown in Fig. 6 (Peckett et al., 2011). Although it is difficult to adequately measure tissue levels of GCs; Masuzaki et al. (2001) have suggested that adipose tissue may have active GCs levels that are 10–15 times higher than the circulating levels due to the actions of HSD1B11. Studies have shown that HSD1B11 was a powerful regulator in modulating WAT metabolism and function (Bujalska, Kumar, Hewison, & Stewart, 1999; Stimson & Walker, 2007). Inhibition of HSD1B11 could prevent the differentiation of white adipocyte, reduce white adipose accumulation, and improve the function of WAT (London & Castonguay, 2009; Staab & Maser, 2010).

Our previous experiments (Quesada-López, González-Dávalos, et al., 2016) demonstrate that the expression of the *HSD1B11* mRNA was increased during the first minute of treatment with 0.1, 1, and 10 μM of cortisone treatment. This response was greater with the 0.1- and 1-μM treatments, showing 4.5- and 4.8-fold induction, respectively. It is noteworthy that the three concentrations exhibited a similar waved-manner response, and it should be noted that the response with 0.1 and 1 μM after 20 min returned to control values, while mRNA expression with 10 μM was decreased ($P < 0.05$). Results found by Kim et al. (2007) characterized a role of HSD1B11 in mediating glucocorticoid-driven adipocyte differentiation. In their studies, they found that 3T3-L1 fibroblasts could be induced to differentiate into adipocytes by five glucocorticoids: dexamethasone, corticosterone, cortisol, dehydrocorticosterone (DHC), and cortisone. Also, these results suggest that HSD1B11 activity is tightly controlled during

Fig. 6 Sterols inter-conversion by *HSD1B11* and *HSD2B11* in adipose tissue.

3T3-L1 differentiation. Fibroblasts show low mRNA levels of *HSD1B11* but the expression levels are enough to enable the conversion of inactive GCs. This activity will cause the initiation of adipocyte differentiation. Endogenous HSD1B11 levels are increased late during adipogenesis, presumably when the differentiating cells have sufficient capacity to generate NADPH and preserve the oxoreductase activity of HSD1B11 and thus prevent glucocorticoid inactivation and suppression of the adipogenic program (Kim et al., 2007).

Concerning our current experiments, cortisone treatment of murine WA showed no significant modifications in the expression of *HSD1B11* during the first 20 min of treatment. However, after 48 h of treatment, 0.1 and 10 μM of cortisone caused an induction of 3 folds in *HSD1B11 mRNA* expression and 1 μM showed a fivefold induction ($P<0.05$, Fig. 7A). This was different from the results obtained in the cell line 3T3-L1 which showed a repression in *HSD1B11* mRNA expression after 48 h of treatment (0.2- and 0.4-fold induction for 0, 1 and 1 μM, respectively). Our results demonstrate that there are differences between isolated precursors and the cell line 3T3-L1, at least regarding the timing in the induction of *HSD1B11* (Quesada-López, González-Dávalos, et al., 2016). Regarding human adipocytes, treatment of SGBS differentiated cells with cortisone showed minor effects in *HSD1B11* mRNA expression. The only significant modification was observed after 5 min of treatment with 1 μM of cortisone which caused an increase in 1.5-folds ($P<0.001$, Fig. 7B).

BAT plays an active role in energy expenditure and fatty acid oxidization for heat dissipation (Cannon & Nedergaard, 2004). Evidence has demonstrated that BAT is critically important for energy equilibrium (Liu et al., 2013). Viengchareun, Penfornis, Zennaro, and Lombès (2001) demonstrate that exposure of human and mouse BAT to chronically elevated GCs levels impairs function by interfering with sympathetic adrenergic signaling and suppression of transcriptional programs that regulate non-shivering thermogenesis. Doig et al. (2017) provide evidence that elevated *HSD1B11* expression within brown adipocytes, as a consequence of GC excess and aging, can impair the BAT thermogenic program, mitochondrial biogenesis, and respiratory capacity. As such, HSD1B11-mediated GCs generation can influence the ability of BAT to execute its primary function of thermogenic regulation.

The results obtained from BA treatment with cortisone show a rapid increase (1.5-fold induction) in *HSD1B11* expression which is stabilized after 10 min. However, the greater registered induction (twofold) was

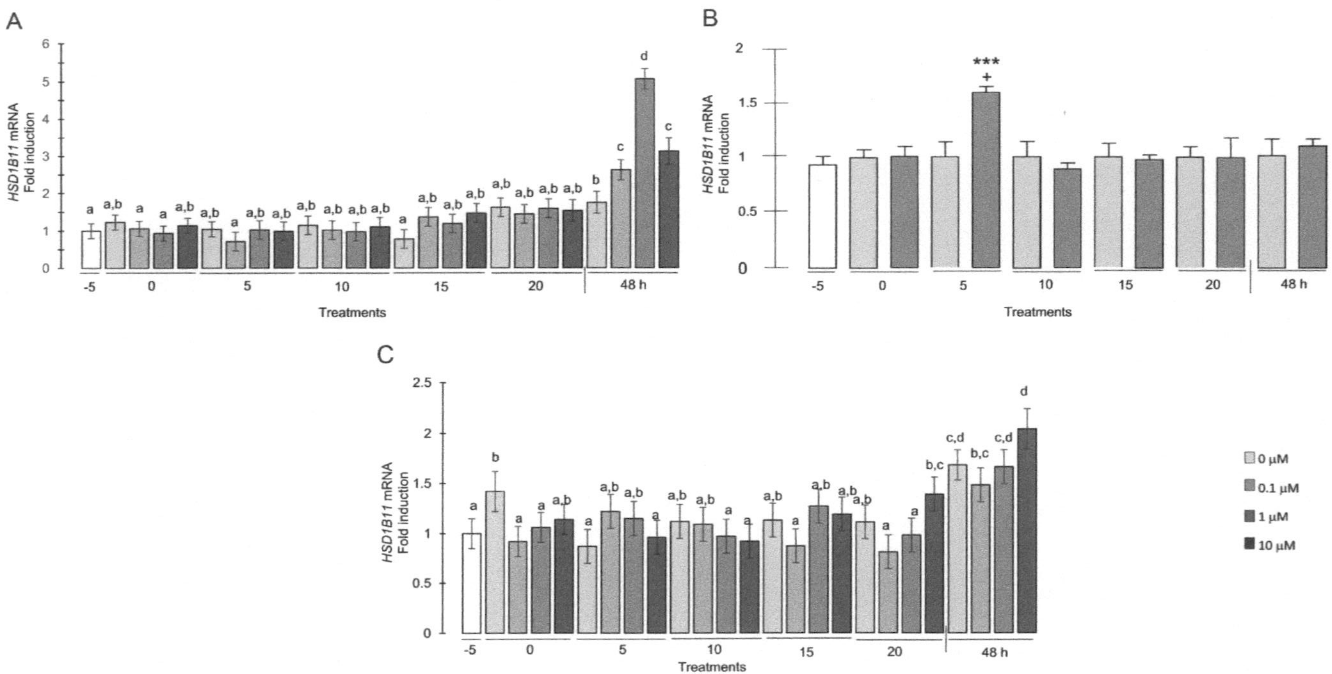

Fig. 7 Effect of cortisone on *HSD1B11* mRNA in adipocytes of mice and human. (A) *HSD1B11* in mice WA. Bars show the relative expression of mRNA from −5 min up to 48 h, as indicated in the figure, for treatments with 0, 0.1, 1 and 10 μM cortisone. (B) *HSD1B11* in SGBS adipocytes. Bars show the relative expression of mRNA from −5 min up to 48 h, as indicated in the figure, for treatments with 0 and 1 μM cortisone. (C) *HSD1B11* in mice BA. Bars show the relative expression of mRNA from −5 min up to 48 h, as indicated in the figure, for treatments with 0, 0.1, 1 and 10 μM cortisone. The intensity of the bars corresponds to the increase in cortisone concentration. All error bars represent mean ± SEM, $P < 0.05$ ($n = 4$, SGBS adipocytes $n = 3$).

observed after 48 h of cortisone treatment at 10 μM ($P < 0.05$, Fig. 7C). This is similar to what Barclay et al. (2015) showed. In these assays, treatment with dexamethasone induced expression of *HSD1B11* in brown and white preadipocytes. However, this effect was significantly greater in brown preadipocytes. Therefore, these results suggest that the differences in *HSD1B11* expression between brown and white adipocytes may probably reflect contrasted effects of GCs between these tissues rather than their local availability. A report by Lee et al. (2008) revealed that dexamethasone did not enhance the expression of *HSD1B11* in white preadipocytes derived from subcutaneous abdominal fat. In contrast, *HSD1B11* from omental fat was upregulated by dexamethasone, indicating that there are also depot-specific differences in responsiveness of *HSD1B11* to GCs (Lee et al., 2008). Nevertheless, the observation that dexamethasone stimulates the expression of *HSD1B11* could be indicative of a novel stimulatory role for GCs in adipogenesis of BA, potentiated via a feed-forward pathway through the induction of *HSD1B11*.

Further work is needed to address matters in the regulation of adipogenesis and thermogenesis in interplay with glucocorticoid metabolism. Specially regarding *HSD1B11* expression and activity direction; localized glucocorticoid activation/inactivation, induction of adipogenesis, and development of insulin resistance.

5. Adrenergic stimulation effects over *AQP7, GK* and *HSD1B11* mRNA expression

The influence of cortisol on the sensitivity to catecholamines could also be compatible with changes in β-adrenoceptor number or functional coupling to its G_s protein. In 3T3-F442A adipocytes dexamethasone downregulates β_3-adrenoceptor protein expression and β_3-adrenoceptor-mediated adenylate cyclase activity (Fève et al., 1992). Observations made by Ottosson, Lönnroth, Björntorp, and Edén (2000) suggest that cortisol affects the cAMP production rate at the adenylate cyclase level or, alternatively, alters the cAMP elimination rate by stimulating phosphodiesterase activity.

In this sense, AQP7 in 3T3-L1 adipocytes is localized at the periphery of the nucleus in steady state. However, AQP7 translocates to the plasma membrane following adrenaline stimulation, which induces lipolysis (Kishida et al., 2000). Adrenaline does not affect *AQP7* gene expression. However, adrenaline elevates intracellular cAMP levels through adrenergic receptor which then activate protein kinase A (PKA) and will facilitate AQP7 translocation to the membrane. Interestingly, similar results are observed for AQP2, which is a key water channel of the kidney. AQP2 exists in the principal cells of the renal collecting duct (Maeda, Hibuse, & Funahashi, 2009; Nielsen et al., 1995).

In the other hand, it is known that thermogenic activation of BAT is also associated with a strong enhancement of intracellular lipolysis, which produces fatty acids that fuel intracellular thermogenesis. Fatty acids might be released from brown adipocytes to mediate relevant signaling processes either locally or at a distance; noradrenergic-mediated stimulation of lipolysis in cultured brown adipocytes can trigger the release of fatty acids to the medium (Cannon & Nedergaard, 2004; Villarroya, Cereijo, Villarroya, & Giralt, 2017). Armengol et al. (2012) showed the action of dexamethasone in BA differentiation and the C/EBPδ–Pref-1 axis, they established that dexamethasone blunted the Pref-1 downregulation in normal BA differentiation.

We used SGBS differentiated adipocytes treated with cAMP and CL316243 (β3-agonist) to measure *AQP7*, *GK*, *HSD1B11* and *UCP1* mRNA expression (Fig. 8). The treatment with cAMP was applied by 6 and 24 h and CL316243 only 24 h. While CL316243 results show no β3-receptor specific effects on the expression of the aforementioned genes, cAMP did show significant modifications in mRNA expression (suggesting a response mediated by other type of β receptors). cAMP treatment of human differentiated SGBS adipocytes shows an induction in the expression of the three genes under study, in parallel with *UCP1* (main thermogenic gene marker); which suggests a response to thermogenic activation of adipocytes and also imply an important regulation in glycerol traffic during thermogenic response. To further elucidate in these results, we decided to check for a possible effect on a murine model. We took the database available for a cold-exposure experiment performed on mice. RNA-seq data were obtained from BAT samples obtained from mice housed under thermoneutral conditions or following a 24 h exposure to 4 °C. The database analysis software developed by Dr. Cereijo (dataset in Quesada-López, Cereijo, et al., 2016) helped us to identify the effects of cold exposure over the expression of *HSD1B11*, *GK* and *AQP7*. In the case of *HSD1B11* and *GK*, the results are concordant with those observed for the RNA-seq. In the case of *AQP7*, although there is a decrease in the induction of *AQP7* upon 24 h of cAMP treatment, in vivo the effects are maintained upon 24 h of cold exposure. This could mean that the peak in *AQP7* mRNA expression in vitro is maintained for short periods of time and what would remain to be elucidated is be the effect over *AQP7* mRNA expression under a chronic cold challenge in vivo.

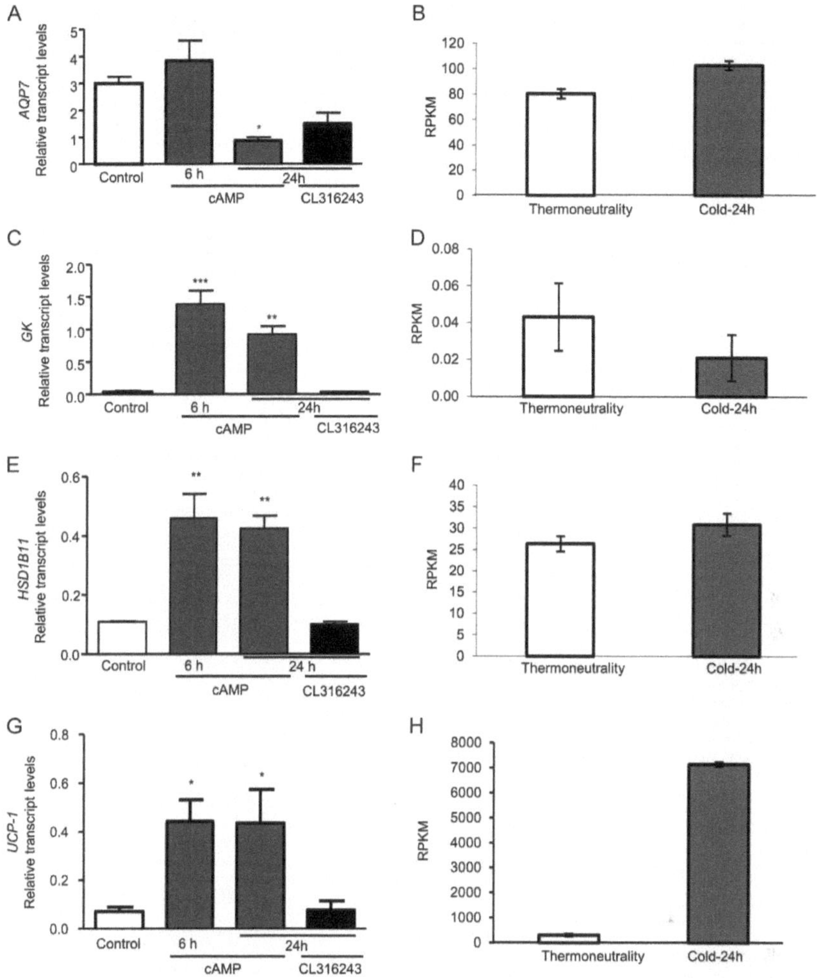

Fig. 8 Effect of cAMP and CL316243 (β3-agonist) on mRNA (A and B) *AQP7*, (C and D) *GK* and (E and F) *HSD1B11*, and *UCP1* (G and H) expression. All error bars represent mean ± SEM, $P<0.01$ ($n=4$). RPKM = Lectures in millions of kilobases. All error bars represent mean ± SEM, $P<0.05$.

6. Conclusions and future directions

To date, the mechanisms by which GCs regulate AQP7 have not been completely elucidated. As for our findings, Fig. 9 summarizes cortisone effects over the mRNA expression of *AQP7*, *GK* and *HSD1B11* in white and brown adipocytes. AQP7 is an aquaglyceroporin which mRNA expression is induced rapidly upon cortisone treatment in adipocytes.

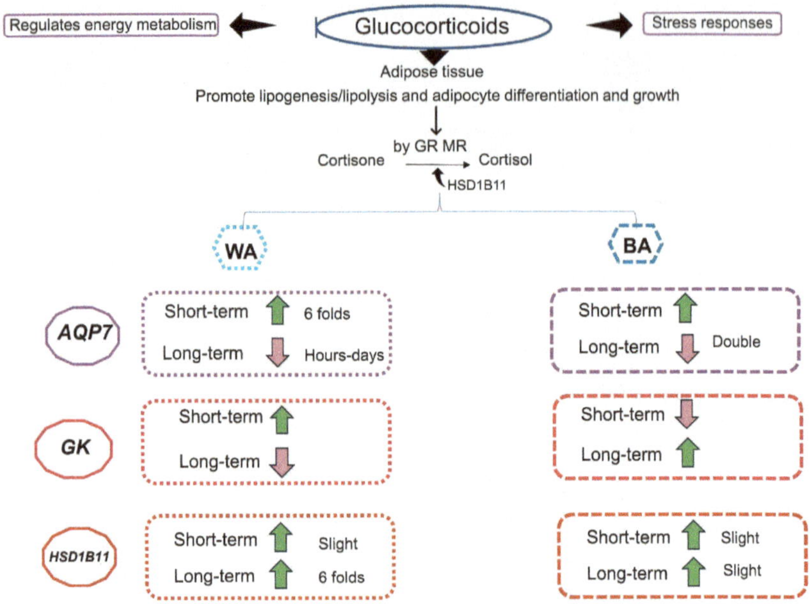

Fig. 9 Summary of the effect of cortisone on the mRNA of *AQP7*, *GK* and *HSD1B11* in white and brown adipocyte cultures. GR, MR = glucocorticoids and mineralocorticoids receptors.

However, after a chronic treatment, *AQP7* mRNA expression in adipose cells is decreased. The regulatory pathways in the distinct adipose depots-derived cells should be better characterized in order to clarify the role of GCs on AQPs and its relevance in the development of metabolic diseases. Also, the mechanisms by which an excessive input of corticosteroids may affect adipose cells' response to metabolic signals (i.e., insulin, catecholamines) should be clarified not only to better understand the pathophysiology of metabolic diseases but also to identify potential therapeutic targets. In addition, taking into consideration the evidence which suggests that GCs may affect adipocyte thermogenic activation should be an important adverse effect to take into account considering the relevance of this activation as a therapeutic target.

Acknowledgments

This study was supported by a grant 211559 from CONACYT (Consejo Nacional de Ciencia y Tecnología, México City, México) and grant SAF2017-85722R, from MINECO, Spain. L.G.-D. and M.A.-P. contributed in the same way in the development of this study. This work is part of the undergraduate thesis that M.A.-P. submitted to the Universidad Nacional Autónoma de México (UNAM). M.A.-P. thanks

SNI-CONACYT for a scholarship at Facultad de Estudios Superiores-Cuautitlán, UNAM. We also thank Martín García-Servín, Alejandra Castilla and Adriana González-Gallardo [INB (Neurobiology Institute)-UNAM, Querétaro, Qro., México] for their technical assistance.

References

Armengol, J., Villena, J. A., Hondares, E., Carmona, M., Sul, H., Iglesias, R., et al. (2012). Pref-1 in brown adipose tissue: Specific involvement in brown adipocyte differentiation and regulatory role of C/EBPδ. *Biochemistry Journal, 443*(3), 799–810. https://doi.org/10.1042/BJ20111714.

Barclay, J. L., Agada, H., Jang, C., Ward, M., Wetzig, N., & Ho, K. K. (2015). Effects of glucocorticoids on human brown adipocytes. *Journal of Endocrinology, 224*(2), 139–147. https://doi.org/10.1530/JOE-14-0538.

Boone, M., Kortenoeven, M. L., Robben, J. H., Tamma, G., & Deen, P. M. (2011). Counteracting vasopressin-mediated water reabsorption by ATP, dopamine, and phorbol esters: Mechanisms of action. *American Journal of Physiology-Renal Physiology, 300*(3), F761–F771. https://doi.org/10.1152/ajprenal.00247.2010.

Bujalska, I. J., Kumar, S., Hewison, M., & Stewart, P. M. (1999). Differentiation of adipose stromal cells: The roles of glucocorticoids and 11beta-hydroxysteroid dehydrogenase. *Endocrinology, 140*(7), 3188–3196. https://doi.org/10.1210/endo.140.7.6868.

Cannon, B., & Nedergaard, J. (2004). Brown adipose tissue: Function and physiological significance. *Physiological Reviews, 84*, 277–359. https://doi.org/10.1152/physrev.00015.2003.

Carrière, A., Jeanson, Y., Berger-Müller, S., André, M., Chenouard, V., Arnaud, E., et al. (2014). Browning of white adipose cells by intermediate metabolites: An adaptive mechanism to alleviate redox pressure. *Diabetes, 63*(10), 3253–3265. https://doi.org/10.2337/db13-1885.

Ceperuelo-Mallafre, V., Miranda, M., Chacon, M. R., Vilarrasa, N., Megia, A., Gutierrez, C., et al. (2007). Adipose tissue expression of the glycerol channel aquaporin-7 gene is altered in severe obesity but not in type 2 diabetes. *The Journal of Clinical Endocrinology & Metabolism, 92*(9), 3640–3645. https://doi.org/10.1210/jc.2077-0531.

Choi, H. J., Jung, H. J., & Kwon, T. H. (2015). Extracellular pH affects phosphorylation and intracellular trafficking of AQP2 in inner medullary collecting duct cells. *American Journal of Physiology-Renal Physiology, 308*(7), F737–F748. https://doi.org/10.1152/ajprenal.00376.2014.

da Silva, I. V., & Soveral, G. (2017). Aquaporins in obesity. *Advances in Experimental Medicine and Biology, 969*, 227–238. https://doi.org/10.1007/978-94-024-1057-0_15.

Day, R. E., Kitchen, P., Owen, D. S., Bland, C., Marshall, L., Conner, A. C., et al. (2014). Human aquaporins: Regulators of transcellular water flow. *Biochimica et Biophysica Acta, 1840*(5), 1492–1506. https://doi.org/10.1016/j.bbagen.2013.09.033.

Delporte, C. (2017). Aquaporins and gland secretion. *Advances in Experimental Medicine and Biology, 969*, 63–79. https://doi.org/10.1007/978-94-024-1057-0_4.

Desbriere, R., Vuaroqueaux, V., Achard, V., Boullu-Ciocca, S., Labuhn, M., Dutour, A., et al. (2006). 11beta-Hydroxysteroid dehydrogenase type 1 mRNA is increased in both visceral and subcutaneous adipose tissue of obese patients. *Obesity, 14*(5), 794–798. https://doi.org/10.1038/oby.2006.92.

Doig, L. C., Fletcher, R. S., Morgan, S. A., McCabe, E. L., Larner, D. P., Tomlinson, J. W., et al. (2017). 11B-HSD1 modulates the set point of brown adipose tissue response to glucocorticoids in male mice. *Endocrinology, 158*(6), 1964–1976. https://doi.org/10.1210/en.2016.1722.

Fasshauer, M., Klein, J., Lossner, U., Klier, M., Kralisch, S., & Paschke, R. (2003). Suppression of aquaporin adipose gene expression by isoproterenol, TNFalpha, and dexamethasone. *Hormone and Metabolic Research*, *35*(4), 222–227. https://doi.org/10.1055/s-2003-39478.

Festuccia, W. T., Guerra-Sá, R., Kawashita, N. H., Garófalo, M. A., Evangelista, E. A., Rodrigues, V., et al. (2003). Expression of glycerokinase in brown adipose tissue is stimulated by the sympathetic nervous system. *American Journal of Physiology. Regulatory, Integrative and Comparative Physiology*, *284*(6), R1536–R1541. https://doi.org/10.1152/ajpregu.00764.2002.

Fève, B., Baude, B., Krief, S., Strosberg, A. D., Pairault, J., & Emorines, L. J. (1992). Inhibition by dexamethasone of β 3-adrenergic receptor responsiveness in 3T3–F442A adipocytes. *Journal of Biological Chemistry*, *267*(22), 15909–15915.

Gazzarrini, S., Kang, M., Epimashko, S., Van Etten, J. L., Dainty, J., Thiel, G., et al. (2006). Chlorella virus MT325 encodes water and potassium channels that interact synergistically. *Proceedings of the National Academy of Sciences of the United States of America*, *103*(14), 5355–5360. https://doi.org/10.1073/pnas.0600848103.

Gunnarson, E., Axehult, G., Baturina, G., Zelenin, S., Zelenina, M., & Aperia, A. (2005). Lead induces increased water permeability in astrocytes expressing aquaporin 4. *Journal of Neuroscience*, *136*(1), 105–114. https://doi.org/10.1016/j.neuroscience.2005.07.027.

Hibuse, T., Maeda, N., Funahashi, T., Yamamoto, K., Nagasawa, A., Mizunoya, W., et al. (2005). Aquaporin 7 deficiency is associated with development of obesity through activation of adipose glycerol kinase. *Proceedings National Academy of Sciences of the United States of America*, *102*(31), 10993–10998. https://doi.org/10.1073/pnas.0503291102.

Iena, F. M., & Lebeck, J. (2018). Implications of aquaglyceroporin 7 in energy metabolism. *International Journal of Molecular Sciences*, *19*(1). E154. https://doi.org/10.3390/ijms19010154.

Ishibashi, K., Morishita, Y., & Tanaka, Y. (2017). The evolutionary aspects of aquaporin family. *Advances in Experimental Medicine and Biology*, *969*, 35–50. https://doi.org/10.1007/978-94-024-1057-0_2.

Ishibashi, K., Tanaka, Y., & Morishita, Y. (2014). The role of mammalian superaquaporins inside the cell. *Biochimica et Biophysica Acta*, *1840*(5), 1507–1512. https://doi.org/10.1016/j.bbagen.2013.10.039.

Kim, J., Temple, K. A., Jones, S. A., Meredith, K. N., Basko, J. L., & Brady, M. J. (2007). Differential modulation of 3T3-L1 adipogenesis mediated by 11β-hydroxysteroid dehydrogenase-1 levels. *Journal of Biological Chemistry*, *282*(15), 11038–11046. https://doi.org/10.1074/jbc.M606197200.

Kishida, K., Kuriyama, H., Funahashi, T., Shimomura, I., Kihara, S., Ouchi, N., et al. (2000). Aquaporin adipose, a putative glycerol channel in adipocytes. *The Journal of Biological Chemistry*, *275*(27), 20896–20902. https://doi.org/10.1074/jbcM001119200.

Kroon, J., Koorneef, L. L., van den Heuvel, J. K., Verzijl, C. R. C., van de Velde, N. M., Mol, I. M., et al. (2018). Selective glucocorticoid receptor antagonist CORT125281 activates brown adipose tissue and alters lipid distribution in male mice. *Endocrinology*, *159*(1), 535–546. https://doi.org/10.1210/en.2017-00512.

Lanaspa, M. A., Andres-Hernando, A., Li, N., Rivard, C. J., Cicerchi, C., Roncal-Jimenez, C., et al. (2010). The expression of aquaporin-1 in the medulla of the kidney is dependent on the transcription factor associated with hypertonicity, TonEBP. *The Journal of Biological Chemistry*, *285*(41), 31694–31703. https://doi.org/10.1074/jbc.M109.093690.

Lee, M. J., Fried, S. K., Mundt, S. S., Wang, Y., Sullivan, S., Stefanni, A., et al. (2008). Depot-specific regulation of the conversion of cortisone to cortisol in human adipose tissue. *Obesity*, *16*(6), 1178–1185. https://doi.org/10.1038/oby.2008.207.

Li, C., & Wang, W. (2017). Molecular biology of aquaporins. In B. Yang (Ed.), *Aquaporins. Advances in experimental medicine and biology* (pp. 1–34). Dordrecht: Springer. https://doi.org/10.1007/978-94-024-1057-0_1.

Lindgren, C. M., Mahtani, M. M., Widen, E., McCarthy, M. I., Daly, M. J., Kirby, A., et al. (2002). Genomewide search for type 2 diabetes mellitus susceptibility loci in Finnish families: The Botnia study. *American Journal of Human Genetics, 70*(2), 509–516. https://doi.org/10.1086/338629.

Liu, J., Kong, X., Wang, L., Qi, H., Di, W., Zhang, X., et al. (2013). Essential roles of 11β-HSD1 in regulating brown adipocyte function. *Journal of Molecular Endocrinology, 50*(1), 103–113. https://doi.org/10.1530/JME-12-0099.

London, E., & Castonguay, T. W. (2009). Diet and the role of 11beta-hydroxysteroid dehydrogenase-1 on obesity. *The Journal of Nutritional Biochemistry, 20*(7), 485–493. https://doi.org/10.1016/j.jnutbio.2009.02.012.

Loos, R. J., Katzmarzyk, P. T., Rao, D. C., Rice, T., Leon, A. S., Skinner, J. S., et al. (2003). Genome-wide linkage scan for the metabolic syndrome in the HERITAGE Family Study. *The Journal of Clinical Endocrinology and Metabolism, 88*(12), 5935–5943. https://doi.org/10.1210/jc.2003-030553.

Maeda, N. (2012). Implications of aquaglyceroporins 7 and 9 in glycerol metabolism and metabolic syndrome. *Molecular Aspects of Medicine, 33*(5–6), 665–675. https://doi.org/10.1016/j.mam.2012.02.004.

Maeda, N., Hibuse, T., & Funahashi, T. (2009). Role of aquaporin-7 and aquaporin-9 in glycerol metabolism; involvement in obesity. In E. Beitz (Ed.), *Aquaporins. Handbook of experimental pharmacology*. Springerhttps://doi.org/10.1007/978-3-540-79885-9_12.

Masuzaki, H., Paterson, J., Shinyama, H., Morton, N. M., Mullins, J. J., Seckl, J. R., et al. (2001). A transgenic model of visceral obesity and the metabolic syndrome. *Science, 294*(5549), 2166–2170. https://doi.org/10.1126/science.1066285.

Miranda, M., Escote, X., Ceperuelo-Mallafre, V., Alcaide, M. J., Simon, I., Vilarrasa, N., et al. (2010). Paired subcutaneous and visceral adipose tissue aquaporin-7 expression in human obesity and type 2 diabetes: Differences and similarities between depots. *The Journal of Clinical Endocrinology & Metabolism, 95*(7), 3470–3479. https://doi.org/10.1210/jc.2009-2655.

Moure, R., Domingo, P., Gallego-Escuredo, J. M., Villarroya, J., Gutierrez Mdel, M., Mateo, M. G., et al. (2016). Impact of elvitegravir on human adipocytes: Alterations in differentiation, gene expression and release of adipokines and cytokines. *Antiviral Research, 132*, 59–65. https://doi.org/10.1016/j.antiviral.2016.05.013.

Nielsen, S., Chou, C. L., Marples, D., Christensen, E. I., Kishore, B. K., & Knepper, M. A. (1995). Vasopressin increases water permeability of kidney collecting duct by inducing translocation of aquaporin-CD water channels to plasma membrane. *Proceedings of the National Academy of Sciences of the United States of America, 92*, 1013–1017. https://doi.org/10.1073/pnas.92.4.1013.

NOM-062-ZOO-1999. (1999). Especificaciones técnicas para la producción, cuidado y uso de los animales domésticos. In *Norma Oficial Mexicana. Secretaría de Agricultura, Ganadería y Desarrollo Rural, Pesca y Alimentación SAGARPA* (pp. 1–19). Estados Unidos Mexicanos: Diario Oficial de la Federación.

Olesen, E. T., Rutzler, M. R., Moeller, H. B., Praetorius, H. A., & Fenton, R. A. (2011). Vasopressin-independent targeting of aquaporin-2 by selective E-prostanoid receptor agonists alleviates nephrogenic diabetes insipidus. *Proceedings of the National Academy of Sciences of the United States of America, 108*(31), 12949–12954. https://doi.org/10.1073/pnas.1104691108.

Ottosson, M., Lönnroth, P., Björntorp, P., & Edén, S. (2000). Effects of cortisol and growth hormone on lipolysis in human adipose tissue. *The Journal of Clinical Endocrinology & Metabolism, 85*(2), 799–803. https://doi.org/10.1210/jcem.85.2.6358.

Peckett, A. J., Wright, D. C., & Riddell, M. C. (2011). The effects of glucocorticoids on adipose tissue lipid metabolism. *Metabolism*, *60*(11), 1500–1510. https://doi.org/10.1016/j.metabol.2011.06.012.

Pérez Di Giorgio, J., Soto, G., Alleva, K., Jozefkowicz, C., Amodeo, G., Muschietti, J. P., et al. (2014). Prediction of aquaporin function by integrating evolutionary and functional analyses. *Journal of Membrane Biology*, *247*(2), 107–125. https://doi.org/10.1007/s00232-013-9618-8.

Prudente, S., Flex, E., Morini, E., Turchi, F., Capponi, D., De Cosmo, S., et al. (2007). A functional variant of the adipocyte glycerol channel aquaporin 7 gene is associated with obesity and related metabolic abnormalities. *Diabetes*, *56*(5), 1468–1474. https://doi.org/10.2337/db06-1389.

Quesada-López, T., Cereijo, R., Turatsinze, J. V., Planavila, A., Cairó, M., Gavaldà-Navarro, A., et al. (2016). The lipid sensor GPR120 promotes brown fat activation and FGF21 release from adipocytes. *Nature Communications*, 7, 13479. https://doi.org/10.1038/ncomms13479.

Quesada-López, T., González-Dávalos, L., Piña, E., & Mora, O. (2016). HSD1 and AQP7 short-term gene regulation by cortisone in 3T3-L1 adipocytes. *Adipocytes*, *5*(3), 298–305. https://doi.org/10.1080/21623945.2016.1187341.

Rodríguez, A., Catalán, V., Gómez-Ambrosi, J., & Frühbeck, G. (2006). Role of aquaporin-7 in the pathophysiological control of fat accumulation in mice. *FEBS Letters*, *580*(20), 4771–4776. https://doi.org/10.1016/j.febslet.2066.07.080.

Rodríguez, A., Catalán, V., Gómez-Ambrosi, J., García-Navarro, S., Rotellar, F., Valentí, V., et al. (2011). Insulin- and leptin-mediated control of aquaglyceroporins in human adipocytes and hepatocytes is mediated via the PI3K/Akt/mTOR signaling cascade. *Journal of Clinical Endocrinology and Metabolism*, *96*(4), E586–E597. https://doi.org/10.1210/jc.2010-1408.

Rodríguez, A., Marinelli, R. A., Tesse, A., Fruhbeck, G., & Calamita, G. (2015). Sexual dimorphism of adipose and hepatic aquaglyceroporins in health and metabolic disorders. *Frontiers in Endocrinology*, *5*(6), 171. https://doi.org/10.3389/fendo.2015.00171.

Rojek, A., Praetorius, J., Frokiaer, J., Nielsen, S., & Fenton, R. A. (2008). A current view of the mammalian aquaglyceroporins. *Annual Review of Physiology*, *70*, 301–327. https://doi.org/10.1146/annurev.physiol.70.113006.100452.

Staab, C. A., & Maser, E. (2010). 11beta-Hydroxysteroid dehydrogenase type 1 is an important regulator at the interface of obesity and inflammation. *The Journal of Steroid Biochemistry and Molecular Biology*, *119*(1–2), 56–72. https://doi.org/10.1016/j.jsbmb.2009.12.013.

Stewart, P. M., & Petersenn, S. (2009). Rationale for treatment and therapeutic options in Cushing's disease. *Best Practice & Research Clinical Endocrinology & Metabolism*, *23*(Suppl 1), S15–S22. https://doi.org/10.1016/S1521-690X(90)70004-1.

Stimson, R. H., & Walker, B. R. (2007). Glucocorticoids and 11beta-hydroxysteroid dehydrogenase type 1 in obesity and the metabolic syndrome. *Minerva Endocrinologica*, *32*(3), 141–159.

Tamma, G., Carmosino, M., Svelto, M., & Valenti, G. (2005). Bradykinin signaling counteracts cAMP-elicited aquaporin 2 translocation in renal cells. *Journal of the American Society of Nephrology*, *16*, 2881–2889. https://doi.org/10.1681/ASN.2005020190.

Verkman, A. S., Anderson, M. O., & Papadopoulos, M. C. (2014). Aquaporins: Important but elusive drug targets. *Nature Reviews Drug Discovery*, *13*, 259–277. https://doi.org/10.1038/nrd4226.

Viengchareun, S., Penfornis, P., Zennaro, M. C., & Lombès, M. (2001). Mineralocorticoid and glucocorticoid receptors inhibit UCP expression and function in brown adipocytes. *American Journal of Physiology Endocrinology and Metabolism*, *280*(4), E640–E649. https://doi.org/10.1152/ajpendo.2001.280.4.E640.

Villarroya, F., Cereijo, R., Villarroya, J., & Giralt, M. (2017). Brown adipose tissue as a secretory organ. *Nature Reviews Endocrinology*, *13*(1), 26–35. https://doi.org/10.1038/nrendo.2016.136.

Yeh, J. I., Charrier, V., Paulo, J., Hou, L., Darbon, E., Claiborne, A., et al. (2004). Structural of enterococcal glycerol kinase in the absence and presence of glycerol: Correlation of conformation to substrate binding and mechanism of activation by phosphorylation. *Biochemistry*, *43*, 362–373. https://doi.org/10.1021/bi034258o.

Zardoya, R., & Villalba, S. (2001). A phylogenetic framework for the aquaporin family in eukaryotes. *Journal of Molecular Evolution*, *52*, 391–404. https://doi.org/10.1007/s002390010169.

Zhang, Y., Peti-Peterdi, J., Muller, C. E., Carlson, N. G., Baqi, Y., Strasburg, D. L., et al. (2015). P2Y12 receptor localizes in the renal collecting duct and its blockade augments arginine vasopressin action and alleviates nephrogenic diabetes insipidus. *Journal of the American Society of Nephrology*, *26*(12), 2978–2987. https://doi.org/10.1681/ASN.2014010118.

Further reading

Cereijo, R., & Sánchez-Infante, D. (2017). Brown adipose tissue and browning: More than just a heating device. *Endocrinología, Diabetes y Nutrición*, *64*(4), 185–187. https://doi.org/10.1016/j.endinu.2017.03.001.

National Research Council. (2011). *Guide for the care and use of laboratory animals* (8th ed.). Washington, DC: The National Academies Presshttps://doi.org/10.17226/12910.

CHAPTER TEN

Modulation by steroid hormones and other factors on the expression of aquaporin-1 and aquaporin-5

A. Skowronska[a,*], D. Tanski[b], L. Jaskiewicz[a], M.T. Skowronski[c]

[a]Department of Human Physiology, School of Medicine, University of Warmia and Mazury in Olsztyn, Olsztyn, Poland

[b]Department of Anatomy and Animal Physiology, Faculty of Biology and Biotechnology, University of Warmia and Mazury in Olsztyn, Olsztyn, Poland

[c]Veterinary Center, University of Nicolaus Copernicus, Torun, Poland

*Corresponding author: e-mail address: agnieszka.skowronska@uwm.edu.pl

Contents

Abstract

Aquaporins are membrane proteins that play crucial roles in water and small solute transport across epithelial and endothelial barriers. This chapter analyzes the involvement of aquaporin 1 and 5 water channels in reproductive tissues, mainly in uterus. AQP1 and 5 are water selective and therefore are members of the classical AQP family.

Vitamins and Hormones, Volume 112
ISSN 0083-6729
https://doi.org/10.1016/bs.vh.2019.08.006

Particular attention was paid to the actions of steroid hormones, and other factors to exhibit how they control/modulate AQP regulation. Significant progress has been made in understanding the roles of AQPs in reproductive physiological processes describing their specific tissue distribution, and their involvement in normal pregnancy, fetal growth, and homeostasis of amniotic fluid volume. This review provides updated information on the underlying mechanisms of these two isoforms AQP1 and 5 in the regulation of reproductive processes in mammals. On the basis of the progress in aquaporins research, consequently, a variety of autocrine and/or paracrine factors have emerged as important modulators of water transport. Further insight into the mechanisms by which uterine AQPs are regulated will be of utmost clinical importance, since perturbed water flow via uterus or ovary AQPs has been implicated in some reproductive diseases. All these studies contribute to understand the physiology of female reproductive tract and open up new research enable to design of compounds targeting specific aquaporins for treatment of human disease. We conclude this chapter by discussing future directions in aquaporin regulation research.

Abbreviations

AA arachidonic acid
AQPs aquaporins
AR androgen receptor
AVP vasopressin
cAMP cyclic adenosine monophosphate
COX cyclooxygenase
DHT dihydrotestosterone
E_2 estradiol
ER estrogen receptor
ERK extracellular-signal-regulated kinase
IMCD inner medullary collecting ducts
JNK c-Jun N-terminal kinase
MAPK mitogen-activated protein kinase
NKCC1 sodium potassium-chloride cotransporter
OT oxytocin
OTR oxytocin receptor
P_4 progesterone
PCOS polycystic ovary syndrome
PGs prostaglandins
PKA cAMP-dependent protein kinase
PLC phospholipase C
V2 vasopressin receptor

1. Aquaporin family of water-specific channels

The passage of water across cell membranes is accomplished by two distinct routes: simple diffusion through the phospholipid bilayer and rapid transit through members of the aquaporin family. The aquaporins (AQPs) are small hydrophobic integral transmembrane proteins and members of a larger family of major intrinsic proteins (MIP), expressed within key osmo-regulatory tissues, where they allow water to move freely and bidirectionally across the cell membrane and in some cases, small nonionic selected compounds, including glycerol (Agre, 2004; Verkman, 2012). These proteins are of significant importance to cell volume control and transcellular water transport. Water movement via AQP allows a more specific regulation of water distribution than water movement via other pathways. They were first described in red blood cells, in the late 1950 (Paganelli & Solomon, 1957). Then in 1987 Agre, Saboori, Asimis, and Smith (1987) isolated and co-purified the antigen from erythrocytes membrane. Further experiments by Zeidel, Ambudkar, Smith, and Agre (1992) and Preston, Caroll, Guggino, and Agre (1992) unveiled characterization of CHIP28 that was later renamed aquaporin-1 (AQP1). AQPs are composed by around 320 amino acid residues with approximately 28 kDa, architected in membranes as tetramers. Each monomer in the tetramer is a functional unit (Preston et al., 1992; Shi, Skach, & Verkman, 1994) and is formed by six transmembrane domains and behaves as an independent pore (King, Kozono, & Agre, 2004). The tetrameric structure has been proposed to be necessary to stabilize the position of individual monomers (Agre, 2006). AQPs are classified into three subfamilies according to their substrate selectivity and sequences homologies. The subfamilies include: (1) the classical AQPs only permeable to water (AQP0, AQP1, AQP2, AQP4, AQP5, AQP6, and AQP8); (2) the aquaglyceroporins permeable to water as well as to small uncharged molecules, such as glycerol and urea (AQP3, AQP7, AQP9, and AQP10), and (3) unorthodox AQPs, whose permeability still remains to be clearly established (AQP11 and AQP12). They were identified with strategic membrane sites in endothelia and a variety of epithelia, with specific organ, tissue most of which have well-defined physiological functions in fluid absorption or secretion (Brown, Katsura, Kawashima, Verkman, & Sabolic, 1995).

There is now substantial evidence that AQPs play a major role in maintaining water balance in many tissues, involved in transcellular and transepithelial water movement, taking part in many biological functions all over the human body and other mammals. A majority of aquaporins have been shown to be post-translationally regulated, either through altered water transport through individual AQPs or by changing the number of AQP molecules present in the plasma membrane through trafficking (Törnroth-Horsefield, Hedfalk, Fischer, Lindkvidt-Petersson, & Richard,

2010). A representative short-term regulatory mechanism is phosphorylation of serine and threonine residues. Binding of vasopressin to its receptor activates protein kinase A (PKA), which phosphorylates the Ser256 residue of AQP2 (Fushimi, Sasaki, & Marumo, 1997; Katsura, Gustafson, Ausiello, & Brown, 1997). AQP1 and AQP5 have been shown to be regulated by reversible phosphorylation. Reversible protein phosphorylation is maybe the most common mechanism by which the function of the protein can be dynamically regulated, i.e., protein kinase A (PKA), protein kinase C (PKC). In addition, heavy metals (mercury, cooper and nickel) are also involved in short-regulation (Zelenina, Bondar, Zelenin, & Aperia, 2003; Zelenina, Tritto, Bondar, Zelenin, & Aperia, 2004) and affect the activity of AQP expression except for AQP4 (Yasui et al., 1999). Much of the short-term regulation is mediated via hormone receptors inserted in the plasma membrane. Long-term regulation can be divided into regulation during the development stage and regulation through interaction between proteins. Furthermore pH changes affect the conformation of AQPs and modify their permeability (Németh-Cahalan, Kalman, & Hall, 2004).

Recently, attention has been focused on the mechanism that underlies regulation of the activity of AQPs in order to clarify the molecular basis of water movement across biological membranes under both physiological and pathological conditions. Emerging evidence suggests that aquaporins play important roles for the dynamic regulation of uterus, water homeostasis and for the regulation of uterine fluid production. AQPs have been found in various organs of the reproductive system, therefore, this chapter addresses the structural and functional features of aquaporin-1 and aquaporin-5, detailing theirs modulation by steroid hormones and other factors, and will discuss the complex roles and overall functions of these AQPs in the female reproductive tissues. Information about the mechanisms by which AQPs can be regulated will be a prerequisite for future development of novel therapeutic tools for diseases associated with disturbances in water homeostasis in reproductive system.

1.1 Structure of AQP1 water channel

AQP1 was the first discovered member of the AQP family, isolated from human red blood cells (King & Agre, 1996), and was originally named channel-forming integral membrane protein of 28 kDa (CHIP28) (Denker, Smith, Kuhajda, & Agree, 1998). Is organized as a tetramer spanning the plasma membrane, consists of six-tilted α-helical domains

(H1–H6), connected by five loops (loops A–E). The water pore is formed by loops containing the two Asn-Pro-Ala (NPA) motifs located in loops B and E (Raina, Preston, Guggino, & Agre, 1995; Wang et al., 2015). The most constricted region of this pore is approximately 2.8 Å in size, which allows the continuous passage of one water molecule at a time (Agre et al., 2002). AQP1 is expressed strongly throughout microvascular endothelial cells (outside of the brain), including microvessels in kidney, lung and airways, secretory glands, skeletal muscle, pleura and peritoneum, gastrointestinal tract. Its physiological activity is especially pronounced in the kidneys, AQP1 is localized in the plasma membranes of proximal tubules, descending thin limbs of Henle and descending vasa recta (Nielsen, Chou, et al., 1995; Nielsen, Pallone, et al., 1995; Zhai, Fenton, Andreasen, Thomsen, & Christensen, 2007). Subsequent studies have demonstrated that is also expressed in endothelial cells in cornea, trabecular meshwork in canal of Schlem and lacteals of the small intestine (Nielsen, Smith, Christensen, & Agre, 1993). Mobasheri and Marples (2004) using tissue microarray technology confirmed the presence of AQP1 in endothelial barriers of almost all tissues and in many epithelial barriers.

1.2 Regulation of AQP1 activity

Mechanism regulating AQP1 channel activity is proposed to involve; cyclic nucleotide and protein kinase. One study demonstrated through an oocyte cell-swelling assay that cyclic adenosine monophosphate (cAMP)-dependent PKA phosphorylation of AQP1 resulted in increased cell swelling due to the increased abundance of functional AQP1 at the cell surface (Han & Patil, 2000). In addition under hyperosmolar conditions, AQP1 expression was enhanced by the agonist of cAMP, vasopressin through increased translocation of AQP1 to the plasma membrane (Jeng, Matheson, Ihara, & Ramirez, 1998). Analysis of human, rat and mouse acid sequence indicates that AQP1 contains potential phosphorylation sites for PKA, PKC and CKII (Kreegipuu, Blom, & Brunak, 1999). In the line with these results, AQP1 water permeability has been shown to be dynamically regulated by several hormones, i.e., vasopressin (AVP), atrial natriuretic peptide (Patil, Han, & Wax, 1997). Two glucocorticoid response elements have been identified in AQP1 promoter region (Moon, King, & Agre, 1997). Exposure to hypertonicity induced AQP1 expression which is mediated by the activation of MAPK pathways (Umenishi & Schrier, 2002). The same authors revealed that the AQP1 promoter contains a hypertonicity response element (Umenishi & Schrier, 2003). So far, several pharmacological agents have been tested as potential regulators of AQP1, i.e., acetazolamide (Ma et al., 2004), and tetraethylammonium (Brooks, Regan, & Yool, 2000).

1.3 Structure of AQP5 water channel

With regard to protein homology, AQP5 is most closely related to AQP2, 63% identity (Raina et al., 1995) within the aquaporin family and shares the core structural features found in other AQP family members (Horsefield et al., 2008). AQP5 is organized as a tetramer spanning the plasma membrane, composed by four identical monomers, each behaving as a water pore. There are three extracellular loops (A, C and E) and two intracellular loops (B and D). Both loops B and E, which contain the Asp-Pro-Ala (NPA) sequence that is highly conserved for AQPs. AQP5 as well as other AQPs (2, 3 and 4) contain an N-glycosylation site in loop C (Raina et al., 1995). The amino acid sequence contains potential phosphorylation sites for protein kinase (PK) A, and casein kinase II. AQP5 also contains a mercury-sensitive cysteine residue at position 182, which is located just upstream of the NPA sequence in loop E (Krane, Towne, & Menon, 1999; Lee et al., 1996; Raina et al., 1995). Within this pore, water passes at almost the diffusion rate along an osmotic gradient. Phosphorylation sites involved in regulation are commonly found in the N- and the C-termini, which are located in the cytoplasm as well as in the cytoplasmic loops B and D (Nesverova & Törnroth-Horsefield, 2019).

1.4 Regulation of AQP5 water channel activity

AQP5 is similar to AQP2, is subject to posttranslational regulation by phosphorylation, at which point it is trafficked between intracellular storage compartments and the plasma membrane. This aquaporin was fist cloned from rat submandibular gland (Raina et al., 1995) and from the human salivary gland (Lee et al., 1996) by homology cloning, and is widely distributed among the human and mammalian species in the apical plasma membrane. AQP5 is known as an exocrine-type water-channel protein, because of its involvement in the excretion of water from exocrine glands, such as the salivary gland, lacrimal gland, and sweat gland (Raina et al., 1995; Yang, Kawedia, & Menon, 2003). Indeed, AQP5 classically prevents excessive water reabsorption and modulates cell shape and volume (Delporte, Bryla, & Perret, 2016). Furthermore, AQP5 is also found in epithelial cells of the lung and the cornea. Apical-basal polarity appears to be important because AQP5 localizes to the apical surface in most of the tissues. AQP5 was shown to translocate from the cytoplasm to the apical membranes in cells of rat salivary glands by stimulation with M3 muscarinic and α1-adrenergic agonists (Ishikawa, Eguchi, Skowronski, & Ishida, 1998; Ishikawa, Skowronski, Inoue, & Ishida, 1999), and has been implicated in trafficking from intracellular membranes to the apical membrane of epithelial cells after C-terminal modifications (Kosugi-Tanaka et al., 2006; Wellner, Hong, Cotrim, Swaim, & Baum, 2005) and in response to the addition of cAMP (Yang et al., 2003) or cevimeline (Ishikawa et al., 2005). Kitchen et al. (2015) proposed three independent regulated mechanisms: one proposed mechanism is phosphorylation of AQP5 in Ser156 resulting in increased targeting or decreased internalization, second; involvement of PKA in basal recycling of AQP5 independently of Ser156 phosphorylation, third; environmental tonicity has been suggested to regulate changes in AQP5 localization (Direito, Madeira, Brito, & Soveral, 2016). The long-term effect of cAMP and PKA involves direct phosphorylation of AQP5 while short-term effect does not (Sidhaye, Hoffert, & King, 2005). Analysis of AQP5 amino acid sequence identified two activator protein 1 (AP-1) binding sequences (Yao et al., 2010), a number of putative GATA-binding sites, specificity protein 1 (Sp1) consensus site (Borok et al., 2000) and an estrogen response element (Kobayashi, Takahashi, Miyagawa, Watanabe, & Iguchi, 2006). Additionally, hyperosmotic stress-mediated AQP5 expression mediates the expression of inflammatory cytokines and cell death through the JNK1/2 MAPK signaling pathway (Ren, Lu, Reinach, et al., 2017). The authors, confirmed that solute-transporting system NKCC1 is physiologically associated with AQP5. NKCC1 and AQP5 co-localized in the lumen of acinar cells and interacted with each other.

2. Physiological relevance of aquaporins in uterus

The functions of AQPs in the uterus physiology may involve water transfer, as well as absorption and secretion of water and even small solutes through the epithelium and stromal blood vessels. The absorption and secretion of transepithelial fluids and solutes are closely related to uterus physiology by maintaining water balance. This balance of uterus fluid movement has been indicated to be regulated by a complex of factors, including steroid hormones, vascular endothelial growth factors, prostaglandins, cAMP, oxytocin and other factors. These regulations may either affect endometrial growth and proliferation or the activation or inhibition of AQPs expression and distribution, which may account for the changes of water transport in the reproductive tract. Furthermore in the preimplantation period, these molecules could help to shuttle water from the myometrial layer, providing for the rapid transfer to the lumen through the stroma, thus resulting the stromal imbibition. The uterus is an essential organ for reproduction, it ensures proper environment for survival and development of the conceptus. Furthermore, precise regulation of uterine fluid volume is crucial for successful reproduction.

Adult human and primate endometria are divided into two functional layers: the upper stratum functional is containing glands surrounded by loose stroma and the lower stratum basal is consisting of branched coiled glands and dense stroma. Histologically, the endometrium consists of two epithelial cell types: luminal epithelium and glandular epithelium. In the endometrium, hydrostatic and osmotic forces regulate water balance across membrane barriers via specific ion pumps, ion channels and solute transport proteins (Zeuthen, 1995). The uterine lumen contains electrolytes/water-based fluid, the volume of which is known to fluctuate throughout the menstrual cycle under the influence of ovarian hormones (Casslen, 1998; Chen et al., 2013; Naftalin, Thiagarajah, Pedley, Pocock, & Milligan, 2002). The function of AQPs in fluid homeostasis in reproductive system has long been suggested, and the presence of aquaporins in the uterus is now well documented by molecular and pharmacological methods in the different species (Zhang, Tan, Qu, Sheng, & Huang, 2012; Zhu et al., 2015). Moreover, their specific expression pattern suggests that they play a role in water movement between the intraluminal, interstitial and capillary compartments (Huang et al., 2006). At least 11 members of the aquaporin protein family are expressed in the reproductive system, where they are distributed in a cell-specific manner (Hua et al., 2013; Huang et al., 2006; Zhang et al., 2012; Zhu et al., 2015). The first reports of the presence of AQPs in the female reproductive system occurred in the human uterus, after cDNA isolation and sequencing (Li, Yu, & Koide, 1994). Over the last 2 decades, a number of AQPs have been found to be expressed in the endometrium of different species, including the mouse, rat, porcine and human. Efforts have been made to investigate the involvement of various endometrial AQPs in different reproductive events, including embryo implantation. This chapter analyses the involvement of aquaporin 1 and 5 water channels in reproductive tissues, mainly uterus. Particular attention was paid to the actions of steroid hormones (P_4, E_2), and other factors to unwell how they control/modulate AQPs regulation (Table 1).

Table 1 The studies of expression, of AQP1 and 5 uterus with steroid hormones, and other factors.

Tissue	Subtypes	Sites of expression	Species	Methods adopted for the AQPs determination	Factors	Authors and references
Uterus	AQP1	Endometrial vessels capillaries during normal cycle	Human	mRNA expression	Mefiperistone	Hildenbrand et al. (2008)
		Endothelial cells (lining the interior surface of blood vessels) in the time implantation		IHC, RT-PCR	Mefiperistone	Zhou et al. (2013)
		Endometrial blood vessels in menorrhage		RT-PCR	–	Mints et al. (2007)
		Inner circular layer of myometrium during peri-implantation period	Mouse	IHC, RT-PCR, in situ hybridization	P_4	Richard, Gao, Brown, and Reese (2003)
		Longitudinal and circulating smooth muscle cells of the myometrium ovariectomized animals		IHC, WB, swelling assay	E_2	Jablonski, McConnell, Hughes Jr., and Huet-Hudson (2003)
		Uterus	Rat	RT-PCR	E_2, Anordiol	Li, Yu, and Koide (1997)
		Inner circular layer of myometrium during early pregnancy		IHC	–	Lindsay and Murphy (2004)
		Endometrial blood vessels ovariectomized animals		IHC	E_2, P_4	Lindsay and Murphy (2006)
		Endothelial cells of the uterine blood vessels during 17–19 days of the estrous cycle	Pig	IHC, WB	–	Skowronski, Kwon, and Nielsen (2009)

Continued

Table 1 The studies of expression, of AQP1 and 5 uterus with steroid hormones, and other factors.—Cont'd

Tissue	Subtypes	Sites of expression	Species	Methods adopted for the AQPs determination	Factors	Authors and references
		Endometrium and myometrium of blood vessels, during different stages of the cycle and early pregnancy—protein		IHC, WB, RT-PCR	–	Skowronski (2010)
		Apical and basal plasma membranes during: mid-luteal phase of the estrous cycle and luteolysis, implantation, early pregnancy, on uterine explants, and separated endometrium and myometrium at the time mid-luteal phase of the estrous cycle and luteolysis		WB, mRNA expression	E_2, P_4, OT, AA, FSK, cAMP	Skowronska, Mlotkowska, Wojciechowicz, et al. (2015) and Skowronska, Mlotkowska, Nielsen, and Skowronski (2015)
		Venous and arterial vessels and in capillaries during phases of the estrous cycle	Equine	IHC, WB, mRNA expression		Skowronska, Mlotkowska, Majewski, Nielsen, and Skowronski (2016) and Skowronska, Mlotkowska, Okrasa, Nielsen, and Skowronski (2016)
		Mesometrial, myometrium and endometrial blood vessels and in the circular and longitudinal layers of myometrium, during phases of estrous cycle	Bitch	IHC, WB, mRNA expression	–	Klein, Troedsson, and Rutllant (2013)

AQP5	Glandular epithelium in the peri-implantation uterus	Mouse	IHC, mRNA expression, in situ hybridization	–	Richard et al. (2003)
	Apical plasma membrane of uterine epithelial cells on ovariectomized and pregnant animals	Rat	IHC	E_2, P_4	Lindsay and Murphy (2006)
			RT-PCR		Lindsay and Murphy (2007)
	Myometrium of non-pregnant and late-pregnancy		RT-PCR	OT	Ducza, Seres, Hajagos-Thot, Falkay, and Gaspar (2014)
	Smooth muscle cells and luminal and glandular epithelial cells	Pig	IHC, WB, RT-PCR	–	Skowronski (2010)
	Uterine epithelial cells and smooth muscle cells, labelling was in the apical (control) and in basolateral plasma membranes of the epithelial cells after treatment with E_2, P_4, FSK, AA and cAMP during mid-luteal phase of the estrous cycle and luteolysis, implantation and placentation		WB, mRNA expression	E_2, P_4, OT, AA, FSK, cAMP	Skowronska, Mlotkowska, Wojciechowicz, et al. (2015) and Skowronska, Mlotkowska, Nielsen, et al. (2015)
			IHC, WB, mRNA expression		Skowronska, Mlotkowska, Majewski, et al. (2016) and Skowronska, Mlotkowska, Okrasa, et al. (2016)

Continued

Table 1 The studies of expression, of AQP1 and 5 uterus with steroid hormones, and other factors.—Cont'd

Tissue	Subtypes	Sites of expression	Species	Methods adopted for the AQPs determination	Factors	Authors and references
		Separated endometrium and myometrium at the time mid-luteal phase of the estrous cycle and luteolysis		WB, mRNA expression	E_2, P_4, OT, AA, FSK, cAMP	Skowronska, Mlotkowska, Wojciechowicz, et al. (2015) and Skowronska, Mlotkowska, Nielsen, et al (2015)
		Luminal and glandular epithelial and stromal cells during estrous cycle and early pregnancy	Equine	IHC, WB, mRNA expression	E_2, P_4	Klein et al. (2013)
		Apical plasma membrane of uterine epithelial cells, during estrous phases	Bitch	IHC, WB	–	Aralla et al. (2009)
		Effect of antibiotics on AQP5 and uterine contractility on 22-day pregnant	Rat	WB, RT-PCR, contractility studies	Amoxicillin, Fosfomycin, Doxycycline	Csányi, Hajagos-Tóth, Kothencz, Gaspar, and Ducza (2018)
		Effect of estrogen and gestagen compounds on the expression of the AQP5 in the late pregnancy		WB, RT-PCR	E_2, clomiphene citrate, tamoxifen citrate, P_4, levonorgestrel, medroxyprogesterone acetate	Csányi, Bota, Falkay, Gaspar, and Ducza (2016)

It is known that AQP1 and AQP5 are present in structures of the female reproductive system (Zhu et al., 2015). AQP1 was the first aquaporin to be identified in a human uterine cDNA library by a one-step polymerase chain reaction (PCR) (Li et al., 1994), subsequently, the presence of AQP1 m RNA was observed in rat uterus (Li et al., 1997) and then was immunolocalized in the myometrium (Jablonski et al., 2003; Richard et al., 2003), uterine vascular smooth muscle cells (Shanahan et al., 1999) and plasma membrane of smooth muscle cells within the inner circular layer (Lindsay & Murphy, 2004) in rodents. The presence of AQP1 in the endometrium has been shown in the uterus of both humans and rats (Lindsay & Murphy, 2004; Mints et al., 2007). The presence of AQP1 in human endometrial blood vessels indicates an involvement in the regulation of edema, and angiogenesis (Hildenbrand et al., 2008). AQP5 is mainly located in the stromal and glandular epithelial cells of the endometrium. The expression of human endometrial AQP5 is menstrual cycle dependent. High levels of AQP5 are found at the proliferative and mid-secretory phases, suggesting that estrogen may regulate the expression of AQP5 in endometrial glandular cells. It is clear that structural features of the AQPs family and the mechanism of selective water passage through the AQP pore are now established (Murata et al., 2000; Törnroth-Horsefield et al., 2010), however less information is available describing how AQPs, regulate cellular water flow.

3. Hormonal regulation of AQP expression

3.1 Progesterone and estradiol

It is well known in many organs that tight control of water homeostasis requires differential expression of water channels in particular by hormones. The development, proliferation and differentiation of the uterus involve the concerted actions of hormones and wide range of protein secretions, including enzymes, transport proteins, growth factors, and cytokines. In addition to these regulatory endocrine factors, normal uterus development requires cellular communication and cell-cell interactions between the stromal, glandular epithelial and myometrial elements of the uterus. The cycling endometrium undergoes recurrent uterine stromal edema in response to hormonal stimuli, and at the time of insemination manifests an inflammatory-like response that is marked by generalized edema and hyperemia. The sex steroids and their cognate receptors dictate the menstrual cycle in mammals and are essential to the establishment and maintenance of pregnancy. Sex steroid receptors such as AR, ERα and ERβ are expressed in many tissues/cells and regulate target gene expression upon ligand binding. ERα is expressed at high levels in the uterus. It has been well established that steroid hormones are synthesized from cholesterol via multiple enzymatic reactions.

AQPs have been shown to either be up or downregulated by sex-steroids namely estrogen, progesterone (Chinigarzadeh, Muniandy, & Salleh, 2016) and testosterone (Salleh, Mokhtar, Kassim, & Giribabu, 2015). It is evident that dramatic changes in the uterus include hyperemia, increased capillary permeability, and edema are seen in response to ovarian steroid hormone, especially estradiol (E_2) (Huang et al., 2006). Estrogen induces fenestration of uterine capillary vessels, and gaps between endothelial cells, contribute to increase vascular permeability, causes secretion of sodium, potassium and water into the lumen of the uterus. Another of the prominent effects of estrogen is water imbibition in the uterine endometrium, whereas minimal edema is seen in the myometrium. This fluid is essential for multiple processes of reproduction including sperm transport, capacitation, fertilization, embryo transport and implantation. In turn, progesterone, in early pregnancy, it is responsible for preparation of endometrium for implantation process, causes reabsorption of those substances, for example to initiate the attachment phase of embryo implantation, and maintenance of gestational sac in uterus, also by modulation of maternal immune system. Undoubtedly, accurate regulation of uterine fluid volume is crucial for successful reproductive processes (Lindsay & Murphy, 2004; Salleh, Baines, Naftalin, & Milligan, 2005), therefore any variation from the normal volume could affect the fertility.

Clearly, the effect of sex hormones on water transport into and within the uterus in mammals is known. An important question yet to be answered about the mechanism, responsible for balancing water concentration within the uterus. Molecular studies have gradually elucidated the physiological functions of aquaporins and their contribution to the mechanism responsible for balancing water concentration within the uterus. The first mRNA and protein study has subsequently demonstrated that E_2 regulate AQP1 in the human and rodents' uterus (Jablonski et al., 2003; Li et al., 1994, 1997; Richard et al., 2003). AQP1 expression to the stromal vasculature in response to estrogen suggests a role for AQP1 in uterine vascular permeability. In the uterine luminal epithelial cells AQP5 mRNA was induced in estrogen-treated, progesterone-primed uteri of mice (Richard et al., 2003), but was absent in the control or estrogen treated mouse uterus (Jablonski et al., 2003). Another interesting study using light and electron microscopy have demonstrated hormonal regulation of AQP1 and 5 protein expression in the ovariectomized and hormone treated rat model (Lindsay & Murphy, 2006). Ovarian hormones, are involved in the coordinated redistribution/expression of AQP5 in the apical plasma membrane of the luminal epithelial cells, and an increase in AQP1 in the inner circular layer of myometrium (Lindsay & Murphy, 2006), strongly suggesting mechanism for the reabsorption of luminal fluid in the progesterone treated rat uterus via the transcellular pathway involving apical AQP5 channels. AQP5 is also expressed in a distinct uterine cell type during the peri-implantation period, when the organ displays edema and hyperemia (Richard et al., 2003). It is well known that progesterone has a crucial role in implantation, and this sex hormone upregulates the AQP1 expression in the inner circular layer of myometrium in ovariectomized rats (Lindsay & Murphy, 2006).

Work using mouse tissue quantitative PCR (ChIP) and DNA microarray has confirmed that AQP5 is activated soon after administration of estrogen in uterus (Kobayashi et al., 2006). These findings ultimately led to the identification of the promoter region of AQP5 contained a functional estrogen response element that was activated directly by estrogen. Taken together, these observations clearly indicated that upon ligand binding, the estrogen receptor could recognize namely estrogen response element in the promoter of AQP5 through their DNA-binding domain. Taking advantage of this information Jiang et al. (2015) identified an ERE-like motif in the promoter region of AQP5 gene, they also observed that increased AQP5 protein expression by E_2 might be a downstream signal for AKT phosphorylation signaling pathways that promote cell proliferation and invasion. Very recently, studies with mice lacking AQP5 have shown that this aquaporin is responsible for E_2 induced abnormal luminal fluid accumulation during implantation (Zhang et al., 2015).

In turn, the presence of AQP1 in blood vessels of the endometrial stroma provides a mechanism for the formation of stromal edema in response to hormonal regimes. Moreover, studies by Huang et al. (2006) clearly indicated progesterone effect in the development of stromal edema at the implantation site. In agreement with these results, the inhibition of the endogenous progesterone after mifepristone treatment (an anti-progestogenic steroid) leads to increase in AQP1 expression suggests that progesterone is connected to AQP1 regulation in the human endometrium (Hildenbrand et al., 2008). The human endometrium and myometrium are both clearly estrogen-responsive tissues. Some result potentially conflict with the findings of some other studies (Jablonski et al., 2003; Li et al., 1997) however, the above series of studies draw a new extensive map for mechanism underlying estrogen-stimulated water transport by aquaporins.

One can assume that during water imbibition into the uterine tissue from the capillaries to the spaces surrounding the stromal cells, in response to E_2, AQPs exerts their function of limiting imbibed water to the stromal layer therefore, protecting the myometrial layer from edema and mediate movement of water into the luminal cavity of the uterus. As expected, presence of tight junctions between the epithelial cells allows water cross this layer through transcellular pathway means through AQP channels, rather than through pericellular routes. Thus these first studies were important in developing the concept of aquaporin regulation by steroids.

Abundant experiments evidence substantiates that AQP5 may play a role by activating ERK or P38/MAPK signaling pathways (Jiang et al., 2015; Sakamoto, Hisatsune, Katsuki, Horie, & Isohama, 2014; Shi et al., 2014). On the one hand, E_2 combined AQP5 directly, on the other hand, E_2 regulates AQP5 by activating AKT pathway. Recent findings confirmed that estrogen enhances expression of AQP1 in the uterine luminal and glandular epithelia that helps in uterine fluid secretion, and AQP1 in stromal blood vessels, which support the role of this isoform in uterine fluid imbibition. Additionally, phyto-estrogen-like compounds including genistein (Chinigarzadeh, Muniandy, & Salleh, 2017) and quercetin could also affect AQP expression (Shahzad et al., 2017).

In vitro studies have extended these results and showed that estradiol and progesterone potentially mediate AQP1 and AQP5 expression in uterine tissue (Fig. 1). Interestingly, E_2 and P_4 caused an emergence of AQP5 from apical to the basolateral plasma membrane of the epithelial cells, which might affect the transcellular water movement (through epithelial cells) between uterine lumen and blood vessels (Skowronska, Mlotkowska, Wojciechowicz, et al., 2015). Furthermore, other studies conducted with separated porcine uterine tissues (*endo-* and myometrium) have confirmed the participation of steroid hormones (E_2 and P_4) in the regulation of AQP1 and 5 (Skowronska, Mlotkowska, Nielsen, et al., 2015). The endometrium and myometrium of porcine uterine tissues exhibit their own profiles of AQPs expression in respond to steroid hormones. Consequently, the action of steroid hormones within the uterus during the estrous cycle also includes their influence on the expression of AQPs, which reacted with quantitative and qualitative changes related to the different processes, such as increasing secretory activity and promoting angiogenesis. Consistent with this observation, during early pregnancy (implantation and placentation)

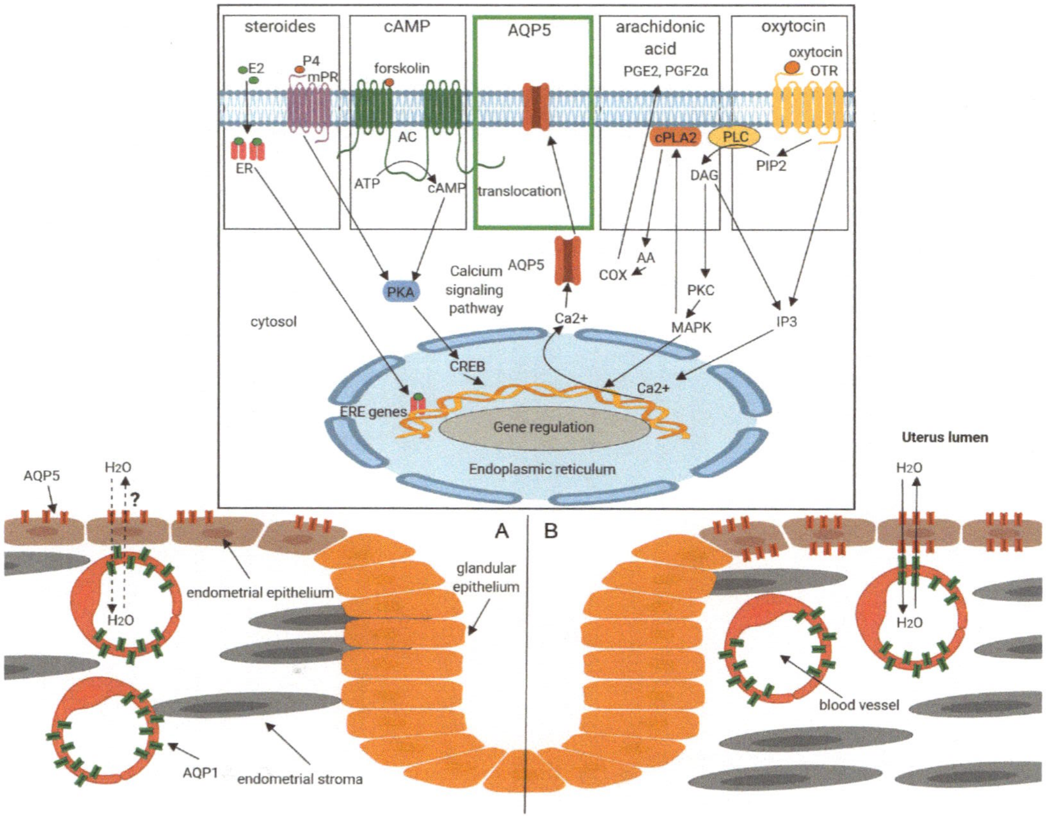

Fig. 1 See figure legend on next page.

in vitro treatment of uterine explants with steroid hormones has caused intense accumulation of AQP5 in the basolateral membrane of the epithelial cells, indicates that this aquaporin is markedly engaged in water efflux and influx within the uterus, and may be important to general fluid homeostasis following implantation and placentation (Skowronska, Mlotkowska, Majewski, et al., 2016; Skowronska, Mlotkowska, Okrasa, et al., 2016). However, the expression patterns of genes encoding these AQPs in response to P_4 and E_2 during implantation and placentation period differ, taking this distinct expression patterns, one could assume that steroids produced by uterus or estrogens secreted by developing placenta may additionally modify the uterine AQPs.

Notably, other studies have shown the effect of estrogen- and gestagen-related compounds on the expression of the AQP5 channel in the late-pregnant rat uterus (Csányi et al., 2016). Thus, for example, AQP5 is highly expressed after progesterone and progesterone analogs treatment in the late-pregnant rat uterus. In addition, estrogen had a significant effect on AQP5 protein expression in the late-pregnant rat uterus. Hormonally-induced model for preterm birth used in this study proved the progesterone effect on AQP5 expression. Taken together, these findings indicate that AQP5 expression is influenced by both female hormones with progesterone predominance in the late-pregnant rat uterus.

3.2 Androgens

Although androgens are typically considered the male hormones, they also play important physiologic roles in females. Active androgens are largely

Fig 1 Hypothetical mechanism of transcellular water movement between the uterine lumen and blood vessels in pigs. Distribution of AQP1 and AQP5 in the pig uterus during the estrous cycle and early pregnancy: (A) under control condition—AQP1 was associated with the apical and basal membranes of endothelial cells, but AQP5 was mainly localized in the apical membranes of uterine epithelial cells and (B) after treatment with P4, E2, AA, cAMP or FSK—the AQP1 localization did not change, but AQP5 labeling additionally appeared in the basolateral membranes of epithelial cells. *AA*, arachidonic acid; *AC*, adenylyl cyclase; *AQP1*, aquaporin 1; *AQP5*, aquaporin 5; *ATP*, adenosine triphosphate; Ca^{2+}, calcium; *cAMP*, cyclic adenosine monophosphate; *COX*, cyclooxygenase; *cPLA2*, Phospholipase A2; *CREB*, Camp response element -binding protein; *DAG*, diacylglycerol; E_2, estradiol; *ER*, estrogen receptor; *ERE*, genes-targeting estrogen responsive elements; H_2O, water; *IP3*, inositol triphosphate; *MAPK*, mitogen-activated protein kinases; *mPR*, membrane progesterone receptor; *OTR*, oxytocin receptor; P_4, progesterone; *PGE2*, prostaglandin E2; *PGF2α*, prostaglandin F2α; *PIP2*, phosphatidylinositol 4,5-bisphosphate; *PKC*, protein kinase C; *PLC*, phosphoinositide phospholipase C. See details in the text.

created from circulating precursors in their target tissues, where they both act and are metabolized. Androgens exert their effects through binding to the androgen receptor (AR), an intracellular key regulatory protein that is necessary for androgenic activity. Androgen precursors present in females include: DHEA sulfate (DHEAS), produced by the zona reticularis of the adrenal glands; DHEA, produced by the zona reticularis, ovarian theca cells, and peripherally from circulating DHEAS; Androstenedione, produced by the zona fasiculata of the adrenal glands, the ovarian stroma, and peripherally from circulating DHEA. Because dihydrotestosterone (DHT) has higher binding affinity for the AR than testosterone, it is the more active androgen. The effects of testosterone are mediated either via the intracellular AR or via the estrogen receptor after the conversion of testosterone to 17β-estradiol.

The data about the presence of androgen-response element in the promoter of AQP5 gene was determined (Moehren, Denayer, Podvinec, Verrijdt, & Claessens, 2008). Analyses, considering the possible mechanism underlying testosterone effects have revealed markedly increase AQP1 and 5 expression in the uteri of ovariectomized rats (Salleh et al., 2015). Testosterone is a predominantly male sex hormone which has been shown to affect several female reproductive processes. Plasma testosterone level in females immediately rises after ovulation and this hormone plays an important role in decidualization. High plasma testosterone in females could interfere with uterine fluid secretions under estrogen influence. Interestingly, under testosterone influence, AQP5 was present to the apical and basolateral membranes of luminal epithelium, it might suggest that AQP5 participate in fluid reabsorption from the uterine lumen. These findings also demonstrated that AQP1 was markedly elevated in the inner stroma and myometrium under influence of estrogen, but not AQP5. One may assume that under testosterone influence, AQP1 might participate in uterine fluid reabsorption resulting in luminal fluid loss (Salleh et al., 2015). Therefore, it was suggested that testosterone effect on uterine AQP expression involved the androgen-receptor-mediated pathway, did not involve the testosterone active metabolites 5a-DHT. Similar effect was reported in rats uteri, which was responsive only toward testosterone and not 5a-DHT via transmembrane regulator (CFTR) (Mohd Mokhtar, Giribabu, Kassim, Muniandy, & Salleh, 2014). Taken together testosterone was able to induce increased in electrolyte secretion into the uterine lumen and might also induce solutes secretion that causes increased in osmolality of uterus. Further, in the kidney, solute secretion into the tubular fluid also increased under testosterone influence (Toot et al., 2008). Interestingly, high level of testosterone which occurs in disease such as polycystic ovaries (PCOS)

(Wawrzkiewicz-Jalowiecka, Kowalczyk, Pluta, Blukacz, & Madej, 2017) could interfere with uterine fluid, therefore these findings are of clinical importance. It is therefore possible that under normal physiological conditions slightly elevated testosterone level following ovulation thereby regulate the precise uterine fluid volume that is needed for embryo implantation.

3.3 Oxytocin

The oxytocin (OT) is released from the neurohypophysis and physiologically plays an important role in myometrial contractility, it is also important regulator of the transition from uterine quiescence to contractility. Oxytocin mediate its biological effect by acting on specific oxytocin receptor (OTR) through G protein-coupled receptor that signals through phospholipase C (PLC). The activated OTR triggers a series of mechanisms resulting in increases myometrial contractions and frequency and increases force by sensitizing the contractile apparatus of the myocytes to calcium (Blanks, Shmygol, & Thornton, 2007). OTR is present in most reproductive organs, and its expression is sometimes regulated by estrogens. OT receptors are also expressed in the brain, the kidney, the heart, bone and in endothelial cells (Gimpl & Fahrenholz, 2001). Oxytocin is also known to possess antidiuretic properties and the use of OT to induce labor in pregnancy has been associated with water retention and hyponatremia (Li et al., 2008). It is known that vasopressin which regulates AQP2 expression in the kidney, differs in sequence from OT by only two amino acids. OT stimulates its own receptor, but also cross-reacts with the vasopressin V1a receptor (Akerlund et al., 1999). Therefore, in vitro observations using OT has been shown to increase osmotic water transport in microdissected renal inner medullary collecting ducts (IMCD) (Chou, DiGiovanni, Luther, & Knepper, 1995). Another experiment but conducted in vivo reveals that OT cause an antidiuresis in vasopressin-deficient Brattleboro rats (Lyness, Robinson, Sheridan, & Gash, 1985; Pouzet et al., 2001). In the line with this prediction, the above effects were reversed by a vasopressin V2 receptor antagonist, suggesting that OT stimulation of vasopressin receptors mediates the antidiuresis. This possibility was supported by another experiments conducted by Li and co-workers (2008) indicting that the antidiuresis of OT is associated with molecular events, including increased AQP2 expression and trafficking, and increased AQP3 expression, which are primarily mediated via the V2 receptor on the basolateral membrane of the collecting duct.

The fact that OT is one of the main hormones responsible for pregnant uterine contractions, made an another question to investigate whether or not OT has any effect on the expression of AQPs in late-pregnancy uteri (Ducza et al., 2014). To assess specificity of action, the authors also evaluated how vasopressin, which regulates AQP2 expression in the kidney affects uterine AQP expression. They determined the effect of oxytocin on the myometrial expression of AQP5 and proved that oxytocin selectively decreases the expression of AQP5 and strongly suggest that the reduction of the oxytocin-regulated AQP5 protein may lead to the initiation of uterine contractions. The same group few years later investigated the effect of antibiotics on AQP5 and uterine contractility on the day of birth in rats (Csányi et al., 2018). They have found further evidence for the crosstalk between AQP5 and oxytocin in the regulation of uterine contraction. More specifically, the changes in the cell volume by AQP5 were induced by antibiotics that influence the intracellular calcium concentration. However, the exact mechanism of this phenomenon is still not clear. With regard to the pharmacological effect, amoxicillin and fosfomycin might sensitize the uterus to oxytocin via the reduction of AQP5 expression and lead to the initiation of uterine contractions. Thus, this findings are of clinical importance during the pharmacotherapy of infection-related preterm birth.

Our group investigated in vitro the effect of oxytocin on AQP1 and AQP5 expression with the uterine explants derived from cyclic and pregnant gilts (Skowronska, Mlotkowska, Majewski, et al., 2016; Skowronska, Mlotkowska, Nielsen, et al., 2015; Skowronska, Mlotkowska, Okrasa, et al., 2016; Skowronska, Mlotkowska, Wojciechowicz, et al., 2015). The main reason for investigating the effect of oxytocin on AQP1 and AQP5 expression was that pig endometrium secretes oxytocin, and its responsible for pulsatile secretion of PGF_{2alpha} and contractions of myometrium during luteolytic period. We observed decreased AQP1 and AQP5 mRNA expression (after 3- and 24-h incubations), without visible changes in protein content. It can be suggested that in the porcine uterus, under physiological conditions, inhibitory action of OT on AQP1 and 5 uterine expression at a transcriptional level is connected with remodeling of endometrial tissues taking place at the end of the luteal phase. Furthermore, during placentation we observed the up-regulation of uterine AQP5 gene expression after 3-h incubation with OT, and similarly to implantation OT affected neither AQP1 nor AQP5 protein expression.

4. The effect of forskolin (AC activator) and cAMP in regulation on aquaporin expression

The second messenger, cyclic adenosine monophosphate (cAMP) is a ubiquitous molecule that is generated from adenosine triphosphate by adenylate cyclase, and plays a crucial role in many biological processes (Astakhova, Kapitskiĭ, Govardovskiĭ, & Firsov, 2012). It is well established that protein kinase A (PKA) is required for cAMP to stimulate transcription. Cyclic nucleotide and protein kinase pathways are two regulatory mechanisms currently proposed to be involved in the activation of AQPs. Exposure of cultured human amnion-derived WISH cells to monobutyryl cyclic adenosine monophosphate or the forskolin (activator of adenylate cyclase, which catalyzes cellular production of cAMP) resulted in a significant increase in AQP8 mRNA level (Wang, Chen, Au, & Ross, 2003). In primary human amnion epithelial cell culture from amnion of normal-term pregnancy AQP1 mRNA expression was also shown to be upregulated by forskolin, an adenylate cyclase activator (Wang, Amidi, Beall, Gui, & Ross, 2006). SP-cAMP, membrane-permeable and phosphodiesterase-resistant cAMP analog, which activates the cellular protein kinase A (PKA), did not have such an effect on AQP1mRNA expression, suggest that cAMP upregulates AQP1 via the protein kinase A independent pathway (Wang, Amidi, Sheng, Beall, & Ross, 2007). Woo, Chae, et al. (2008) reported that PKA may participate in regulating the expression of AQP5 in cultivated human bronchial epithelial cells via the phosphorylation of serine 156. Additionally, S156, was shown to be preferentially phosphorylated in tumor cells (Woo, Lee, et al., 2008). Increasing concentrations of cAMP have a stimulatory effect on AQP1mRNA and protein in trophoblasts cell lines (Belkacemi, Beall, Magee, Pourtemour, & Ross, 2008). Furthermore, in line with authors predictions, obtained data provide strong evidence that arginine vasopressin (AVP) at the physiological concentration may play a role in water balance in placental trophoblasts cells by acting through a cAMP signaling pathway to increase AQP1 expression. However, it remains to be elucidated the role of PKA phosphorylation in AQP1 regulation, to identify the specific residue at which this phosphorylation occurs.

The possible role of cAMP through the PKA pathway in the regulation of AQP5 (i.e., expression, intracellular trafficking and localization in the epithelial cells) has been investigated (Hasegawa et al., 2011; Sidhaye et al., 2005; Wang & Zheng, 2011; Yang et al., 2003). Thus, AQP5 has several potential phosphorylation sites. Two consensus PKA/PKG-activated phosphorylation motifs are present at residues RRTSP at 153–157 and RKKT at 256–259, in loop D and at the COOH-terminal domain, respectively. Interestingly, cAMP-induced biphasic effect on AQP5 expression and membrane localization when cells were exposed to cAMP short-term exposure (minutes) to cAMP results in AQP5 internalization from the plasma membrane while long-term (8-h) exposure resulted in increased abundance and an increase in trafficking of AQP5 to the plasma membrane in mouse lung epithelial cells. Both are dependent on PKA, but where only the long-term effect involves direct AQP5 phosphorylation (Sidhaye et al., 2005). In turn, Hasegawa et al. (2011) reported that AQP5 is phosphorylated at Thr259 by PKA through cAMP. Moreover, in vitro and ex-vivo experiments using MDCK (Madin-Darby Canine Kidney) cells, and ex vivo using isolated corneas from wild-type mice revealed PKA-induced AQP5 internalization (Kumari, Varadaraj, Yerramilli, Menon, & Varadaraj, 2012). When discussing the regulation pathway proposed for AQP5, it would be important to note that Kitchen et al. (2015) proposed three independent mechanisms: phosphorylation of AQP5 in Ser156 resulting in either increased targeting or decreased internalization or both; involvement of PKA; and environmental tonicity.

We observed the stimulatory effects of forskolin and sp-cAMP analog on the expression of AQP1/AQP5 mRNAs and proteins in the endometrium and myometrium clearly dominated during luteolysis. In contrast, results concerning the expression of AQPs during the mid-luteal phase of the cycle were differentiated, e.g., in the endometrium there was no response or inhibition at the mRNA level, but no response or stimulation at the protein level. Our data indicate the participation of the adenylate cyclase/cAMP pathway in the intracellular transduction of signals connected with the regulation of AQP1 and AQP5 expression in the porcine uterine tissues, which seems to be augmented during luteolysis (Skowronska, Mlotkowska, Wojciechowicz, et al., 2015). The above regulations seem to be important for cyclic changes connected with the preparation of an intrauterine environment for proper development of embryos/fetuses and successful reproduction. Moreover, other in vitro studies (Skowronska, Mlotkowska, Majewski, et al., 2016; Skowronska, Mlotkowska, Okrasa, et al., 2016) have shown that mainly AQP5 was submitted to regulatory effect of cAMP and forskolin, since these compounds markedly increased AQP5 mRNA during 3-h exposure and protein abundance after 24-h incubation. This indicates that elevated AQP protein abundance most likely resulted from increased number of its transcripts, which was accompanied by subsequent decrease in AQP5 mRNA level (24h). Similar relationship was found by other researchers (Sidhaye et al., 2005; Yang et al., 2003). In contrast, during implantation, cAMP and FSK increased both AQP1 mRNA and protein expression (Skowronska, Mlotkowska, Majewski, et al., 2016). Altogether, these in vitro studies indicate that participation of cAMP in the regulation of AQP1 expression in the uterine tissues is modified depending on the stage of gestation. It is worth mentioning that, cAMP and forskolin have caused a prominent accumulation of AQP5 in the basolateral membrane of the uterine epithelial cells. This finding raised the possibility that this aquaporin is markedly engaged in water efflux and influx within uterus. It is therefore likely that these changes seem to be connected with high secretory activity of endometrial glands during early pregnancy, which release multiple substances that promote uterine receptivity and embryonic development.

5. Prostaglandins

On the basis of the progress in aquaporins research, consequently, a variety of autocrine and/or paracrine factors have emerged as important modulators of water transport including prostaglandins. Prostaglandins belong to a group of prostanoids, derived from arachidonic acid (AA), which is converted to PGH_2 by cyclooxygenase-1 and -2, and then to various forms of PGs. The association between PGs and AQPs expression has been suggested by several researchers. The effect of PGE_2 on renal water absorption has been extensively investigated. Most of studies pointed to a significant effect of PGE_2 on collecting duct water transport (Sakairi, Jacobson, Noland, & Breyer, 1995; Zelenina et al., 2000). Prostaglandins have direct actions on tubular sodium, chloride, and water transport in a number of nephron segments (Li, Wei, Zheng, Guan, & Zhang, 2017). Under particular physiological conditions, PGE_2 induces water excretion/reabsorption (Olesen & Fenton, 2013). It has been reported that: PGE_2 interacts with AVP and regulates distribution of AQP2 in the rat medulla (Zelenina et al., 2000), and that mice lacking cPLA2, the initial enzyme in prostaglandin synthesis, have abnormal AQP1 expression (Downey et al., 2001) and ureteral obstruction stimulates renal COX-2 expression and lowers AQP2 and AQP3 abundance. Moreover, the authors present strong evidence that cyclooxygenase (COX)-PG pathway may contribute to renal functional changes (Nilsson et al., 2012).

Prostaglandins (PGs), especially PGE_2 and $PGF_{2\alpha}$, are implicated in the regulation of reproductive processes at the utero-ovarian level, including the crucial role of PGE_2 in maintenance of pregnancy and conceptus development. Both prostaglandins are synthesized in the porcine uterus mainly by endometrial tissue. The developmental connection between aquaporins and prostaglandins, in particular effect of PGE_2, $PGF_{2\alpha}$, on water permeability in uterus still remain to be examine. Interestingly, the prostaglandin E_2 receptor EP_3 has the same distribution as AQP1 in the uterine myometrium (Yang et al., 1997), suggesting that AQP and prostaglandin signaling pathways may interact to produce uterine edema before the onset of implantation in mouse (Richard et al., 2003). It has been discussed that prostaglandins produced by cyclooxygenase (COX) are implicated in the production of peri-implantation edema because they are inflammatory in nature, and they are increased at implantation sites (Richard et al., 2003). Implication of the COX-1 and COX-2 isoforms has also been investigated in mice knockouts. COX-1 isoform, which is normally expressed in the uterine luminal epithelium before attachment, whereas the inducible COX-2 isoform is expressed at the time of implantation but restricted to the implantation site (Chakraborty, Das, Wang, & Dey, 1996). Because COX-1- and COX-2-deficient mice have reduced uterine weight and vascular permeability (Lim et al., 1997; Reese, Brown, Paria, Morrow, & Dey, 1999), prostaglandin signaling seems to be required for the implantation process. Endometrial PGE_2 and $PGF_{2\alpha}$ are known for their luteotrophic and luteolytic activities, respectively, in many species, including pigs (Waclawik, 2011). In pigs, PGF_{2a} exhibits luteolytic properties (De Rensis, Saleri, Tummaruk, Techakumphu, & Kirkwood, 2012), whereas PGE_2 is luteoprotective (Stefańczyk-Krzymowska et al., 2006). Arachidonic acid is metabolized in the uterus to prostaglandins (PGE_2 and $PGF_{2\alpha}$), which are involved in many reproductive activities. Prostaglandin synthesis is thus dependent upon accessibility of AA (Irvine, 1982) and the activity of enzymes involved in its metabolism (Blitek et al., 2006). Moreover, an in vitro studies performed on uterine explants have shown that the administration of AA stimulated AQP1 and 5 expression at the protein level only during luteolysis (Skowronska, Mlotkowska, Wojciechowicz, et al., 2015) This well corresponds with physiological situation, since PGF_{2alpha} is released by endometrium in a pulsatile manner to cause corpus luteum regression. Other in vitro studies have shown that AA did lead to upregulation of AQP1/AQP5 mRNAs and proteins in the endometrial slices during luteolysis (Skowronska, Mlotkowska, Nielsen, et al., 2015). Thus, these in vitro studies strongly suggest the stimulatory effect of $PGF_{2\alpha}$ on water permeability that may affect the endometrial expression of AQPs during luteolysis and early pregnancy (Skowronska, Mlotkowska, Majewski, et al., 2016; Skowronska, Mlotkowska, Okrasa, et al., 2016).

Considering the regulatory actions of prostaglandins discuss above, these observations indirectly indicate that prostaglandins may exert regulatory effects on AQP1 and 5 in uterine tissue. All these studies have gradually elucidated the association between PGs and AQPs expression however, there is no colocalization study of examined uterine AQPs with major enzymes of prostaglandin pathway, COX-1 or COX-2, which would provide deeper insight into the role of prostaglandins in the uterine AQPs.

6. Conclusion and future directions

This review highlights some of the current findings about uterine AQPs and their potential regulation. In terms of the structure, AQPs constitute one of the best characterized families of integral membrane proteins, however much remains to be learnt about how different AQPs can be specifically regulated. Furthermore, limited information is available about their pharmacological properties due to the lack of "non-toxic tissue"- and subtype-selective agonists, antagonists or blockers. Moreover, studies on the specific role of AQPs much rely upon the availability of AQP knockout mice models. Despite the great progress in aquaporins research, in the reproductive system, few compounds have been proposed to modulate both classical and aquaglyceroporin types of AQPs. It is evident that AQP1 and 5 play a major role in water regulation in the uterus but differential expression of AQPs in different cell types suggests multiple functions that require greater understanding. From a pathophysiological perspective, there is a growing evidence supports their involvement in the development and progression of many types of cancer in reproductive tissues. Promising, AQP-selective ligands may contribute to the development of new drug candidates. Nonetheless, further work is required to determine whether aquaporins are viable therapeutic targets and eventual disease therapy.

Acknowledgments

A.S. wrote the manuscript. D.T. and L.J. drafted the figure and table, search the data. A.S. and M.T.S. searched and analyzed the publication data. A.S. and M.T.S. critically revised the manuscript.

Conflict of interest

The authors declare no conflict of interest.

References

Agre, P., Saboori, A. M., Asimis, A., & Smith, B. L. (1987). Purification and partial characterization of the Mr. 30,000 integral membrane protein associated with the erythrocyte Rh (D) antigen. *Journal of Biological Chemistry*, *262*(36), 17497–17503.

Agre, P., King, L. S., Yasui, M., Guggino, W. B., Ottersen, O. P., Fujiyoshi, Y., et al. (2002). Aquaporin water channels—From atomic structure to clinical medicine. *Journal of Physiology*, *542*, 3–16.

Agre, P. (2004). Nobel lecture. Aquaporin water channels. *Bioscience Reports*, *24*, 127–163.

Agre, P. (2006). The aquaporin water channels. *Proceedings of the American Thoracic Society*, *3*, 5–13.

Akerlund, M., Bossmar, T., Brouard, R., Kostrzewska, A., Laudanski, T., Lemancewicz, A., et al. (1999). Receptor binding of oxytocin and vasopressin antagonists and inhibitory effects on isolated myometrium from preterm and term pregnant women. *Journal of Obstetrics and Gynaecology*, *106*, 1047–1053.

Aralla, M., Borromeo, V., Groppetti, D., Secchi, C., Cremonesi, F., & Arrighi, S. (2009). A collaboration of aquaporins handles water transport in relation to the estrous cycle in the bitch uterus. *Theriogenology*, *72*, 310–321.

Astakhova, L. A., Kapitskiĭ, S. V., Govardovskiĭ, V. I., & Firsov, M. L. (2012). cAMP as a regulator of the phototransduction cascade. *Rossiĭskii Fiziologicheskiĭ Zhurnal Imeni I.M. Sechenova*, *98*, 1273–1285.

Belkacemi, L., Beall, M. H., Magee, T. R., Pourtemour, M., & Ross, M. G. (2008). AQP1 gene expression is upregulated by arginine vasopressin and cyclic AMP agonists in trophoblast cells. *Life Sciences*, *82*, 1272–1280.

Blanks, A. M., Shmygol, A., & Thornton, S. (2007). Regulation of oxytocin receptors and oxytocin receptor signaling. *Seminars in Reproductive Medicine*, *25*(1), 052–059.

Blitek, A., Waclawik, A., Kaczmarek, M. M., Stadejek, T., Pejsak, Z., & Ziecik, A. J. (2006). Expression of cyclooxygenase-1 and -2 in the porcine endometrium during the oestrous cycle and early pregnancy. *Reproduction in Domestic Animals*, *41*, 251–257.

Borok, Z., Li, X., Fernandes, V. F., Zhou, B., Ann, D. K., & Crandall, E. D. (2000). Differential regulation of rat aquaporin-5 promoter/enhancer activities in lung and salivary epithelial cells. *Journal of Biological Chemistry*, *275*(34), 26,507–26,514.

Brooks, H. L., Regan, J. W., & Yool, A. J. (2000). Inhibition of aquaporin-1water permeability by tetraethylammonium: Involvement of the loop E pore region. *Molecular Pharmacology*, *57*, 1021–1026.

Brown, D., Katsura, T., Kawashima, M., Verkman, A. S., & Sabolic, I. (1995). Cellular distribution of the aquaporins: A family of water channel proteins. *Histochemistry and Cell Biology*, *104*(1), 1–9.

Csányi, A., Bota, J., Falkay, G., Gaspar, R., & Ducza, E. (2016). The effects of female sexual hormones on the expression of aquaporin 5 in the late-pregnant rat uterus. *International Journal of Molecular Sciences*, *17*, 1300.

Csányi, A., Hajagos-Tóth, J., Kothencz, A., Gaspar, R., & Ducza, E. (2018). Effects of different antibiotics on the uterine contraction and the expression of aquaporin 5 in term pregnant rat. *Reproductive Toxycology*, *81*, 64–70.

Casslen, B. (1998). Uterine fluid volume. Cyclic variations and possible extra-uterine contributions. *Journal of Reproductive Medicine*, *31*, 506–510.

Chakraborty, I., Das, S. K., Wang, J., & Dey, S. K. (1996). Developmental expression of the cyclo-oxygenase-1 and cyclo-oxygenase-2 genes in the peri-implantation mouse uterus and their differential regulation by the blastocyst and ovarian steroids. *Journal of Molecular Endocrinology*, *16*, 107–122.

Chen, Q., Zhang, Y., Elad, D., Jaffa, A. J., Cayo, Y., Ye, X., et al. (2013). Navigating the site for embryo implantation: Biomechanical and molecular regulation of intrauterine embryo distribution. *Molecular Aspects of Medicine, 34*, 1024–1042.

Chinigarzadeh, A., Muniandy, S., & Salleh, N. (2016). Combination effect of genistein and female sex-steroids on uterine fluid volume and secretion rate and aquaporins (AQP)-1,2, 5, and 7 expression in the uterus in rats. *Environmental Toxicology, 32*, 832–844.

Chinigarzadeh, A., Muniandy, S., & Salleh, N. (2017). Combinatorial effect of genistein and female sex-steroids on uterine fluid volume and secretion rate and aquaporin (AQP)-1, 2, 5, and 7 expression in the uterus in rats. *Environmental Toxycology, 32*(3), 832–844, Epub 2016 May 28. https://doi.org/10.1002/tox.22283.

Chou, C. L., DiGiovanni, S. R., Luther, A., & Knepper, M. A. (1995). Oxytocin as an antidiuretic hormone: II, role of V2 vasopressin receptor. *American Journal of Physiology, 269*, F78–F85.

Delporte, C., Bryla, A., & Perret, J. (2016). Aquaporins in salivary glands: From basic research to clinical applications. *International Journal of Molecular Sciences, 17*, 166.

Denker, B. M., Smith, B. K., Kuhajda, F. P., & Agree, P. (1998). Identification, purification, and partial characterization of a novel MR 28,000 integral membrane protein form erythrocytes and renal tubules. *Journal of Biological Chemistry, 263*, 15634–15642.

De Rensis, F., Saleri, R., Tummaruk, P., Techakumphu, M., & Kirkwood, R. N. (2012). Prostaglandin F2α and control of reproduction in female swine: A review. *Theriogenology*, 77(1), 1–11.

Direito, I., Madeira, A., Brito, M. A., & Soveral, G. (2016). Aquaporin-5: From structure to function and dysfunction in cancer. *Cellular and Molecular Life Sciences, 73*(8), 1623–1640.

Downey, P., Sapirstein, A., O'Leary, E., Sun, T. X., Brown, D., & Bonventre, J. V. (2001). Renal concentrating defect in mice lacking group IV cytosolic phospholipaseA(2). *American Journal of Physiology. Renal Physiology, 270*, F388–F394.

Ducza, E., Seres, A. B., Hajagos-Thot, J., Falkay, G., & Gaspar, R. (2014). Oxytocin regulates the expression of aquaporin 5 in the late-pregnant rat uterus. *Molecular Reproduction & Development, 81*, 524–530.

Fushimi, K., Sasaki, S., & Marumo, F. (1997). Phosphorylation of serine 256 is required for cAMP-dependent regulatory exocytosis of the aquaporin-2water channel. *Journal of Biological Chemistry, 272*, 14800–14804.

Gimpl, G., & Fahrenholz, F. (2001). The oxytocin receptor system: Structure function and regulation. *Physiological Reviews, 81*, 629–683.

Han, Z., & Patil, R. V. (2000). Protein kinase A-dependent phosphorylation of aquaporin-1. *Biochemical and Biophysical Research Communications, 273*, 328–332.

Hasegawa, T., Azlina, A., Javkhlan, P., Yao, C., Akamatsu, T., & Hosoi, K. (2011). Novel phosphorylation of aquaporin-5 at its threonine 259 through cAMP signaling in salivary gland cells. *American Journal of Physiology—Cell Physiology, 301*, C667–C678.

Hua, Y., Jiang, W., Zhang, W., Shen, Q., Chen, M., & Zhu, X. (2013). Expression and significance of aquaporins during pregnancy. *Frontiers in Bioscience (Landmark Ed.), 1*(18), 1373–1383.

Huang, H. F., He, R. H., Sun, C. C., Zhang, Y., Meng, Q. X., & Ma, Y. Y. (2006). Function of aquaporins in female and male reproductive systems. *Human Reproduction Update, 12*, 785–795.

Hildenbrand, A., Stavreus-Evers, A., Lalitkumar, P. G. L., Nielsen, S., Mints, M., & Gemzell-Danielsson, K. (2008). Aquaporin 1 is expressed in the human endometrium during normal cycle and increases after mifepristone treatment. *International Journal of Molecular Medicine, 22*, 49–53.

Horsefield, R., Norde, K., Fellert, M., Backmark, A., Tornroth-Horsefield, S., Terwisscha van Scheltinga, A. C., et al. (2008). High-resolution X-ray structure of human aquaporin 5. *Proceedings of the National Academy of Sciences of the United States of America, 105*(36), 13327–13332.

Ishikawa, Y., Eguchi, T., Skowronski, M. T., & Ishida, H. (1998). Acetylocholine acts on M_3 muscarinic receptors and induces the translocation of aquaporin5 water channel via cytosolic Ca^{2+} elevation in rat parotid glands. *Biochemical and Biophysical Research Communications, 245*, 835–874.

Ishikawa, Y., Skowronski, M. T., Inoue, N., & Ishida, H. (1999). alpha(1) adrenoreceptor-induced traffiking of aquaporin-5 to the apical plasma membrane of rat parotid cells. *Biochemical and Biophysical Research Communications, 265*, 94–100.

Ishikawa, Y., Yuan, Z., Inoue, N., Skowronski, M. T., Nakae, Y., Shono, M., et al. (2005). Identification of AQP5 in lipid rafts and its translocation to apical membranes by activation of M3 mAChRs in interlobular ducts of rat parotid gland. *American Journal of Physiology—Cell Physiology, 289*, C1303–C1311.

Irvine, R. F. (1982). How is the level of free arachidonic acid controlled in cells? *Biochemical Journal, 204*, 3–16.

Jablonski, E. M., McConnell, N. A., Hughes, F. M., Jr., & Huet-Hudson, Y. M. (2003). Estrogen regulation of aquaporins in the mouse uterus: Potential roles in uterine water movement. *Biology of Reproduction, 69*, 1481–1487.

Jeng, W., Matheson, I. M., Ihara, W., & Ramirez, G. (1998). Aquaporin-1: An osmoinducible water channel in cultured mIMCD-3 cells. *Biochemical and Biophysical Research Communications, 245*, 804–809.

Jiang, X. X., Wei, F. X., Zhao, L., Ye, X. L., Xin, L. B., Qu, Y., et al. (2015). Aquaporin 5 plays a role in estrogen-induced ectopic implantation of endometrial stromal cells in endometriosis. *PLoS ONE, 10*(12), e0145290. https://doi.org/10.1371/journal.

Katsura, T., Gustafson, C. E., Ausiello, D. A., & Brown, D. (1997). Protein kinase A phosphorylation is involved inregulated exocytosis of aquaporin-2 in transfected LLC-PK1 cells. *American Journal of Physiology, 272*, F817–F822.

King, L. S., & Agre, P. (1996). Pathophysiology of the aquaporin water channels. *Annual Review of Physiology, 58*, 619–648.

King, L. S., Kozono, D., & Agre, P. (2004). From structure to disease: The evolving tale of aquaporin biology. *Nature Reviews. Molecular Cell Biology, 5*, 687–698.

Kitchen, P., Oberg, F., Sjohamn, J., Hedfalk, K., Bill, R. M., Conner, A. C., et al. (2015). Plasma membrane abundance of human aquaporin 5 is dynamically regulated by multiple pathways. *PLoS One, 10*(11), e0143027.

Klein, C., Troedsson, M. H., & Rutllant, J. (2013). Expression of aquaporin water channels in equine endometrium is differentially regulated during the oestrous cycle and early pregnancy. *Reproduction in Domestic Animals, 48*(4), 529–537, Epub 2012 Oct 26. https://doi.org/10.1111/rda.12116.

Kobayashi, M., Takahashi, E., Miyagawa, S., Watanabe, H., & Iguchi, T. (2006). Chromatin immunoprecipitation-mediated target identification proved aquaporin 5 is regulated directly by estrogen in the uterus. *Genes to Cells, 11*, 1133–1143.

Kosugi-Tanaka, C., Li, X., Yao, C., Akamatsu, T., Kanamori, N., & Hosoi, K. (2006). Protein kinase A-regulated membrane trafficking of a green fluorescent protein-aquaporin 5 chimera in MDCK cells. *Biochimica et Biophysica Acta, 1763*, 337–344.

Krane, C. M., Towne, J. E., & Menon, A. G. (1999). Cloning and characterization of murine Aqp5: Evidence for a conserved aquaporins gene cluster. *Mammalian Genome, 10*, 498–505.

Kreegipuu, A., Blom, N., & Brunak, S. (1999). PhosphoBase, a database of phosphorylation sites: Release 2.0. *Nucleic Acids Research, 27*, 237–239.

Kumari, S. S., Varadaraj, M., Yerramilli, V. S., Menon, A. G., & Varadaraj, K. (2012). Spatial expression of aquaporin 5 in mammalian cornea and lens, and regulation of Its localization by phosphokinase A. *Molecular Vision, 18*, 957–967.

Lee, M. D., Bhakta, K. Y., Raina, S., Yonescu, R., Griffin, C. A., Copeland, N. G., et al. (1996). The human aquaporin-5 gene. Molecular characterization and chromosomal localization. *Journal of Biological Chemistry, 271*, 8599–8604.

Li, X., Yu, H., & Koide, S. S. (1994). The water channel gene in human uterus. *International Journal of Biochemistry and Molecular Biology*, *32*, 371–377.

Li, X., Yu, H., & Koide, S. (1997). Regulation of the water channel gene (AQP-CHIP) expressionby estradiol and anordiol in the rat uterus. *Yao Xue Xue Bao*, *32*, 586–592.

Li, C., Wang, W., Summer, S. N., Westfall, T. D., Brooks, D. P., Falk, S., et al. (2008). Molecular mechanisms of antidiuretic effect of oxytocin. *Journal of the American Society of Nephrology*, *19*, 225–232.

Li, Y., Wei, Y., Zheng, F., Guan, Y., & Zhang, X. (2017). Prostaglandin E2 in the regulation of water transport in renal collecting ducts. *International Journal of Molecular Sciences*, *18*, 2539.

Lim, H., Paria, B. C., Das, S. K., Dinchuk, J. E., Langenbach, R., Trzaskos, J. M., et al. (1997). Multiple female reproductive failures in cyclooxygenase 2-deficient mice. *Cell*, *91*, 197–208.

Lindsay, L. A., & Murphy, C. R. (2004). Redistribution of aquaporins in uterine epithelial cells at the time of implantation in the rat. *Acta Histochemica*, *106*, 299–307.

Lindsay, L. A., & Murphy, C. R. (2006). Redistribution of aquaporins 1 and 5 in the rat uterus is dependent on progesterone: A study with light and electron microscopy. *Reproduction*, *131*, 369–378.

Lindsay, L. A., & Murphy, C. R. (2007). Aquaporins are upregulated in glandular epithelium at the time of implantation in the rat. *Journal of Molecular Histology*, *38*, 87–95.

Lyness, J., Robinson, A. G., Sheridan, M. N., & Gash, D. M. (1985). Antidiuretic effects of oxytocin in the Brattleboro rat. *Experientia*, *41*, 1444–1446.

Ma, B., Xiang, Y., Mu, S. M., Li, T., Yu, H. M., & Li, X. J. (2004). Effects of acetazolamide and anordiol on osmotic water permeability in AQP1-cRNAinjected Xenopus oocyte. *Acta Pharamcologica Sinica*, *25*, 90–97.

Mints, M., Hildenbrand, A., Lalitkumar, L. P., Andersson, S., Nielsen, S., Gemzell-Danielsson, K., et al. (2007). Expression of aquaporin-1 in endometrial blood vessels in menorrhagia. *International Journal of Molecular Sciences*, *19*, 407–411.

Mobasheri, A., & Marples, D. (2004). Expression of the AQP-1 water channel in normal human tissues: A semiquantitative study using tissue microarray technology. *American Journal of Physiology—Cell Physiology*, *286*, C529–C537.

Mohd Mokhtar, H., Giribabu, N., Kassim, N., Muniandy, S., & Salleh, N. (2014). Testosterone decreases fluid and chloride secretion in the uterus of adult female rats via down regulating cystic fibrosis transmembrane regulator(CFTR) expression and functional activity. *Journal of Biochemistry and Molecular Biology*, *144*, 361–372.

Moon, C., King, L. S., & Agre, P. (1997). Aqp1 expression in erythroleukemia cells: Genetic regulation of glucocorticoid and chemical induction. *American Journal of Physiology*, *273*, C1562–C1570.

Moehren, U., Denayer, S., Podvinec, M., Verrijdt, G., & Claessens, F. (2008). Identification of androgen-selective androgen-response elements in the human aquaporin-5 and Rad9 genes. *Biochemical Journal*, *411*(3), 679–686. https://doi.org/10.1042/BJ20071352.

Murata, K., Mitsuoka, K., Hirai, T., Walz, T., Agre, P., Heymann, J. B., et al. (2000). Structural determinants of water permeation through aquaporins-1. *Nature*, *407*, 599–605.

Naftalin, R. J., Thiagarajah, J. R., Pedley, K. C., Pocock, V. J., & Milligan, S. R. (2002). Progesterone stimulation of fluid absorption. *Reproduction*, *123*(5), 633–638.

Németh-Cahalan, K. L., Kalman, K., & Hall, J. E. (2004). Molecular basis of pH and Ca2+ regulation of aquaporin water permeability. *Journal of General Physiology*, *123*, 573–580.

Nesverova, V., & Törnroth-Horsefield, S. (2019). Phosphorylation-dependent regulation of mammalian aquaporins. *Cells*, *8*, 82. https://doi.org/10.3390/cells8020082.

Nielsen, S., Smith, B. L., Christensen, E. I., & Agre, P. (1993). Distribution of the aquaporin CHIP in secretory and resorptive epithelia and capillary endothelia. *Proceedings of the National Academy of Sciences of the United States of America, 90*, 7275–7279.
Nielsen, S., Pallone, T., Smith, B. L., Christensen, E. I., Agre, P., & Maunsbach, A. B. (1995). Aquaporin-1 water channels in short and long loop descending thin limbs and in descending vasa recta in rat kidney. *American Journal of Physiology, 268*, F1023–F1037.
Nielsen, S., Chou, C. L., Marples, D., Christensen, E. I., Kishore, B. K., & Knepper, M. A. (1995). Vasopressin increases water permeability of kidney collecting duct by inducing translocation of aquaporin-CD water channels to plasma membrane. *Proceedings of the National Academy of Sciences of the United States of America, 92*(4), 1013–1017.
Nilsson, L., Madsen, K., Topcu, S. O., Jensen, B. L., Frøkiær, J., & Nørregaard, R. (2012). Disruption of cyclooxygenase-2 prevents downregulation of cortical AQP2 and AQP3 in response to bilateral ureteral obstruction in the mouse. *American Journal of Physiology—Renal Physiology, 302*(11), F1430–F1439.
Olesen, E. T., & Fenton, R. A. (2013). Is there a role for PGE_2 in urinary concentration? *Journal of the American Society of Nephrology, 24*(2), 169–178.
Paganelli, C. V., & Solomon, A. K. (1957). The rate of exchange of tritiated water across the human red cell membrane. *Journal of General Physiology, 41*(2), 259–277.
Patil, R. V., Han, Z., & Wax, M. B. (1997). Regulation of water channel activity of aquaporin1 by arginine vasopressin and natriuretic peptide. *Biochemical and Biophysical Research Communications, 238*, 392–396.
Pouzet, B., Serradeil-Le, G. C., Bouby, N., Maffrand, J. P., Le, F. G., & Bankir, L. (2001). Selective blockade of vasopressin V2 receptors reveals significant V2-mediated water reabsorption in Brattleboro rats with diabetes insipidus. *Nephrology, Dialysis, Transplantation, 16*, 725–734.
Preston, G. M., Caroll, T. P., Guggino, W. B., & Agre, P. (1992). Apperance of water channels in Xenopus oocytes expressing red cell CHIP28 protein. *Science, 256*(5055), 385–387.
Raina, S., Preston, G. M., Guggino, W. B., & Agre, P. (1995). Molecular cloning and characterization of an aquaporin cDNA from salivary, lacrimal, and respiratory tissues. *Journal of Biological Chemistry, 270*(4), 1908–1912.
Ren, Y., Lu, H., Reinach, P. S., et al. (2017). Hyperosmolarity-induced aqp5 upregulation promotes inflammation and cell death via jnk1/2 activation in human corneal epithelial cells. *Scientific Reports, 7*, 4727.
Reese, J., Brown, N., Paria, B. C., Morrow, J., & Dey, S. K. (1999). COX-2 compensation in the uterus of COX-1 deficient mice during the pre-implantation period. *Molecular and Cellular Endocrinology, 150*, 23–31.
Richard, C., Gao, J. U., Brown, N., & Reese, J. (2003). Aquaporin water channel genes are differentially expressed and regulated by ovarian steroids during the periimplantation period in the mouse. *Endocrinology, 144*, 1533–1541.
Salleh, N., Baines, D., Naftalin, R., & Milligan, S. (2005). The hormonal control of uterine luminal fluid secretion and absorption. *Journal of Membrane Biology, 206*(1), 17–28.
Salleh, N., Mokhtar, H. M., Kassim, N. M., & Giribabu, N. (2015). Testosterone induces increase in aquaporin (AQP)-1, 5 and 7 expression in the uteri of ovariectomized rats. *Journal of Membrane Biology, 248*(6), 1097–1105.
Sakamoto, Y., Hisatsune, A., Katsuki, H., Horie, I., & Isohama, Y. (2014). Aquaporin 5 increases keratinocyte-derived chemokine expression and NF-κB activity through ERK activation. *Biochemical and Biophysical Research Communications, 448*, 355–360.
Sakairi, Y., Jacobson, H. R., Noland, T. D., & Breyer, M. D. (1995). Luminal prostaglandin E receptors regulate salt and water transport in rabbit cortical collecting duct. *The American Journal of Physiology, 269*(2 Pt. 2), F257–F265.

Shanahan, C. M., Connolly, D. L., Tyson, K. L., Cary, N. R., Osbourn, J. K., Agre, P., et al. (1999). Aquaporin-1 is expressed by vascular smooth muscle cells and mediates rapid water transport across vascular cell membranes. *Journal of Vascular Research, 36*, 353–362.

Shahzad, H., Giribabu, N., Karim, K., Muniandy, S., Kassim, N. M., & Salleh, N. (2017). Quercetin alters uterine fluid volume and aquaporin (AQP) subunits (AQP-1, 2, 5 & 7) expression in the uterus in the presence of sex-steroids in rats. *Reproductive Toxicology, 69*, 276–285, Epub 2017 Mar 22. https://doi.org/10.1016/j.reprotox.2017.03.012.

Shi, L. B., Skach, W. R., & Verkman, A. S. (1994). Functional independence of monomeric CHIP28 water channels revealed by expression of wild-type mutant heterodimers. *Journal of Biological Chemistry, 269*, 10417–10422.

Shi, X., Wu, S., Yang, Y., Tang, L., Wang, Y., Zhao, W., et al. (2014). AQP5 silencing suppresses p38 MAPK signaling and improves drug resistance in colon cancer cells. *Tumour Biology, 35*, 7035–7045.

Sidhaye, V., Hoffert, J. D., & King, L. S. (2005). cAMP has a distinct acute and chronic effect on aquaporin-5 in lung epithelial cells. *Journal of Biological Chemistry, 280*(5), 3590–3596.

Skowronska, A., Mlotkowska, P., Wojciechowicz, B., Okrasa, S., Nielsen, S., & Skowronski, M. T. (2015). Progesterone, estradiol, arachidonic acid, oxytocin, forskolin and cAMP influence on aquaporin 1 and 5 expression in porcine uterine explants during the mid-luteal phase of the estrous cycle and luteolysis: An in vitro study. *Reproductive Biology and Endocrinology, 18*, 13–17. https://doi.org/10.1186/s12958-015-0004-5.

Skowronska, A., Mlotkowska, P., Nielsen, S., & Skowronski, M. T. (2015). Difference in expression between AQP1 and AQP5 in porcine endometrium and myometrium in response to steroid hormones, oxytocin, arachidonic acid, forskolin and cAMP during the mid-luteal phase of the estrous cycle and luteolysis. *Reproductive Biology and Endocrinology, 1*, 13–131. https://doi.org/10.1186/s12958-015-0128-7.

Skowronska, A., Mlotkowska, P., Majewski, M., Nielsen, S., & Skowronski, M. T. (2016). Expression of aquaporin 1 and 5 and their regulation by ovarian hormones, arachidonic acid, forskolin and cAMP during implantation in pigs. *Physiological Research, 65*(4), 637–650.

Skowronska, A., Mlotkowska, P., Okrasa, S., Nielsen, S., & Skowronski, M. T. (2016). Modulatory effects of steroid hormones, oxytocin, arachidonic acid, forskolin and cyclic AMP on the expression of aquaporin 1 and aquaporin 5 in the porcine uterus during placentation. *Journal of Physiology and Pharmacology, 67*(2), 311–319.

Skowronski, M. T. (2010). Distribution and quantitative changes in amounts of aquaporin 1, 5 and 9 in the pig uterus during the estrous cycle and early pregnancy. *Reproductive Biology and Endocrinology, 8*, 109. https://doi.org/10.1186/1477-7827-8-109.

Skowronski, M. T., Kwon, T. H., & Nielsen, S. (2009). Immunolocalization of aquaporin 1, 5, and 9 in the female pig reproductive system. *Journal of Histochemistry and Cytochemistry, 57*(1), 61–67. https://doi.org/10.1369/jhc.2008.952499.

Stefańczyk-Krzymowska, S., Wasowska, B., Chłopek, J., Gilun, P., Grzegorzewski, W., & Radomski, M. (2006). Retrograde and local destination transfer of uterine prostaglandin E_2 in early pregnant sow and its physiological consequences. *Prostaglandins and Other Lipid Mediators, 81*(1–2), 71–79.

Toot, J., Jenkins, C., Dunphy, G., Boehme, S., Hart, M., Milsted, A., et al. (2008). Testosterone influences renal electrolyte excretion in SHR/y and WKY males. *BMC Physiology, 8*, 5. https://doi.org/10.1186/1472-6793-8-5.

Törnroth-Horsefield, S., Hedfalk, K., Fischer, G., Lindkvidt-Petersson, K., & Richard, N. (2010). Structural insights into eukaryotic aquaporin regulation. *FEBS Letters, 584*, 2580–2588.

Umenishi, F., & Schrier, R. W. (2002). Identification and characterization of a novel hypertonicity-responsive element in the human aquaporin-1 gene. *Biochemical and Biophysical Research Communications, 292*, 771–775.

Umenishi, F., & Schrier, R. W. (2003). Hypertonicity-induced aquparin-1 (AQP1) expression is mediated by the activation of MAPK pathways and hypertonicity-responsive element in the AQP1 gene. *Journal of Biological Chemistry*, *278*, 15765–15770.
Yang, Z. M., Das, S. K., Wang, J., Sugimoto, Y., Ichikawa, A., & Dey, S. K. (1997). Potential sites of prostaglandin actions in the periimplantation mouse uterus: Differential expression ad regulation of prostaglandin receptor genes. *Biology of Reproduction*, *56*, 368–379.
Yang, F., Kawedia, J. D., & Menon, A. G. (2003). Cyclic AMP regulates aquaporin 5 expression at both transcriptional and post-transcriptional levels through a protein kinase A pathway. *Journal of Biological Chemistry*, *278*, 32,173–32,180.
Yao, C., Purwant, N., Karabasil, M. R., Azlina, A., Javkhlan, P., Hasegawa, T., et al. (2010). Potentail down-regulation of salivary gland AQP5 by LPS via cross-coupling of NK-kappaB and p-c-Jun/c-Fos. *American Journal of Pathology*, *177*(2).
Yasui, M., Hazama, A., Kwon, T. H., Nielsen, S., Guggino, W. B., & Agre, P. (1999). Rapid gating and anion permeability of an intracellular aquaporin. *Nature*, *402*, 184–187.
Verkman, A. S. (2012). Aquaporins in clinical medicine. *Annual Review of Medicine*, *63*, 303–316.
Waclawik, A. (2011). Novel insights into the mechanisms of pregnancy establishment: Regulation of prostaglandin synthesis and signalling in the pig. *Reproduction*, *142*(3), 389–399.
Wang, S., Chen, J., Au, K. T., & Ross, M. G. (2003). Expression of aquaporin 8 and its up-regulation by cyclic adenosine monophosphate in human WISH cells. *American Journal of Obstetrics and Gynecology*, *188*, 997–1001.
Wang, S., Amidi, F., Beall, M., Gui, L., & Ross, M. G. (2006). Aquaporin 3 expression in human fetal membranes and its up-regulation by cyclic adenosine monophosphate in amnion epithelial cell culture. *Journal of the Society for Gynecology Investigation*, *13*, 181–185.
Wang, S. B., Amidi, F., Sheng, L. Y., Beall, M., & Ross, M. G. (2007). Cyclic adenosine monophosphate regulation of aquaporin gene expression in human amnion epithelia. *Reproductive Sciences*, *14*, 234–240.
Wang, W., & Zheng, M. (2011). Role of cAMP-PKA/CREB pathway in regulation of AQP 5 production in rat nasal epithelium. *Rhinology*, *49*(4), 464–469.
Wang, J., Feng, L., Zhu, Z., Zheng, M., Wang, D., Chen, Z., et al. (2015). Aquaporins as diagnostic and therapeutic targets in cancer: How far we are? *Journal of Translational Medicine*, *21*, 13–96. https://doi.org/10.1186/s12967-015-0439-7.
Wawrzkiewicz-Jalowiecka, A., Kowalczyk, K., Pluta, D., Blukacz, L., & Madej, P. (2017). The role of aquaporins in polycystic ovary syndrome-a way towards a novel drug target in PCOS. *Medical Hypotheses*, *102*, 23–27.
Wellner, R. B., Hong, S., Cotrim, A. P., Swaim, W. D., & Baum, B. J. (2005). Modifying the NH2 and COOH termini of aquaporin-5: Effects on localization in polarized epithelial cells. *Tissue Engineering*, *11*, 1449–1458.
Woo, J., Chae, Y. K., Jang, S. J., Kim, M. S., et al. (2008). Membrane trafficking of AQP5 and cAMP dependent phosphorylation in bronchial epithelium. *Biochemical and Biophysical Research Communications*, *366*, 321–327.
Woo, J., Lee, J., Chae, Y. K., Kim, M. S., Baek, J. H., Park, J. C., et al. (2008). Overexpression of AQP5, a putative oncogene, promotes cell growth and transformation. *Cancer Letters*, *264*, 54–62.
Zelenina, M., Christensen, B. M., Palmér, J., Nairn, A. C., Nielsen, S., & Aperia, A. (2000). Prostaglandin E(2) interaction with AVP: Effects on AQP2 phosphorylation and distribution. *American Journal of Physiology—Renal Physiology*, *278*(3), F388–F394.
Zelenina, M., Bondar, A. A., Zelenin, S., & Aperia, A. (2003). Nickel and extracellular acidification inhibit the water permeability of human aquaporin-3 in lung epithelial cells. *Journal of Biological Chemistry*, *278*, 30,037–30,043.

Zelenina, M., Tritto, S., Bondar, A. A., Zelenin, S., & Aperia, A. (2004). Copper inhibits the water and glycerol permeability of aquaporin-3. *Journal of Biological Chemistry*, *279*, 51939–51,943.

Zeidel, M. L., Ambudkar, S. V., Smith, B. L., & Agre, P. (1992). Reconstitution of functional water channels in liposomes containing purified red cell CHIP28 protein. *Biochemistry*, *31*(33), 7436–7440.

Zhang, D., Tan, Y. J., Qu, F., Sheng, J. Z., & Huang, H. F. (2012). Functions of water channels in male and female reproductive systems. *Molecular Aspects of Medicine*, *33*, 676–690.

Zhang, Y., Chen, Q., Zhang, H., Wang, Q., Li, R., Jin, Y., et al. (2015). Aquaporin-dependent excessive intrauterine fluid accumulation is a major contributor in hyper-estrogen induced aberrant embryo implantation. *Cell Research*, *25*(1), 139–142.

Zhou, F., Liang, Y., Qian, Z. D., Zhuang, Y. L., Chen, Y. Z., Lv, B. J., et al. (2013). Low-dose mifepristone increases uterine expression of aquaporin 1/aquaporin 2 at the time of implantation. *Contraception*, *87*(6), 844–849.

Zhu, C., Jiang, Z., Bazer, F. W., Johnson, G. A., Burghardt, R. C., & Wu, G. (2015). Aquaporins in the female reproductive system of mammals. *Frontiers in Bioscience (Landmark Ed.)*, *1*, 838–871.

Zeuthen, T. (1995). Molecular mechanisms for passive and active transport of water. *International Review of Cytology*, *160*, 99–161.

Zhai, X. Y., Fenton, R. A., Andreasen, A., Thomsen, J. S., & Christensen, E. I. (2007). Aquaporin-1 is not expressed in descending thin limbs of short-loop nephrons. *Journal of the American Society of Nephrology*, *18*, 2937–2944.

Further reading

Escobar, J., Gormaz, M., Arduini, A., Gosens, K., Martinez, A., Perales, A., et al. (2012). Expression of aquaporins early in human pregnancy. *Early Human Development*, *88*(8), 589–594.

King, L. S., Nielsen, S., & Agre, P. (1996). Aquaporin-1 water channel protein in lung: Ontogeny, steroid-induced expression, and distribution in rat. *The Journal of Clinical Investigation*, *97*, 2183–2191.

Nishimoto, G., Zelenina, M., Li, D., Yasui, M., Aperia, A., Nielsen, S., et al. (1999). Arginine vasopressin stimulates phosphorylation of AQP-2 in rat renal tissue. *American Journal of Physiology*, *276*(2 Pt. 2), F254–F259.

CHAPTER ELEVEN

Estrogen regulates aquaporin-2 expression in the kidney

Stine Julie Tingskov, Henricus A.M. Mutsaers, Rikke Nørregaard*
Department of Clinical Medicine, Aarhus University, Aarhus, Denmark
*Corresponding author: e-mail address: rn@clin.au.dk

Contents

Abstract

Estrogens are primarily identified as sex hormones that, for a long time, have been known as important regulators of female reproductive physiology. However, our understanding of the role of estrogens has changed over the past years. It is now well accepted that estrogens are also involved in other physiological and pathological processes in both genders. This is due to the fact that estrogen can act both local as well as on a systemic level. Next to its role in reproductive physiology, there is accumulating evidence that estrogen influences multiple systems involved in water homeostasis. This chapter will delineate the regulatory effects of estrogen on the water channel aquaporin-2 (AQP2) found in the renal collecting duct. We will first provide an introduction to estrogen, the estrogen receptors and their role in renal physiology as well as describe the effect of selective estrogen receptor modulators (SERMs) on the kidney. Subsequently, we will focus on how estrogen and SERMs influence water balance and regulate AQP2 expression in principal cells of the collecting duct. Finally, we will describe how estrogen regulates AQP2 functionality in other organ systems.

Vitamins and Hormones, Volume 112
ISSN 0083-6729
https://doi.org/10.1016/bs.vh.2019.08.016

1. Introduction

Estrogen was discovered in the 1920s (Allen & Doisy, 1923; Marrian & Butenandt, 1931) and belongs to the group of steroid hormones. Estrogen is the main female sex hormone present in woman of reproductive age, and it is mainly responsible for the development and regulation of the female reproductive system. In premenopausal women, the ovaries are the principal source of circulating estrogens, although during pregnancy, the placenta also secrets substantial amounts of estrogen into the circulation (Hall, 2011). Next to the effects on the female reproductive system, estrogens also play a crucial role in the regulation of lipid and carbohydrate metabolism, skin physiology, skeletal homeostasis, the cardiovascular system, electrolyte balance and the central nervous system (Hall, 2011). Sex hormones influence not only the gender-specific organs but also the structure and function of nearly every tissue and organ in the mammalian body, including the kidneys (Sabolic et al., 2007). Due to this widespread regulatory role, it is not surprising that estrogens also play a significant role in male physiology and cannot be exclusively considered as a female sex hormone. Men as well as postmenopausal women, who cease to produce estrogen because of a decline in ovary function, are dependent on local production of estrogens in extragonadal tissues, including mesenchymal cells of adipose tissue, the vascular endothelium, aortic smooth muscle cells and the brain. Due to this local synthesis, estrogens not only have an endocrine effect but also function in an autocrine and paracrine manner (Simpson, 2003).

2. Estrogen signaling

It is assumed that estrogens exert their effects in target tissues by binding to their designated receptors. In spite of the significance of estrogen signaling there are only three estrogen receptors (ERs), including ERα, ERβ and G protein-coupled estrogen receptor 1 (GPER1) that mediate all the effects of estrogens. These receptors can be localized in the cell membrane and cytoplasm as well as in the nucleus of target cells (Sabolic et al., 2007; Simpson, 2003). ERα was the first estrogen receptor to be discovered and is therefore investigated the most (Jensen, 1962). ERβ, which has a significant sequence homology to ERα, was identified about four decades later and showed to have different, nonredundant roles (Kuiper, Enmark, Pelto-Huikko, Nilsson, & Gustafsson, 1996). In 1997, a cDNA screen identified the GPER1, which was then also linked to estrogen signaling

(Carmeci, Thompson, Ring, Francke, & Weigel, 1997). ERα has at least three different isoforms and ERβ has five. Both ERα and ERβ are nuclear estrogen receptors, whereas GPER1 is a plasma membrane receptor that is genetically and structurally distinct from ERα and ERβ. The function and expression of GPER1 is independent of the two others ERs (Vrtacnik, Ostanek, Mencej-Bedrac, & Marc, 2014). In comparison to the two nuclear ERs, the GPER1 binding affinity for estradiol—the most potent estrogen produced in the body—and other estrogens is lower and receptor-ligand interaction is very brief (Vrtacnik et al., 2014).

Estrogen-dependent signaling can be divided into genomic regulation of transcriptional processes through estrogen response elements and non-genomic pathways via activation of cell-signaling cascades (Sabolic et al., 2007). Different mechanisms of genomic and nongenomic estrogen signaling are represented in Fig. 1. Genomic signaling is the classic mechanism of

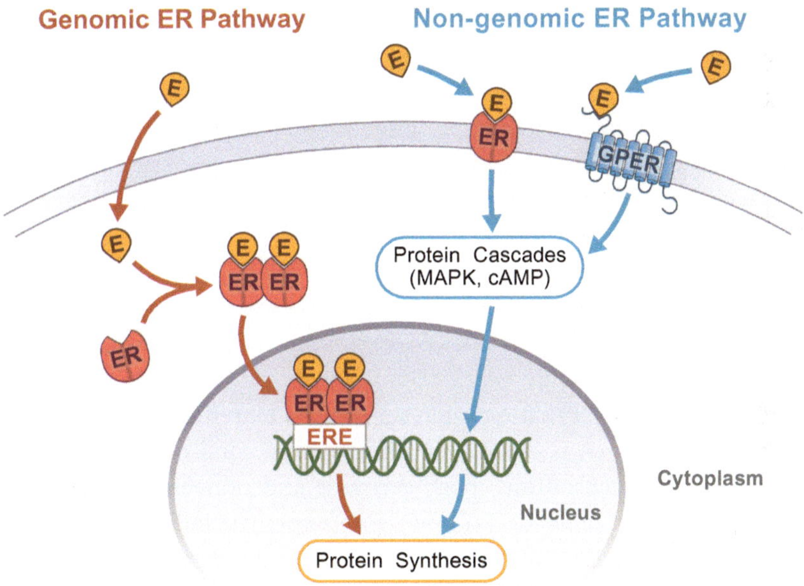

Fig. 1 Estrogen signaling pathway. Schematic illustration of the genomic estrogen receptor (ER) and nongenomic ER signaling pathways. Genomic ER pathway (red): Estrogen binds to one of the ER located in the cytoplasm. This complex binds to the estrogen response element (ERE) in the promoter of the target gene in nucleus resulting in protein synthesis. Nongenomic ER pathway (blue): Estrogen can activate the G protein-coupled estrogen receptor-1 (GPER1) or one of the ER isoforms located in the cell membrane. This can result in stimulation of different protein-kinase cascade, including cAMP and MAPK activation, leading to indirect changes in gene transcription and protein synthesis. *E*, estrogen; *ER*, estrogen receptor; *GPER1*, G protein-coupled estrogen receptor; *MAPK*, mitogen-activated protein kinases.

estrogen signaling. Binding of estrogen to ERα or ERβ in the cytoplasm triggers receptor dimerization and nuclear translocation where these complexes bind to estrogen response elements (ERE) in the promoter of the target gene (Klinge, 2001). Nonetheless, many estrogen responsive genes lack ERE-like promoter sequences. In the case of ERE-independent genomic signaling, the ER complex does not bind directly to the target gene, but induces translation via protein-protein interaction with other transcription factors at the response elements (Göttlicher, Heck, & Herrlich, 1998).

The nongenomic pathway is too rapid to involve gene transcription, but is common among steroid hormones. Nongenomic signaling is associated with the membrane bound GPER1 and certain isoforms of ERα and ERβ (Losel & Wehling, 2003), and can involve mobilization of intracellular calcium, stimulation of adenylate cyclase activity and cAMP production, as well as lead to activation of mitogen-activated protein kinase (MAPK) signaling pathway (Aronica, Kraus, & Katzenellenbogen, 1994; Improta-Brears et al., 1999). These events can indirectly affect gene expression by changing the phosphorylation of transcription factors.

In addition to genomic and nongenomic ligand-dependent signaling, ERs can also be activated in the absence of estrogen or other ligands, which is termed ligand-independent signaling. This type of signaling events occur when the ER gets phosphorylated on specific residues, mainly serine and tyrosine, or when ER-associated coregulators get phosphorylated. Factors that can cause ligand-independent signaling include protein kinase A (PKA), protein kinase C (PKC) and peptide growth factors. The latter includes epidermal growth factor (EGF), insulin, and transforming growth factor β (TGFβ) (Nilsson et al., 2001).

It has become clear that estrogen signaling is extremely complex as it is assumed that estrogen is capable to regulate expression of the same target gene via several mechanisms, both genomic and nongenomic. The final response therefore depends on multiple factors, including transcription factors present on the gene promoter, expression levels and cellular localization of the ERs, their coregulators and signaling factors as well as the type of stimuli.

3. Estrogen receptors in the kidney

3.1 Estrogen receptor α and estrogen receptor β

ERα was discovered in the late 1950s, and the first ERα knockout mouse was created in 1993 by Lubahn et al. (1993). At that time, it was surprising that life was possible without ERα, because it was assumed to be the only mediator of estrogen signaling. Later, the ERβ knockout mouse was developed (Krege et al., 1998) and this was followed by the double ERαβ knockout model (Couse et al., 1999). These knockout mice revealed that life is possible without ERs, but reproductive functions are heavily impaired (Couse & Korach, 1999). Studies with female ERα knockout mice have demonstrated that these mice have larger kidneys compared to wild type mice. Furthermore, these ERα knockout females have elevated plasma levels of androgens, similar to the levels observed in wild type males. Both androgens and luteinizing hormone can promote renal growth which might explain the increase in kidney weight observed in female ERα knockout mice (Lovegrove et al., 2004).

The kidney has been described as the most estrogen responsive organ after gonadal tissue. All three ERs are expressed in the mammalian kidney, both in males and females (Fig. 2) (Jelinsky et al., 2003). However, it has been

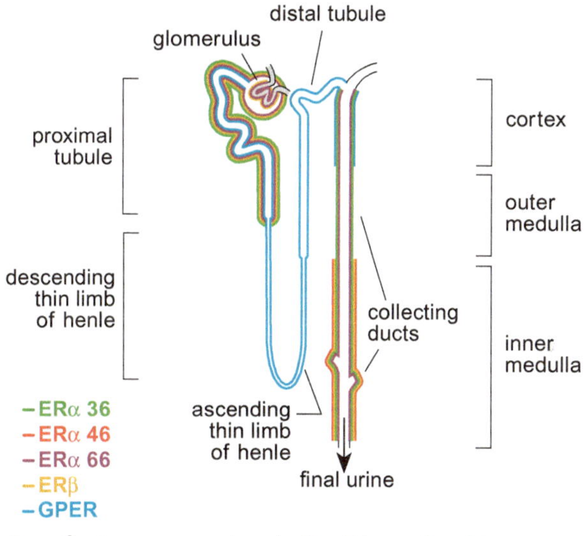

Fig. 2 Expression of estrogen receptors in the kidney. Graphic representation of the localization of different estrogen receptors (ER) in the nephron and collecting duct system. ERα36 (green) is present in glomerulus, proximal tubules in the cortical and inner medullary collecting ducts. ERα46 (red) is expressed in inner medullary collecting ducts. ERα66 (purple) is expressed in the cortical and inner medullary collecting ducts as well as the vasculature and the glomerulus, but also in the brush border of the proximal tubules. ERβ (yellow) is located in proximal tubules, in the podocytes and in the vasculature but also in inner medullary collecting ducts. GPER (blue) is predominantly expressed in the proximal tubules, distal convoluted tubules and the loop of Henle, additionally it is present in the cortical collecting duct.

suggested that ERα isoforms contribute the most to estrogen signaling in the kidney (Irsik, Carmines, & Lane, 2013; Jelinsky et al., 2003). Expression of all three ERα isoforms (ERα66, ERα46, and ERα36) was observed in the renal cortex from rats, while mice studies revealed the presence of ERα66 and ERα36 in both the cortical and inner medullary collecting ducts. In addition, our group demonstrated that ERα46 is expressed in isolated inner medullary collecting ducts (Cheema et al., 2015). Furthermore, using the kidneys of female mice, Irsik and colleagues demonstrated that ERα66 was mainly localized in the vasculature and the glomerulus as well as the brush border of the proximal tubule and the apical membrane of collecting duct cells (Irsik et al., 2013). Furthermore, they demonstrated that ERα36 was localized in glomerular mesangial cells and proximal tubular epithelial cells.

The renal expression profile of ERβ has been more difficult to identify and therefore it is suggested that the receptor is less abundantly present in the kidney (Wells et al., 2005). However, some rat studies revealed the presence of ERβ in the renal cortex at least at the transcriptional level. In mice, ERβ was detected in both the proximal tubule and collecting duct (Cheema et al., 2015). Additionally, it has been demonstrated that ERβ is more expressed in proximal tubular cells as compared to cells from distal nephron segments such as the collecting duct. ERβ was also detected in podocytes and endothelial cells (Irsik et al., 2013).

Taken together this demonstrate that both ERα and ERβ are expressed in the different nephron segments and additional studies are needed to further clarify the role of these estrogen receptors in renal physiology.

3.2 G protein-coupled estrogen receptor 1

GPER1, formerly known as orphan G protein-coupled receptor 30 (GPR30), is located on the cell surface and is mainly responsible for rapid estrogen signaling. GPER1 is expressed in the inner medulla of both rats and mice (Hazell et al., 2009), whereas other studies have showed GPER1 expression in tubular epithelial cells (Lindsey, Yamaleyeva, Brosnihan, Gallagher, & Chappell, 2011). In addition, Cheng and colleagues reported that GPER1 is predominantly expressed in proximal tubules, distal convoluted tubules and the loop of Henle, but they did not observe any expression in the collecting duct (Cheng et al., 2014). However, our group has demonstrated that GPER1 is expressed, at the transcriptional level, in cortical collecting duct mpkCCD cells (Cheema et al., 2015), consisted with studies by Hofmeister et al. who reported that GPER1 mRNA was present in isolated cortical collecting ducts (Hofmeister et al., 2011). The localization of GPER1 depends on cell type and the receptor is found intracellular or on the plasma membrane. In the kidney, GPER1 is predominantly localized to the basolateral membrane of both medullary and cortical renal epithelial cells (Cheng, Graeber, Quinn, & Filardo, 2011). Interestingly, it has been shown that the expression and subcellular localization of GPER1 is modulated during the estrus cycle. In the secretory phases of the estrus cycle, GPER1 expression is increased in cortical epithelia and localized to the basolateral membrane domains during proestrus, and redistributed intracellularly during estrus. In contrast, during metestrus and diestrus, when circulating levels of estrogen are low, GPER1 expression is significantly decreased in both cortical and medullary renal epithelial cell, possibly due to endocytic degradation (Cheng et al., 2014). Thus, there is clear evidence supporting the notion that GPER1 expression is modulated by the amount of circulating estrogens during the estrus cycle. Further studies are needed to demonstrate whether the different renal expression pattern of GPER1 during the estrus cycle plays a role in the regulation of renal functions.

4. Renoprotective effects of estrogen

Observations in mammals indicate differences between males and females regarding renal (patho)physiology. To illustrate, in most experimental models of renal disease, male animals progress more rapidly than females (Neugarten & Golestaneh, 2013; Silbiger & Neugarten, 1995). In addition, clinical studies have shown that chronic kidney disease (CKD) progresses more slowly in women as compared to men (Neugarten, Acharya, & Silbiger, 2000). These gender differences may be related to sex-hormones, which affect a number of renal cellular processes by influencing both the synthesis and activity of various cytokines, growth factors and vasoactive agents (Neugarten & Golestaneh, 2013). Estrogen plays an important role in the regulation of the renin-angiotensin system (Kang & Miller, 2002), transforming growth factor-β (TGFβ) signaling (Zdunek, Silbiger, Lei, & Neugarten, 2001) and extracellular matrix metabolism (Guccione, Silbiger, Lei, & Neugarten, 2002; Neugarten, Medve, Lei, & Silbiger, 1999), resulting in renoprotective effects of estrogen.

In both human and animal studies, estrogen has a vasodilatory effect by stimulating nitric oxide (NO) synthesis as well as decreasing the production of vasoconstrictor agents (Caulin-Glaser, García-Cardeña, Sarrel, Sessa William, & Bender Jeffrey, 1997; Huang, Sun, Koller, & Kaley, 2000).

Inner medullary endothelial nitric oxide synthase (eNOS) protein levels were higher in female rats as compared to male rats, and protein levels declined following ovariectomy (OVX), which is surgical removal of the ovaries. Renal medullary expression of eNOS was increased by estrogen administration in OVX rats (Neugarten, Ding, Friedman, Lei, & Silbiger, 1997). This ability of estrogen to upregulate eNOS may describe gender-related differences in renal medullary hemodynamics and contribute to the higher papillary blood flow seen in females (Neugarten et al., 1997; Silbiger & Neugarten, 1995). The production of NO is affected in renal failure and endothelial dysfunction occurs in CKD, which decreases the production of NO (Blum et al., 1998). This means that the lower level of NO production in patients with renal failure contribute to the progression of CKD and this progression is higher in males than females. Targeting the NO system using estrogen might provide therapeutic options for slowing the progression of CKD.

Estrogen has been shown to have a protective effect against renal damage (Ahmed & Ramesh, 2016; Gluhovschi et al., 2012). As mentioned before, renal injury is more pronounced in male compared to female rats, and these sex differences in renal disease reflect clinical studies indicating that progression of renal disease is more rapid in men than in women (Neugarten & Golestaneh, 2013). Estrogen deficiency can accelerate the progression of glomerulosclerosis in susceptible mice. In a study by Elliot and colleagues it was shown that ovariectomized sclerosis-prone ROP OS/+ mice developed more severe renal dysfunction than intact estrogen sufficient female mice (Elliot et al., 2003). Additionally, estrogen is known to preserve renal function in a chronic unilateral ureteral obstruction (UUO) model, a well-known model of CKD. Mao and colleagues showed improved GFR of the obstructed kidney in female rats treated with estrogen as compared to OVX rats with estrogen deficiency (Mao et al., 2014). They also demonstrated that OVX induced the expression of the fibrosis markers TGF-β and α-SMA, which could be mitigated by treatment with estrogen (Mao et al., 2014). Other rat models have also demonstrated positive effects of estrogen administration; e.g., estradiol therapy after OVX has been shown to normalize the renal functional reserve in conscious oophorectomized rats (Nielsen, Flyvbjerg, et al., 2003), to reduce glomerular damage in uninephrectomized, spontaneously hypertensive rats (Gross et al., 2004) and to diminish glomerulosclerosis and tubulointerstitial fibrosis in aging Dahl salt-sensitive rats (Maric, Sandberg, & Hinojosa-Laborde, 2004). Furthermore, Ji et al. demonstrated that estrogen protects against the development of glomerular and tubular damage in a renal wrap model of experimental hypertension (Ji et al., 2005). In addition, the same laboratory has shown that estrogen reduces hypertension-induced renal injury by suppressing the activity of nicotinamide adenine dinucleotide phosphate (NADPH) oxidase, which stimulates reactive oxygen species (ROS) production (Ji et al., 2007). Taken together, these findings demonstrate an important role for estrogen in the reduction of renal damage.

5. Selective estrogen receptor modulators and the kidney

Selective estrogen receptor modulators (SERMs) are synthetic non-steroidal compounds that can interact with estrogen receptors. These compounds have the capability of acting as either ER agonist or antagonist depending on the site of action, most likely due to the receptor confirmation structural changes associated with SERMs binding and the subsequent effect on transcription (Lonard & Smith, 2002). Preclinical and clinical studies indicate that each SERM has unique tissue-selective effects depending of specific ER agonist or antagonist effects (Dutertre & Smith, 2000; Love et al., 1992). SERMs have been proven to be effective against various diseases, including osteoporosis and breast cancer (Ellis, Hendrick, Williams, & Komm, 2015). FDA approved SERMs include tamoxifen and toremifene, which are used for the prevention and treatment of breast cancer; raloxifene approved for the prevention and treatment of osteoporosis as well as prevention of invasive breast cancer; and ospemifene approved for the treatment of dyspareunia from menopausal vaginal atrophy (Elkinson & Yang, 2013; Fisher et al., 1998; Gams, 1997; Vogel et al., 2010). To illustrate the duality of SERMs, it is known that tamoxifen has antiestrogenic effects on the mammary gland but exhibits estrogenic effects on the cardiovascular and skeletal system (Lonard & Smith, 2002; Sugerman, 2013). Tamoxifen exerts its effect via binding to the ERs at the same site as estradiol (Catalano et al., 2014) and it acts both as an agonist and antagonist for ERα, but only as an antagonist for ERβ (Barkhem et al., 1998; Hall & McDonnell, 1999; Matthews & Gustafsson, 2003).

Both tamoxifen and raloxifene have been demonstrated to have renoprotective effects in different in vitro and in vivo models (Lee et al., 2013). Tamoxifen significantly reduced UUO-induced renal tubular injury, interstitial matrix deposition and inflammatory processes, possibly via an ERα-dependent TGF-β/Smad signaling pathway (Kim et al., 2014).

Tamoxifen also showed antifibrotic effects in a L-NAME model of hypertensive nephrosclerosis (Delle et al., 2012). In addition, it has been shown that raloxifene ameliorates renal damage in db/db mice, a model of diabetic nephropathy (Chin et al., 2005). Likewise, a post hoc analysis of the MORE (Multiple Outcomes of Raloxifene Evaluation) trail, involving 7705 postmenopausal women, showed that the glomerular filtration rate (GFR) of women on raloxifene decreased at a slower rate over a 3-year period than those in the placebo group, suggesting that raloxifene might have renoprotective effects in postmenopausal women (Melamed et al., 2011). Together these finding demonstrate that SERMs can be used to modulate renal function and renal injury.

6. Regulation of water and salt homeostasis by estrogen

Several studies have reported an important role for estrogen in the regulation of water and salt homeostasis (Brunette & Leclerc, 2001; Stachenfeld, 2014; Stachenfeld, Taylor, Leone, & Keefe, 2003). High estrogen levels are associated with water and salt retention, edema and hypertension (Stachenfeld, 2014). Water transport in the collecting ducts is mainly regulated by arginine vasopressin (AVP) which stimulates the V2 receptor to enhance AQP2-mediated water reabsorption (Nielsen et al., 2002). Several studies have focused on the role of AVP and the V2 receptor in estrogen-mediated regulation of water and salt homeostasis and it has been demonstrated that estrogen can influence the threshold of osmoreceptors, and either stimulate or inhibit AVP release from the hypothalamus (Sladek & Somponpun, 2008). In addition to water retention observed during pregnancy and menopause, several studies in humans have described that women are more sensitive than men to the antidiuretic effects of AVP (Hvistendahl, Frokiaer, Nielsen, & Djurhuus, 2007; Stachenfeld, Dipietro, Palter, & Nadel, 1998; Stachenfeld, Splenser, Calzone, Taylor, & Keefe, 2001). This might also be the reason why women are more at risk to develop hyponatremia, for example, in response to administration of desmopressin—an AVP analog (Juul, Klein, Sandström, Erichsen, & Nørgaard, 2011). In line with this observation, it has been demonstrated that whole kidney V2 receptor expression is almost twofold higher in female rats as compared to male rats (Liu et al., 2011), providing a likely explanation for the increased sensitivity of females to V2 receptor agonism.

In addition, it has been reported that estrogen induces AVP secretion in postmenopausal women. Progesterone alone had no effect on AVP levels and it counteracts estrogen-induced AVP secretion in postmenopausal women given both estrogen and progesterone (Forsling, Stromberg, & Akerlund, 1982) indicating that progesterone can influence estrogen-induced AVP secretion. Graugaard-Jensen and colleagues showed that high and low levels of estradiol, as seen in the follicular phase of the menstrual cycle, had no influence on the circadian rhythm of AVP or subsequent urine production in young women (Graugaard-Jensen, Hvistendahl, Frokiaer, Bie, & Djurhuus, 2008). An explanation for the variation in AVP concentration could be related to which ER subtypes are activated in the supraoptic nucleus of the hypothalamus. It has been shown that young premenopausal women mainly express ERβ, whereas postmenopausal women express ERα (Ishunina, Kruijver, Balesar, & Swaab, 2000). ERβ inhibits AVP activity in the supraoptic nucleus whereas ERα has a stimulatory effect (Paech et al., 1997). This might explain the elevated AVP levels detected in postmenopausal women in response to estrogen administration and the absence of a response in young premenopausal women. The studies discussed above primarily focused on AVP and the V2 receptor, yet the impact of estrogen on the AVP-AQP2 axis is less widely studied and needs further attention.

7. Estrogen-mediated regulation of AQP2 in the kidney

Water retention is common during pregnancy and it has been demonstrated that inner medullary AQP2 mRNA and protein levels are increased during this period, resulting in V2 receptor-mediated water retention (Ohara et al., 1998). During pregnancy, the extracellular fluid, plasma and blood volume increases 30%–50% in humans and other mammalian species (Phippard et al., 1986; Schrier & Briner, 1991). Additionally, there is increased arterial vasodilation, which, together with sodium and water retention, results in a lower plasma osmolality. Studies have suggested that a change in vasopressin secretion during pregnancy increases the expression of AQP2, which can also cause the observed decrease in plasma osmolality (Ohara et al., 1998).

Menopause can likewise cause water retention. The loss of estrogen during menopause is associated with increased salt-sensitivity and reduced GFR (Pechere-Bertschi & Burnier, 2004). In postmenopause, sex hormone levels are altered, which also changes the regulation of body water (Stachenfeld et al., 1998). The postmenopausal condition can be mimicked in female OVX animal models. Our group has demonstrated that OVX resulted in increased urine osmolality and increased AQP2 phosphorylation at Serine

256 (p256-AQP2), which is critical for AQP2 trafficking to the apical membrane of collecting duct principal cells and water reabsorption. Interestingly, estradiol treatment reduced OVX-induced expression of AQP2, Ser256 phosphorylation and AQP2 trafficking in collecting ducts. Consistent with these findings, estradiol administration restored urine output and urinary osmolality in OVX animals (Cheema et al., 2015), indicating that estradiol might be responsible for changes in renal water balance via regulation of AQP2.

In vitro systems can be used to investigate the isolated effect of estrogen on AQP2 regulation without interference of circulating hormones. We demonstrated that exposure of mpkCCD cells to estradiol, in the absence of deamino-8-D-AVP (dDAVP), reduced the gene and protein expression of AQP2. Moreover, estradiol treatment significantly diminished dDAVP-induced AQP2 protein expression (Cheema et al., 2015). These data indicate a direct inhibitory effect of estradiol on AQP2 gene transcription, but also suggest that estradiol attenuates AQP2 protein expression in the presence of AVP. Interestingly, Zou and colleagues identified a functional ERE in the AQP2 promoter in endometrial carcinoma cells (Zou et al., 2011). This ERE is a 35 bp sequence, containing a short palindromic sequence as essential element. A similar palindrome-containing ERE has been identified in the liver of frogs and chickens in genes that encode for yolk and egg white, respectively, and the production of these proteins is known to be regulated by estrogen (Klein-Hitpaß, Schorpp, Wagner, & Ryffel, 1986; Lai, Riser, & O'Malley, 1983; Zou et al., 2011). ERE can mediate the regulation of AQP2 by estradiol in normal endometrium as well as in endometrial carcinoma cells (Zou et al., 2011), and a similar molecular mechanisms might underlie estrogen-mediated regulation of AQP2 in the renal collecting duct as well.

Precisely how the ERE in the AQP2 promoter contributes to the regulation of AQP2 transcription is somewhat complex as ERs bind to their cognate ERE as a dimer. However, studies from our laboratory have demonstrated that the splice variants ERα46 and ERα36, as well as ERβ, colocalize with AQP2 in both male and female inner medullary collecting ducts and mpkCCD cells (Cheema et al., 2015). Irsik and colleagues have likewise demonstrated that both ERα66 and ERα36 co-localize with AQP2 in the cortical collecting duct of female mice (Irsik et al., 2013). Taken together, these findings support a direct role of estrogen in the regulation of AQP2 expression in the collecting duct. In line with this, we have demonstrated increased AQP2 protein expression in mice lacking ERα as

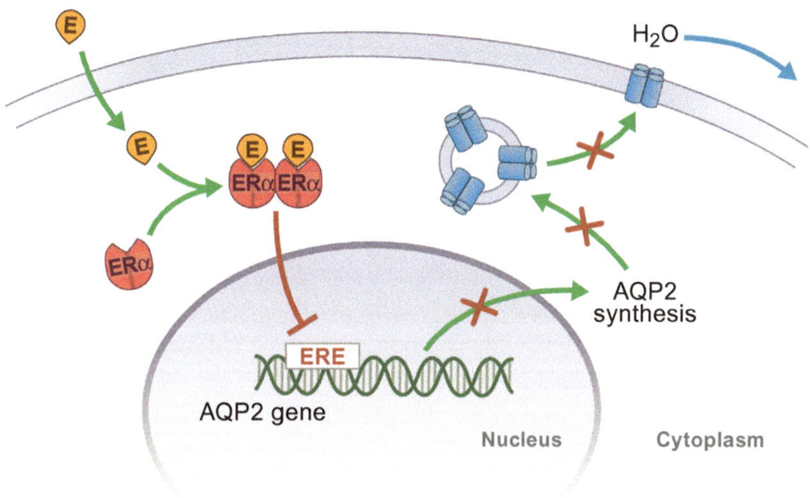

Fig. 3 Proposed model for estrogen-mediated regulation of AQP2 in collecting duct principal cell. Estrogen binds to the estrogen receptor α subunit (ERα) in the cytoplasm which mediate an inhibitory effect of estrogen on gene transcription of AQP2. This will result in decreased synthesis of AQP2 protein and thereby mitigate trafficking to the apical membrane, leading to reduced water reabsorption. See text for further details. *E*, estrogen; *ERα*, estrogen receptor α; *ERE*, estrogen response element.

compared to wild type controls, indicating that the inhibitory effect of estrogen on AQP2 expression is mediated by ERα (Cheema et al., 2015). A proposed model for estrogen-mediated regulation of AQP2 in the collecting duct is represented in Fig. 3.

8. Selective estrogen receptor modulators affect renal AQP2

SERMs have several pharmacological effects on ERs, including both agonist and antagonist actions. The SERM tamoxifen has been reported to be both an agonist and antagonist for ERα, but only an antagonist for ERβ (Barkhem et al., 1998; Hall & McDonnell, 1999; Matthews & Gustafsson, 2003). Recent studies from our laboratory have shown that tamoxifen affects renal water and salt homeostasis (Tingskov, Hu, et al., 2018; Tingskov, Kwon, Frøkiær, & Nørregaard, 2018). Using a lithium-induced nephrogenic diabetes insipidus (NDI) model in rats, we showed that tamoxifen can counteract the adverse effects of lithium-induced polyuria after the establishment of lithium-induced NDI. Lithium is used for the treatment of bipolar disorders and is the predominant cause of acquired NDI (Boton, Gaviria, & Batlle, 1987).

NDI is associated with polyuria caused by a marked downregulation of AQP2 (Kwon et al., 2000; Marples, Christensen, Christensen, Ottosen, & Nielsen, 1995; Nielsen, Kwon, et al., 2003), and our data demonstrate that tamoxifen attenuates the downregulation of both AQP2 mRNA and protein levels in rats subjected to lithium-induced NDI (Tingskov, Hu, et al., 2018). Additionally, we demonstrated that tamoxifen increased AQP2 expression in freshly isolated inner medullary collecting duct (IMCD) tubule suspension from lithium-treated rats, indicating that tamoxifen can affect AQP2 expression directly in the collecting duct (Tingskov, Hu, et al., 2018). cAMP response element-binding protein (CREB) is a key component in the cAMP signaling pathway, and it plays an important role in the regulation of AQP2. Numerous studies have demonstrated that increased phosphorylation and activation of CREB causes stimulation of AQP2 expression (Matsumura, Uchida, Rai, Sasaki, & Marumo, 1997; Umenishi, Narikiyo, Vandewalle, & Schrier, 2006). Our ex vivo studies demonstrated that tamoxifen increased the pCREB/CREB ratio in the IMCD suspension prepared from lithium-treated rats, indicating that tamoxifen might induce AQP2 expression via activation of the CREB pathway (Tingskov, Hu, et al., 2018). In line with this, it has previously been shown that tamoxifen can activate cAMP-dependent pCREB signaling in rat astrocytes (Karki et al., 2013) and the hippocampus (Sharma, Mehra, Dhar, & Vij, 2007), indicating a direct role of tamoxifen in regulation of the cAMP/CREB pathway. Even though tamoxifen can interact with all three ERs, it is still unclear whether the impact we observed in lithium-treated NDI rats is due to an agonistic or antagonistic effect of tamoxifen (Tingskov, Hu, et al., 2018). Future studies are needed to identify whether tamoxifen has agonistic or antagonistic effects and to evaluate whether chronic tamoxifen administration could be a useful therapeutic approach for patients with lithium-induced NDI.

9. Estrogen-mediated regulation of AQP2 in other organ systems

AQP2 is also present in other organs than the kidney. A recent study has shown that estrogen affects AQP2 expression in glioma cells by interacting with ERα and ERβ. It has been suggested that female sex hormones play an important role in the development of gliomas in women (Qi, Shao, Zhang, Hui, & Wang, 2013). No studies have previously demonstrated AQP2 expression in gliomas, but Wan and colleagues detected AQP2 with differential sublocalization in gliomas. In addition, they found that AQP2 was located in all parts of glioma cells or glial cells in the tissues and was located mainly in the nuclei in cell lines. They found that estradiol was able to decrease AQP2 expression in the cell lines, particularly in the nucleus, suggesting that the molecular mechanisms by which estradiol regulates AQP2 localization occurs at the nuclear and membrane levels mediated by ERs. Furthermore, their study shows that estradiol is able to inhibit the phosphorylation of AQP2 (ser256) and pAQP2 (ser264) and increase the amount of pAQP2 (ser261), which may reduce AQP2 accumulation on the membrane (Wan et al., 2018). Taken together, these findings suggest that estrogen reduces AQP2 membrane accumulation by inhibiting phosphorylation of AQP2 and reducing the quantity of AQP2 in the nucleus.

Another study showed that AQP2 mRNA in human ectocervical-vaginal epithelial cells (hECEs) was highest in normally menstruating women and lowest in postmenopausal women, indicating that AQP2 expression correlated with circulating estrogen levels. These findings also suggest that estrogen and AQP2 might play a role in the water balance of the uterus (He et al., 2005). In line with this, it was demonstrated that hECEs were sensitive to estrogen and that it modulates vaginal secretion. This study also established that estradiol enhanced AQP2 expression in hECEs leading to a higher vaginal secretion (Zhao, Lai, & Dong, 2014). The production of vaginal endocrine fluid varies with age. The vagina is moist during the growth period, during perimenopause the vaginal wall gradually reduces in size, decreases secretions, and exhibits dryness, and in menopause almost no vaginal secretion occurs (Semmens & Wagner, 1982). Although AQP2 expression is affected by a number of factors, the analysis of AQP2 mRNA and protein expression in hECEs suggests a close relationship with age. One of the consequences of reduced estrogen levels, as observed in postmenopausal women, is vaginal dryness. This is likely due to the decreased expression of AQP2 in hECEs, leading to water transport barriers and environmental imbalance in the vagina. Thus, there is a theoretical basis to use estrogen as a treatment for impaired water transport in hECEs (Zhao et al., 2014). The reduced excretion of estrogen due to menopause or OVX affects AQP2 expression in the bladder and uterus of rodents. One study, using OVX rats, showed that AQP2 expression in the bladder was regulated by estrogen (Kim et al., 2012). Additionally, it was also demonstrated that estrogen regulates AQP2 expression in the uterus of OVX mice (Jablonski, McConnell, Hughes, & Huet-Hudson, 2003). Together these studies demonstrate that estrogen does also play an important role for the regulation of AQP2 in other organ systems and these studies provide several novel pieces of information regarding mechanisms of AQP2 function.

10. Conclusions and perspectives

AQP2 was originally found in the collecting ducts of the kidney, and it functions as an AVP-sensitive water channel in principal cells. Recently, it has been demonstrated that estrogen plays an important role in regulating AQP2 abundance and membrane translocation in renal collecting duct cells. These findings add to the current knowledge about estrogen-mediated water homeostasis. In terms of wider implications, these findings indicate that changes in estrogen levels during the menstrual cycle may have an impact on renal water handling. Moreover, decreased levels of estrogen may contribute to the dysfunctional renal water handling associated with menopause. Yet, additional studies are required to fully elucidate the molecular mechanisms underlying estrogen-mediated regulation of AQP2 expression and functionality in the kidney and other organs.

Acknowledgments

We thank Professor Jens Christian Djurhuus for critical reading of this manuscript. This work was supported by the Danish Council for Independent Research | Medical Sciences, Lundbeck Foundation and Aarhus University Foundation.

References

Ahmed, S. B., & Ramesh, S. (2016). Sex hormones in women with kidney disease. *Nephrology, Dialysis, Transplantation*, *31*, 1787–1795.

Allen, E., & Doisy, E. A. (1923). An ovarian hormone: Preliminary report on its localization, extraction and partial purification, and action in test animals. *Journal of the American Medical Association*, *81*, 819–821.

Aronica, S. M., Kraus, W. L., & Katzenellenbogen, B. S. (1994). Estrogen action via the camp signaling pathway: Stimulation of adenylate cyclase and cAMP-regulated gene transcription. *Proceedings of the National Academy of Sciences of the United States of America*, *91*, 8517–8521.

Barkhem, T., Carlsson, B., Nilsson, Y., Enmark, E., Gustafsson, J.-Å., & Nilsson, S. (1998). Differential response of estrogen receptor α and estrogen receptor β to partial estrogen agonists/antagonists. *Molecular Pharmacology*, *54*, 105–112.

Blum, M., Yachnin, T., Wollman, Y., Chernihovsky, T., Peer, G., Grosskopf, I., et al. (1998). Low nitric oxide production in patients with chronic renal failure. *Nephron*, *79*, 265–268.

Boton, R., Gaviria, M., & Batlle, D. C. (1987). Prevalence, pathogenesis, and treatment of renal dysfunction associated with chronic lithium therapy. *American Journal of Kidney Diseases*, *10*, 329–345.

Brunette, M. G., & Leclerc, M. (2001). Effect of estrogen on calcium and sodium transport by the nephron luminal membranes. *Journal of Endocrinology*, *170*, 441–450.

Carmeci, C., Thompson, D. A., Ring, H. Z., Francke, U., & Weigel, R. J. (1997). Identification of a gene (GPR30) with homology to the G-protein-coupled receptor superfamily associated with estrogen receptor expression in breast cancer. *Genomics*, *45*, 607–617.

Catalano, S., Giordano, C., Panza, S., Chemi, F., Bonofiglio, D., Lanzino, M., et al. (2014). Tamoxifen through GPER upregulates aromatase expression: A novel mechanism sustaining tamoxifen-resistant breast cancer cell growth. *Breast Cancer Research and Treatment*, *146*, 273–285.
Caulin-Glaser, T., García-Cardeña, G., Sarrel, P., Sessa William, C., & Bender Jeffrey, R. (1997). 17β-estradiol regulation of human endothelial cell basal nitric oxide release, independent of cytosolic Ca^{2+} mobilization. *Circulation Research*, *81*, 885–892.
Cheema, M. U., Irsik, D. L., Wang, Y., Miller-Little, W., Hyndman, K. A., Marks, E. S., et al. (2015). Estradiol regulates AQP2 expression in the collecting duct: A novel inhibitory role for estrogen receptor alpha. *American Journal of Physiology. Renal Physiology*, *309*, F305–F317.
Cheng, S. B., Dong, J., Pang, Y., LaRocca, J., Hixon, M., Thomas, P., et al. (2014). Anatomical location and redistribution of G-protein-coupled estrogen receptor-1 during the estrus cycle in mouse kidney and specific binding to estrogens but not aldosterone. *Molecular and Cellular Endocrinology*, *382*, 950–959.
Cheng, S. B., Graeber, C. T., Quinn, J. A., & Filardo, E. J. (2011). Retrograde transport of the transmembrane estrogen receptor, G-protein-coupled-receptor-30 (GPR30/GPER) from the plasma membrane towards the nucleus. *Steroids*, *76*, 892–896.
Chin, M., Isono, M., Isshiki, K., Araki, S., Sugimoto, T., Guo, B., et al. (2005). Estrogen and raloxifene, a selective estrogen receptor modulator, ameliorate renal damage in db/db mice. *American Journal of Pathology*, *166*, 1629–1636.
Couse, J. F., Hewitt, S. C., Bunch, D. O., Sar, M., Walker, V. R., Davis, B. J., et al. (1999). Postnatal sex reversal of the ovaries in mice lacking estrogen receptors α and β. *Science*, *286*, 2328–2331.
Couse, J. F., & Korach, K. S. (1999). Estrogen receptor null mice: What have we learned and where will they lead us? *Endocrine Reviews*, *20*, 358–417.
Delle, H., Rocha, J. R., Cavaglieri, R. C., Vieira, J. M., Jr., Malheiros, D. M., & Noronha, I. L. (2012). Antifibrotic effect of tamoxifen in a model of progressive renal disease. *Journal of the American Society of Nephrology*, *23*, 37–48.
Dutertre, M., & Smith, C. L. (2000). Molecular mechanisms of selective estrogen receptor modulator (SERM) action. *Journal of Pharmacology and Experimental Therapeutics*, *295*, 431–437.
Elkinson, S., & Yang, L. P. H. (2013). Ospemifene: First global approval. *Drugs*, *73*, 605–612.
Elliot, S. J., Karl, M., Berho, M., Potier, M., Zheng, F., Leclercq, B., et al. (2003). Estrogen deficiency accelerates progression of glomerulosclerosis in susceptible mice. *American Journal of Pathology*, *162*, 1441–1448.
Ellis, A. J., Hendrick, V. M., Williams, R., & Komm, B. S. (2015). Selective estrogen receptor modulators in clinical practice: A safety overview. *Expert Opinion on Drug Safety*, *14*, 921–934.
Fisher, B., Costantino, J. P., Wickerham, D. L., Redmond, C. K., Kavanah, M., Cronin, W. M., et al. (1998). Tamoxifen for prevention of breast cancer: Report of the National Surgical Adjuvant Breast and Bowel Project P-1 Study. *Journal of the National Cancer Institute*, *90*, 1371–1388.
Forsling, M. L., Stromberg, P., & Akerlund, M. (1982). Effect of ovarian steroids on vasopressin secretion. *Journal of Endocrinology*, *95*, 147–151.
Gams, R. (1997). Phase III trials of toremifene vs tamoxifen. *Oncology (Williston Park, NY)*, *11*, 23–28.
Gluhovschi, G., Gluhovschi, A., Anastasiu, D., Petrica, L., Gluhovschi, C., & Velciov, S. (2012). Chronic kidney disease and the involvement of estrogen hormones in its pathogenesis and progression. *Romanian Journal of Internal Medicine*, *50*, 135–144.
Göttlicher, M., Heck, S., & Herrlich, P. (1998). Transcriptional cross-talk, the second mode of steroid hormone receptor action. *Journal of Molecular Medicine*, *76*, 480–489.

Graugaard-Jensen, C., Hvistendahl, G. M., Frokiaer, J., Bie, P., & Djurhuus, J. C. (2008). The influence of high and low levels of estrogen on diurnal urine regulation in young women. *BioMed Central Urology*, *8*, 16.

Gross, M.-L., Adamczak, M., Rabe, T., Harbi, N. A., Krtil, J., Koch, A., et al. (2004). Beneficial effects of estrogens on indices of renal damage in uninephrectomized SHRsp rats. *Journal of the American Society of Nephrology*, *15*, 348–358.

Guccione, M., Silbiger, S. R., Lei, J., & Neugarten, J. (2002). Estradiol upregulates mesangial cell MMP-2 activity via the transcription factor AP-2. *American Journal of Physiology. Renal Physiology*, *282*, F164–F169.

Hall, J. E. (2011). *Guyton and hall textbook of medical physiology* (12th ed.). . Elsevier Health Sciences.

Hall, J. M., & McDonnell, D. P. (1999). The estrogen receptor β-isoform (ERβ) of the human estrogen receptor modulates ERα transcriptional activity and is a key regulator of the cellular response to estrogens and antiestrogens. *Endocrinology*, *140*, 5566–5578.

Hazell, G. G., Yao, S. T., Roper, J. A., Prossnitz, E. R., O'Carroll, A.-M., & Lolait, S. J. (2009). Localisation of GPR30, a novel G-protein-coupled oestrogen receptor, suggests multiple functions in rodent brain and peripheral tissues. *Journal of Endocrinology*, *202*, 223–236.

He, R. H., Zhou, C. Y., Wang, B., Sun, H. L., Jin, F., & Huang, H. F. (2005). Expression of aquaporin 2, a water channel, in human endometrium during the menstrual cycle. *Zhonghua Yi Xue Za Zhi*, *85*, 2420–2424.

Hofmeister, M. V., Damkier, H. H., Christensen, B. M., Olde, B., Fredrik Leeb-Lundberg, L., Fenton, R. A., et al. (2011). 17β-estradiol induces nongenomic effects in renal intercalated cells through G-protein-coupled estrogen receptor 1. *American Journal of Physiology. Renal Physiology*, *302*, F358–F368.

Huang, A., Sun, D., Koller, A., & Kaley, G. (2000). 17β-estradiol restores endothelial nitric oxide release to shear stress in arterioles of male hypertensive rats. *Circulation*, *101*, 94–100.

Hvistendahl, G. M., Frokiaer, J., Nielsen, S., & Djurhuus, J. C. (2007). Gender differences in nighttime plasma arginine vasopressin and delayed compensatory urine output in the elderly population after desmopressin. *Journal of Urology*, *178*, 2671–2676.

Improta-Brears, T., Whorton, A. R., Codazzi, F., York, J. D., Meyer, T., & McDonnell, D. P. (1999). Estrogen-induced activation of mitogen-activated protein kinase requires mobilization of intracellular calcium. *Proceedings of the National Academy of Sciences of the United States of America*, *96*, 4686–4691.

Irsik, D. L., Carmines, P. K., & Lane, P. H. (2013). Classical estrogen receptors and ERα splice variants in the mouse. *PLoS One*, *8*, e70926.

Ishunina, T. A., Kruijver, F. P., Balesar, R., & Swaab, D. F. (2000). Differential expression of estrogen receptor alpha and beta immunoreactivity in the human supraoptic nucleus in relation to sex and aging. *Journal of Clinical Endocrinology and Metabolism*, *85*, 3283–3291.

Jablonski, E. M., McConnell, N. A., Hughes, F. M., Jr., & Huet-Hudson, Y. M. (2003). Estrogen regulation of aquaporins in the mouse uterus: Potential roles in uterine water movement 1. *Biology of Reproduction*, *69*, 1481–1487.

Jelinsky, S. A., Harris, H. A., Brown, E. L., Flanagan, K., Zhang, X., Tunkey, C., et al. (2003). Global transcription profiling of estrogen activity: Estrogen receptor α regulates gene expression in the kidney. *Endocrinology*, *144*, 701–710.

Jensen, E. V. (1962). Basic guides to the mechanism of estrogen action. *Recent Progress in Hormone Research*, *18*, 387–414.

Ji, H., Menini, S., Mok, K., Zheng, W., Pesce, C., Kim, J., et al. (2005). Gonadal steroid regulation of renal injury in renal wrap hypertension. *American Journal of Physiology. Renal Physiology*, *288*, F513–F520.

Ji, H., Zheng, W., Menini, S., Pesce, C., Kim, J., Wu, X., et al. (2007). Female protection in progressive renal disease is associated with estradiol attenuation of superoxide production. *Gender Medicine*, *4*, 56–71.

Juul, K. V., Klein, B. M., Sandström, R., Erichsen, L., & Nørgaard, J. P. (2011). Gender difference in antidiuretic response to desmopressin. *American Journal of Physiology. Renal Physiology*, *300*, F1116–F1122.

Kang, A. K., & Miller, J. A. (2002). Effects of gender on the renin-angiotensin system, blood pressure, and renal function. *Current Hypertension Reports*, *4*, 143–151.

Karki, P., Webb, A., Smith, K., Lee, K., Son, D. S., Aschner, M., et al. (2013). Camp response element-binding protein (CREB) and nuclear factor κB mediate the tamoxifen-induced up-regulation of glutamate transporter 1 (GLT-1) in rat astrocytes. *Journal of Biological Chemistry*, *288*, 28975–28986.

Kim, D., Lee, A. S., Jung, Y. J., Yang, K. H., Lee, S., Park, S. K., et al. (2014). Tamoxifen ameliorates renal tubulointerstitial fibrosis by modulation of estrogen receptor alpha-mediated transforming growth factor-beta 1/Smad signaling pathway. *Nephrology, Dialysis, Transplantation*, *29*, 2043–2053.

Kim, S.-O., Song, S. H., Hwang, E. C., Oh, K. J., Ahn, K., Jung, S. I., et al. (2012). Changes in aquaporin (AQP)2 and AQP3 expression in ovariectomized rat urinary bladder: Potential implication of water permeability in urinary bladder. *World Journal of Urology*, *30*, 207–212.

Klein-Hitpaß, L., Schorpp, M., Wagner, U., & Ryffel, G. U. (1986). An estrogen-responsive element derived from the 5′ flanking region of the Xenopus vitellogenin A2 gene functions in transfected human cells. *Cell*, *46*, 1053–1061.

Klinge, C. M. (2001). Estrogen receptor interaction with estrogen response elements. *Nucleic Acids Research*, *29*, 2905–2919.

Krege, J. H., Hodgin, J. B., Couse, J. F., Enmark, E., Warner, M., Mahler, J. F., et al. (1998). Generation and reproductive phenotypes of mice lacking estrogen receptor β. *Proceedings of the National Academy of Sciences of the United States of America*, *95*, 15677–15682.

Kuiper, G. G., Enmark, E., Pelto-Huikko, M., Nilsson, S., & Gustafsson, J. A. (1996). Cloning of a novel receptor expressed in rat prostate and ovary. *Proceedings of the National Academy of Sciences of the United States of America*, *93*, 5925–5930.

Kwon, T.-H., Laursen, U. H., Marples, D., Maunsbach, A. B., Knepper, M. A., Frøkiær, J., et al. (2000). Altered expression of renal AQPs and Na^+ transporters in rats with lithium-induced NDI. *American Journal of Physiology. Renal Physiology*, *279*, F552–F564.

Lai, E. C., Riser, M. E., & O'Malley, B. W. (1983). Regulated expression of the chicken ovalbumin gene in a human estrogen-responsive cell line. *Journal of Biological Chemistry*, *258*, 12693–12701.

Lee, W. L., Cheng, M. H., Tarng, D. C., Yang, W. C., Lee, F. K., & Wang, P. H. (2013). The benefits of estrogen or selective estrogen receptor modulator on kidney and its related disease-chronic kidney disease-mineral and bone disorder: Osteoporosis. *Journal of the Chinese Medical Association*, *76*, 365–371.

Lindsey, S. H., Yamaleyeva, L. M., Brosnihan, K. B., Gallagher, P. E., & Chappell, M. C. (2011). Estrogen receptor GPR30 reduces oxidative stress and proteinuria in the salt-sensitive female mRen2.Lewis rat. *Hypertension*, *58*, 665–671.

Liu, J., Sharma, N., Zheng, W., Ji, H., Tam, H., Wu, X., et al. (2011). Sex differences in vasopressin V(2) receptor expression and vasopressin-induced antidiuresis. *American Journal of Physiology. Renal Physiology*, *300*, F433–F440.

Lonard, D. M., & Smith, C. L. (2002). Molecular perspectives on selective estrogen receptor modulators (SERMs): Progress in understanding their tissue-specific agonist and antagonist actions. *Steroids*, *67*, 15–24.

Losel, R., & Wehling, M. (2003). Nongenomic actions of steroid hormones. *Nature Reviews Molecular Cell Biology*, *4*, 46–56.

Love, R. R., Mazess, R. B., Barden, H. S., Epstein, S., Newcomb, P. A., Jordan, V. C., et al. (1992). Effects of tamoxifen on bone mineral density in postmenopausal women with breast cancer. *New England Journal of Medicine, 326*, 852–856.

Lovegrove, A. S., Sun, J., Gould, K. A., Lubahn, D. B., Korach, K. S., & Lane, P. H. (2004). Estrogen receptor α-mediated events promote sex-specific diabetic glomerular hypertrophy. *American Journal of Physiology. Renal Physiology, 287*, F586–F591.

Lubahn, D. B., Moyer, J. S., Golding, T. S., Couse, J. F., Korach, K. S., & Smithies, O. (1993). Alteration of reproductive function but not prenatal sexual development after insertional disruption of the mouse estrogen receptor gene. *Proceedings of the National Academy of Sciences of the United States of America, 90*, 11162–11166.

Mao, S., Xu, H., Zou, L., Xu, G., Wu, Z., Ding, Q., et al. (2014). Estrogen preserves split renal function in a chronic complete unilateral ureteral obstruction animal model. *Experimental and Therapeutic Medicine, 7*, 1555–1562.

Maric, C., Sandberg, K., & Hinojosa-Laborde, C. (2004). Glomerulosclerosis and tubulointerstitial fibrosis are attenuated with 17β-estradiol in the aging Dahl salt sensitive rat. *Journal of the American Society of Nephrology, 15*, 1546–1556.

Marples, D., Christensen, S., Christensen, E. I., Ottosen, P. D., & Nielsen, S. (1995). Lithium-induced downregulation of aquaporin-2 water channel expression in rat kidney medulla. *Journal of Clinical Investigation, 95*, 1838.

Marrian, G. F., & Butenandt, A. (1931). Oestrus-producing hormones. *Nature, 128*, 305.

Matsumura, Y., Uchida, S., Rai, T., Sasaki, S., & Marumo, F. (1997). Transcriptional regulation of aquaporin-2 water channel gene by cAMP. *Journal of the American Society of Nephrology, 8*, 861–867.

Matthews, J., & Gustafsson, J.-Å. (2003). Estrogen signaling: A subtle balance between ERα and ERβ. *Molecular Interventions, 3*, 281.

Melamed, M. L., Blackwell, T., Neugarten, J., Arnsten, J. H., Ensrud, K. E., Ishani, A., et al. (2011). Raloxifene, a selective estrogen receptor modulator, is renoprotective: A post-hoc analysis. *Kidney International, 79*, 241–249.

Neugarten, J., Acharya, A., & Silbiger, S. R. (2000). Effect of gender on the progression of nondiabetic renal disease: A meta-analysis. *Journal of the American Society of Nephrology, 11*, 319–329.

Neugarten, J., Ding, Q., Friedman, A., Lei, J., & Silbiger, S. (1997). Sex hormones and renal nitric oxide synthases. *Journal of the American Society of Nephrology, 8*, 1240–1246.

Neugarten, J., & Golestaneh, L. (2013). Gender and the prevalence and progression of renal disease. *Advances in Chronic Kidney Disease, 20*, 390–395.

Neugarten, J., Medve, I., Lei, J., & Silbiger, S. R. (1999). Estradiol suppresses mesangial cell type I collagen synthesis via activation of the map kinase cascade. *American Journal of Physiology Renal Physiology, 277*, F875–F881.

Nielsen, C. B., Flyvbjerg, A., Bruun, J. M., Forman, A., Wogensen, L., & Thomsen, K. (2003). Decreases in renal functional reserve and proximal tubular fluid output in conscious oophorectomized rats: Normalization with sex hormone substitution. *Journal of the American Society of Nephrology, 14*, 3102–3110.

Nielsen, S., Frokiaer, J., Marples, D., Kwon, T. H., Agre, P., & Knepper, M. A. (2002). Aquaporins in the kidney: From molecules to medicine. *Physiological Reviews, 82*, 205–244.

Nielsen, J., Kwon, T. H., Praetorius, J., Kim, Y. H., Frokiaer, J., Knepper, M. A., et al. (2003). Segment-specific ENaC downregulation in kidney of rats with lithium-induced NDI. *American Journal of Physiology. Renal Physiology, 285*, F1198–F1209.

Nilsson, S., Mäkelä, S., Treuter, E., Tujague, M., Thomsen, J., Andersson, G., et al. (2001). Mechanisms of estrogen action. *Physiological Reviews, 81*, 1535–1565.

Ohara, M., Martin, P. Y., Xu, D. L., St John, J., Pattison, T. A., Kim, J. K., et al. (1998). Upregulation of aquaporin 2 water channel expression in pregnant rats. *The Journal of Clinical Investigation, 101*, 1076–1083.

Paech, K., Webb, P., Kuiper, G. G., Nilsson, S., Gustafsson, J., Kushner, P. J., et al. (1997). Differential ligand activation of estrogen receptors ERα and ERβ at AP1 sites. *Science*, *277*, 1508–1510.

Pechere-Bertschi, A., & Burnier, M. (2004). Female sex hormones, salt, and blood pressure regulation. *American Journal of Hypertension*, *17*, 994–1001.

Phippard, A. F., Horvath, J. S., Glynn, E. M., Garner, M. G., Fletcher, P. J., Duggin, G. G., et al. (1986). Circulatory adaptation to pregnancy—Serial studies of haemodynamics, blood volume, renin and aldosterone in the baboon (*Papio hamadryas*). *Journal of Hypertension*, *4*, 773–779.

Qi, Z.-Y., Shao, C., Zhang, X., Hui, G.-Z., & Wang, Z. (2013). Exogenous and endogenous hormones in relation to glioma in women: A meta-analysis of 11 case-control studies. *PLoS One*, *8*, e68695.

Sabolic, I., Asif, A. R., Budach, W. E., Wanke, C., Bahn, A., & Burckhardt, G. (2007). Gender differences in kidney function. *Pflugers Archiv—European Journal of Physiology*, *455*, 397–429.

Schrier, R. W., & Briner, V. A. (1991). Peripheral arterial vasodilation hypothesis of sodium and water retention in pregnancy: Implications for pathogenesis of preeclampsia-eclampsia. *Obstetrics & Gynecology*, *77*, 632–639.

Semmens, J. P., & Wagner, G. (1982). Estrogen deprivation and vaginal function in postmenopausal women. *Journal of the American Medical Association*, *248*, 445–448.

Sharma, K., Mehra, R. D., Dhar, P., & Vij, U. (2007). Chronic exposure to estrogen and tamoxifen regulates synaptophysin and phosphorylated camp response element-binding (CREB) protein expression in CA1 of ovariectomized rat hippocampus. *Brain Research*, *1132*, 10–19.

Silbiger, S. R., & Neugarten, J. (1995). The impact of gender on the progression of chronic renal disease. *American Journal of Kidney Diseases*, *25*, 515–533.

Simpson, E. R. (2003). Sources of estrogen and their importance. *Journal of Steroid Biochemistry and Molecular Biology*, *86*, 225–230.

Sladek, C. D., & Somponpun, S. J. (2008). Estrogen receptors: Their roles in regulation of vasopressin release for maintenance of fluid and electrolyte homeostasis. *Frontiers in Neuroendocrinology*, *29*, 114–127.

Stachenfeld, N. S. (2014). Hormonal changes during menopause and the impact on fluid regulation. *Reproductive Sciences*, *21*, 555–561.

Stachenfeld, N. S., Dipietro, L., Palter, S. F., & Nadel, E. R. (1998). Estrogen influences osmotic secretion of AVP and body water balance in postmenopausal women. *American Journal of Physiology. Regulatory, Integrative and Comparative Physiology*, *274*, R187–R195.

Stachenfeld, N. S., Splenser, A. E., Calzone, W. L., Taylor, M. P., & Keefe, D. L. (2001). Selected contribution: Sex differences in osmotic regulation of AVP and renal sodium handling. *Journal of Applied Physiology*, *91*, 1893–1901.

Stachenfeld, N. S., Taylor, H. S., Leone, C. A., & Keefe, D. L. (2003). Oestrogen effects on urine concentrating response in young women. *Journal of Physiology*, *552*, 869–880.

Sugerman, D. T. (2013). Tamoxifen update. *Journal of the American Medical Association*, *310*, 866.

Tingskov, S. J., Hu, S., Frøkiær, J., Kwon, T.-H., Wang, W., & Norregaard, R. (2018). Tamoxifen attenuates development of lithium-induced nephrogenic diabetes insipidus in rats. *American Journal of Physiology. Renal Physiology*, *314*, F1020–F1025.

Tingskov, S. J., Kwon, T.-H., Frøkiær, J., & Nørregaard, R. (2018). Tamoxifen decreases lithium-induced natriuresis in rats with nephrogenic diabetes insipidus. *Frontiers in Physiology*, *9*. https://doi.org/10.3389/fphys.2018.00903.

Umenishi, F., Narikiyo, T., Vandewalle, A., & Schrier, R. W. (2006). cAMP regulates vasopressin-induced AQP2 expression via protein kinase A-independent pathway. *Biochimica et Biophysica Acta (BBA)—Biomembranes*, *1758*, 1100–1105.

Vogel, V. G., Costantino, J. P., Wickerham, D. L., Cronin, W. M., Cecchini, R. S., Atkins, J. N., et al. (2010). Update of the National Surgical Adjuvant Breast and Bowel Project Study of tamoxifen and raloxifene (STAR) p-2 trial: Preventing breast cancer. *Cancer Prevention Research*, 1940–6207.

Vrtacnik, P., Ostanek, B., Mencej-Bedrac, S., & Marc, J. (2014). The many faces of estrogen signaling. *Journal of Croatian Society of Medical Biochemistry and Laboratory Medicine, 24*, 329–342.

Wan, S., Jiang, J., Zheng, C., Wang, N., Zhai, X., Fei, X., et al. (2018). Estrogen nuclear receptors affect cell migration by altering sublocalization of AQP2 in glioma cell lines. *Cell Death Discovery, 4*, 49.

Wells, C. C., Riazi, S., Mankhey, R. W., Bhatti, F., Ecelbarger, C., & Maric, C. (2005). Diabetic nephropathy is associated with decreased circulating estradiol levels and imbalance in the expression of renal estrogen receptors. *Gender Medicine, 2*, 227–237.

Zdunek, M., Silbiger, S. R., Lei, J., & Neugarten, J. (2001). Protein kinase CK2 mediates TGF-β; 1-stimulated type IV collagen gene transcription and its reversal by estradiol. *Kidney International, 60*, 2097–2108.

Zhao, Y., Lai, A., & Dong, W. (2014). Expression and significance of aquaporin-2 in human ectocervical-vaginal epithelial cells. *Gynecologic and Obstetric Investigation, 78*, 186–193.

Zou, L. B., Zhang, R. J., Tan, Y. J., Ding, G. L., Shi, S., Zhang, D., et al. (2011). Identification of estrogen response element in the aquaporin-2 gene that mediates estrogen-induced cell migration and invasion in human endometrial carcinoma. *Journal of Clinical Endocrinology and Metabolism, 96*, E1399–E1408.

CHAPTER TWELVE

Hormonal regulation of aquaporins in fishes

Jason P. Breves*
Department of Biology, Skidmore College, Saratoga Springs, NY, United States
*Corresponding author: e-mail address: jbreves@skidmore.edu

Contents

Abstract

Comparative studies over the last two decades have revealed that fishes leverage aquaporins (AQPs) to facilitate the movements of water, small non-ionic solutes, and gases across cell membranes. Accordingly, AQPs are expressed in tissues responsible for maintaining hydromineral balance of the whole organism. In teleost fishes, threats to hydromineral balance imposed by fluctuations in environmental salinity are met with the activation of multiple endocrine axes. This chapter first discusses recent advances in our understanding of how hormones control the expression of AQPs in tissues that support hydromineral balance, namely the gastrointestinal tract, kidney, and gill. The second objective of this chapter is to review how hormones regulate teleost AQPs in support of fluid transport processes underlying the production of gametes specialized for release into aquatic environments.

Vitamins and Hormones, Volume 112
ISSN 0083-6729
https://doi.org/10.1016/bs.vh.2019.10.002

1. Introduction

The study of how organisms control the movements of solutes and water at the macromolecular, cellular, and tissue levels constitutes a fundamental line of biological inquiry. To reach a nuanced understanding of how vertebrate osmoregulatory systems operate in adaptive, as well as pathological states, physiologists must illuminate how intrinsic regulators, such as hormones of the endocrine system, control the activities of effectors of solute and water transport. Since fishes inhabit a varied array of aquatic (and semi-aquatic) habitats that pose formidable challenges to hydromineral balance, they represent powerful systems from which to develop mechanistic models of how homeostasis of the internal milieu is maintained. From an ecological perspective, salinity is among the most important physical characteristics of aquatic environments that govern the distribution of species (Whitehead, Roach, Zhang, & Galvez, 2011). Thus, the operation and control of osmoregulatory systems within fishes is inextricable from their patterns of evolution and ecological specialization.

Aquaporins (AQPs) constitute a superfamily of integral membrane proteins that facilitate passive movements of water, small non-ionic compounds, and gases across cell membranes (Cerdà & Finn, 2010). Vertebrate AQPs are assigned into distinct groups based upon their permeation preferences: classical aquaporins (AQP0, -1, -2, -4, -5, and -6), aquaglyceroporins (AQP3, -7, -9, and -10), aquaporin 8 (AQP8), and unorthodox aquaporins (AQP11 and -12). While fishes do not have orthologs of mammalian AQP2, -5, and -6, they possess a larger number of total AQPs in comparison to mammals as a result of tandem and genomic duplication events that occurred during their evolution (Finn & Cerdà, 2011; Finn, Chauvigné, Hlidberg, Cutler, & Cerdà, 2014; Madsen, Engelund, & Cutler, 2015). The evolution, structure-function relationships, and tissue-expression patterns of piscine AQPs have been extensively reviewed previously (Cerdà & Finn, 2010; Finn & Cerdà, 2011; Hillyard, 2015; Madsen et al., 2015; Tingaud-Sequeira et al., 2010). This chapter intends to complement these reviews by outlining our current understanding of the endocrine regulation of piscine AQPs.

Fishes are traditionally classified into three classes: Agnatha (jawless fishes), Chondrichthyes (cartilaginous fishes), and Osteichthyes (bony fishes). The vast majority of investigations into the hormonal regulation of piscine AQPs have utilized teleost models (class Osteichthyes; subclass

Actinopterygii; infraclass Neopterygii; division Teleostei) (Moyle & Cech, 2004). The ensuing discussion reflects this fact and primarily describes the endocrine regulation of teleost AQPs. This chapter begins by broadly introducing the tissues that support hydromineral balance in teleost fishes prior to describing how particular hormones control their activities. This is followed by a discussion of how hormones control AQPs expressed within epithelia of the esophagus, intestine, kidney, and gill. This chapter concludes with an outline of how endocrine control of gonadal AQPs underlies the production of teleost eggs and sperm before proposing avenues for future investigation.

2. Teleost strategies to maintain hydromineral balance

The majority of teleost fishes maintain internal conditions at roughly one-third the osmotic strength (270–400 mOsm) of seawater (SW) (Evans & Claiborne, 2008). Teleosts residing in freshwater (FW) are therefore in constant risk of both excessive hydration and salt loss across body surfaces. FW-acclimated teleosts counter this situation by excreting water via dilute urine while actively absorbing ions (Na^+, Cl^-, Ca^{2+}) from the external environment and the diet across branchial and gastrointestinal epithelia, respectively (Evans, Piermarini, & Choe, 2005; Guh, Lin, & Hwang, 2015; Kaneko, Watanabe, & Lee, 2008). Marine teleosts, on the other hand, must excrete ions gained by passive diffusion from the surrounding SW. Teleosts cannot produce urine that is hyperosmotic to body fluids as a means to conserve water because their renal tubules lack the Loop of Henle (Nishimura & Fan, 2003). To fend off dehydration, marine teleosts drink SW to replace water that is lost via osmosis to the external environment. Imbibed SW is desalinated by the esophagus to produce a fluid closer to the osmolality of plasma (Hirano & Mayer-Gostan, 1976). Then, the intestine partakes in fluid uptake through solute-linked transport processes (Grosell, 2014). Water is conserved by reabsorption across the urinary bladder (Hirano, Johnson, Bern, & Utida, 1973). Finally, the gills and kidney actively secrete monovalent (Na^+, Cl^-) and divalent ions (Mg^{2+}, Ca^{2+}, and SO_4^{2-}), respectively (Kaneko et al., 2008).

While the vast majority of teleosts inhabit a single aquatic environment characteristic of either FW or SW, a relatively small percentage of species (~5%) can thrive in waters that range between FW (≤0.5‰) and SW (30–40‰) conditions (Schultz & McCormick, 2013). These "euryhaline" teleosts tolerate a wide range of salinities because of their evolved capacities to modulate or reverse salt and water transport pathways in the gill, kidney, gut, and urinary bladder (Kültz, 2015). This plasticity is mediated, in part, by the endocrine system that orchestrates the activities of these various tissues (Takei & McCormick, 2013). In turn, euryhaline fishes are intensively studied to resolve how hormones modulate the functions of osmoregulatory tissues at the molecular level. Nearly all of the studies investigating the hormonal control of piscine AQPs have been conducted with euryhaline species.

3. Hormonal control of osmoregulation in teleost fishes

Hormones are central players in the homeostatic regulation of salt and water balance. Perturbations of internal osmotic conditions, caused by changes in environmental salinity, elicit the secretion of hormones that orchestrate the expression, localization, and function of sub-cellular effectors of solute and water transport (Seale, Watanabe, & Grau, 2012). These effectors include AQPs, ion channels, co-transporters, exchangers, and Na^+/K^+-ATPases (Breves, McCormick, & Karlstrom, 2014; McCormick, 2001; Takei, Hiroi, Takahashi, & Sakamoto, 2014). The actions of established "osmoregulatory hormones" are outlined below to provide context for the ensuing discussion of their demonstrated or putative control of AQPs.

3.1 Prolactin

Among the hormones secreted from the teleost pituitary, prolactin is considered the "FW-adapting hormone" because it promotes survival in dilute waters by stimulating ion-conserving and water-excreting processes operating in the gill, kidney, gut, and urinary bladder (Hirano, 1986; Pickford & Phillips, 1959). Accordingly, both *prolactin* gene expression and prolactin hormone release are directly stimulated by reductions in extracellular/plasma osmolality (Fuentes, Brinca, Guerreiro, & Power, 2010; Yada, Hirano, & Grau, 1994); the signal transduction mechanisms underlying these responses have been extensively studied (Seale et al., 2012). Based on the broad range of tissues known to respond to prolactin across teleosts, it is widely believed that prolactin is a highly-conserved stimulator of adaptive processes in dilute environments (Hirano, 1986; McCormick, 2001). Only more recently, however, actual molecular targets of prolactin have been identified which underlie discreet solute transporting processes (Breves et al., 2014).

3.2 Growth hormone

Growth hormone (GH) secreted from the anterior pituitary is regarded as a "SW-adapting hormone" because it promotes the survival of teleosts in hyperosmotic environments. This activity is well established in salmonids (Sakamoto & McCormick, 2006; Sakamoto, McCormick, & Hirano, 1993), whereas evidence in non-salmonid fishes is far more limited. Contrasting with the regulation of prolactin, the release of GH from the pituitary is stimulated by increases in extracellular/plasma osmolality (Nilsen et al., 2008; Seale, Fiess, Hirano, Cooke, & Grau, 2006; Seale et al., 2002). GH directly controls ion extrusion through branchial ionocytes expressing its cognate receptor, and indirectly through systemic (liver derived) and/or locally produced insulin-like growth-factor 1 (IGF1) (Seidelin, Madsen, Byrialsen, & Kristiansen, 1999). Information in non-salmonids regarding the direct effects of IGF1 on hyposmoregulatory mechanisms is especially scarce (Mancera & McCormick, 1998). While GH is likely pleiotropic in its support of SW acclimation, it is largely unknown whether it regulates intestinal and renal functions despite the robust expression of its receptor in these tissues (Björnsson, 1997; Fukada et al., 2004).

3.3 Cortisol

Cortisol, which operates as both a mineralocorticoid and a glucocorticoid in teleosts, is produced by interrenal tissue under the control of adrenocorticotropic hormone (Mommsen, Vijayan, & Moon, 1999). Cortisol is typically deemed a "SW-adapting hormone" because it directly stimulates the activities and/or expression of Na^+/K^+-ATPases and ion transporters tied to ion extrusion pathways in the gill; cortisol also acts indirectly by synergizing with GH/IGF1 signaling (McCormick, 2001). Synergism with other factors is a theme of cortisol action in fishes (Mommsen et al., 1999), and interestingly, cortisol can promote FW acclimation by acting alone or in concert with prolactin (McCormick, 2001). Collectively, the patterns of circulating prolactin, GH, and cortisol in response to changes in salinity (Kajimura et al., 2004; Nilsen et al., 2008; Yada et al., 1994), and the expression of their receptors in osmoregulatory tissues, are in strong agreement with their indispensability to survival in FW and/or SW.

3.4 Additional "osmoregulatory hormones"

There is no question that a far broader collection of hormones beyond the "classical" osmoregulatory hormones outlined above underlies hydromineral balance in teleosts (Guh et al., 2015; Takei et al., 2014). Very few of these factors have been implicated in the regulation of teleost AQPs, and therefore, only a small number of them will appear in this chapter. The neurohypophyseal hormones, arginine vasotocin (AVT) and isotocin, are the teleost homologs to mammalian vasopressin and oxytocin, respectively. AVT has been considered a possible mediator of hyposmoregulation by marine fishes given the antidiuretic action of vasopressin in mammals. Moreover, cortisol may enhance AVT signaling by stimulating the expression of its receptor in target tissues (Cádiz et al., 2015). However, hypothalamic *avt* gene expression and plasma AVT levels exhibit species-dependent responses to changes in salinity, making it difficult to ascribe a universal osmoregulatory function to AVT in teleosts (Takei et al., 2014). A similar scenario has emerged for isotocin because it stimulates either hyposmoregulation or hyperosmoregulation in species-dependent fashions (Chou, Hung, Wu, Hwang, & Hwang, 2011; Martos-Sitcha et al., 2013; Takei et al., 2014). Parathyroid hormone-related protein (PTHrP), a calciotropic hormone expressed in multiple tissues (Guerreiro, Renfro, Power, & Canario, 2007), regulates intestinal water absorption in at least one euryhaline model, the sea bream (*Sparus aurata*), and may therefore regulate intestinal AQPs (Carvalho, Gregório, Canário, Power, & Fuentes, 2015).

4. Endocrine control of AQPs in osmoregulatory tissues

4.1 AQPs in the gastrointestinal tract

4.1.1 Esophagus

To fend off dehydration in SW environments, segments of the gastrointestinal tract must work in concert to sustain solute-linked water absorption. Teleosts inhabiting marine environments drink SW as a necessary source of water. However, to enable water absorption, the imbibed SW (~1000 mOsm) must be desalinated to ~500 mOsm prior to its passage to the stomach (Grosell, 2014). Na^+ and Cl^- are moved from the luminal fluid into blood plasma by active and passive transport; Na^+ and Cl^- are subsequently excreted by branchial ionocytes (Hirano & Mayer-Gostan, 1976; Kaneko et al., 2008; Takei et al., 2017). To effectively reduce the osmolality of the luminal fluid, the osmotic permeability of the esophageal epithelium must be low. Accordingly, in euryhaline Japanese eel (*Anguilla japonica*), esophageal *aqp1* and *-3* mRNA levels are down regulated in SW-acclimated animals (Takei et al., 2017). Whether AQP1 and -3 are expressed in apical and/or basolateral membrane of epithelial cells stands unresolved; nonetheless, the regulated expression of these AQPs could underlie the low osmotic permeability in the esophagus that enables the desalination of SW (Hirano & Mayer-Gostan, 1976). Cortisol seems to be a plausible regulator of AQP1 and -3 given its response to salinity challenges and promotion of SW tolerance (McCormick, 2001). However, in the only study to date that investigated esophageal AQP responses to cortisol, cortisol actually stimulated *aqp1a* mRNA in silver European eels (*A. anguilla*) preparing for downstream migration (Martinez et al., 2005). Interestingly, SW exposure induced *GH receptor* gene expression in Atlantic salmon (*Salmo salar*) esophagus (J.P. Breves, unpublished), a response paralleling elevated plasma GH levels (Nilsen et al., 2008). While highly speculative, GH may support SW tolerance by modulating solute and/or water handling processes in the esophagus. In any case, while the regulation of *aqps* following SW exposure suggest they underpin adaptive functions of the esophagus, a clearer picture of their endocrine control awaits further studies.

4.1.2 Intestine

In marine environments, desalinated SW from the esophagus passes through the stomach where it subsequently enters the anterior intestine at ~400 mOsm (Grosell, 2014). Upon entering the intestine, monovalent ions and water are absorbed from the luminal fluid. Both transcellular and paracellular routes contribute to net fluid transport; AQPs offer a means by which transcellular osmotic permeability can be enhanced (Madsen et al., 2015; Sundell & Sundh, 2012). This notion was supported by the identification of intestinal AQPs/*aqps* that were up regulated in euryhaline fishes, such as sea bream (*S. aurata* and *S. sarba*), sea bass (*Dicentrarchus labrax*), European eel, Japanese eel, Atlantic salmon, and mummichog (*Fundulus heteroclitus*), during SW acclimation (Aoki et al., 2003; Deane, Luk, & Woo, 2011; Engelund et al., 2013; Giffard-Mena et al., 2007; Jung et al., 2015; Kim, Watanabe, Kaneko, Huh, & Park, 2010; Lignot, Cutler, Hazon, & Cramb, 2002; Madsen et al., 2011; Martinez et al., 2005b; Raldúa, Otero, Fabra, & Cerdà, 2008; Tipsmark, Sorensen, & Madsen, 2010). In support of acclimation to SW, cortisol stimulates intestinal fluid uptake (Hirano & Utida, 1968; Utida et al., 1972). A link between cortisol and AQP1/*aqp1* was first shown in both yellow and silver European eels when 8-day cortisol implants stimulated AQP1 levels in mucosal scrapes of whole intestine (Martinez et al., 2005b). Similarly, cortisol-implanted sea bream showed elevated *aqp1* levels in the intestine (Deane et al., 2011). It is noteworthy that in these two instances cortisol stimulated AQP1/*aqp1* in animals inhabiting hyposmotic environments. In other words, a hyperosmotic external environment coupled with the presence of desalinated SW in the gastrointestinal tract was not required for cortisol to stimulate AQP1. In some instances (see the discussion of branchial AQP3/*aqp3*), environmental salinity can exert an overriding effect on hormonal regulators of AQPs.

While intestinal AQP1/*aqp1* is generally increased following SW exposure in teleosts (Madsen et al., 2015), it is curious that black porgy (*Acanthopagrus schlegeli*) and Japanese medaka (*Oryzias latipes*) do not fit this pattern (An, Kim, & Choi, 2008; Madsen, Bujak, & Tipsmark, 2014). These species challenge the "dogma" of intestinal water transport such that paracellular, rather than transcellular, pathways may dominate (Madsen et al., 2014). It is probable that AQP1/*aqp1* sensitivity to cortisol, and perhaps to other hormones, in porgy and medaka would contrast with the patterns reported in eel and sea bream.

In preparation for their migration from FW to marine habitats, many salmonids undergo smoltification, a developmental phase that entails the acquisition of phenotypes supporting hyposmoregulation (Hoar, 1988). Coinciding with springtime elevations in plasma cortisol, Atlantic salmon smolts increase their capacity for intestinal fluid absorption (Specker, 1982; Sundell et al., 2003; Veillette, White, & Specker, 1993). When considering the stimulatory action of cortisol on fluid transport capacities (Cornell, Portesi, Veillette, Sundell, & Specker, 1994; Veillette, Sundell, & Specker, 1995) along with the dynamic *aqp1a, -1b, -8b,* and *-10* expression patterns which occur seasonally and/or following SW exposure in smolts (Tipsmark, Sorensen, & Madsen, 2010), Atlantic salmon are a fitting model for identifying how connections between cortisol and AQPs underlie life-stage transitions. Moreover, Atlantic salmon provide an opportunity to probe whether the elevated plasma GH levels in smolts (Björnsson, 1997) play a role in "pre-adapting" the intestine for fluid uptake by controlling AQPs. While Tipsmark, Sorensen, Hulgard, and Madsen (2010) proposed that stimulation of *claudin-25b* mRNA levels by GH may favor solute and water absorption, no published studies have described any effects of GH on AQPs.

There is some evidence that AVT may regulate AQPs in support of intestinal fluid uptake. AQP1a/*aqp1a* and AQP1b/*aqp1b* are highly expressed in sea bream anterior intestine and rectum, respectively (Raldúa et al., 2008). AVT stimulated swelling by *Xenopus* oocytes expressing the AVT receptor

(V2-type) in combination with AQP1a or -1b (Martos-Sitcha, Campinho, Mancera, Martínez-Rodríguez, & Fuentes, 2015). Isotocin stimulated swelling via AQP1a, but not AQP1b, in oocytes expressing the isotocin receptor. While regulatory links between AVT, isotocin, and AQP1a/-1b await confirmation in intact sea bream, the *in vitro* activities of AVT and isotocin indicate capacities for stimulating AQP-mediated fluid absorption.

Euryhaline teleosts must down regulate processes appropriate to marine environments (e.g., intestinal fluid uptake) upon entering FW environments. In SW, HCO_3^- is secreted into the lumen of the intestine by enterocytes to produce Ca^{2+} and Mg^{2+} carbonate aggregates. The formation of these aggregates enhances water absorption by lowering the osmolality of the luminal fluid (Grosell, 2014). *In vitro* treatment with PTHrP inhibits HCO_3^- secretion and water absorption by sea bream intestine (Carvalho et al., 2015). Intact sea bream injected with the PTHrP receptor agonist, PTHrP(1–34), showed reduced *aqp1b* mRNA levels in the anterior intestine. Accordingly, *aqp1b* levels were stimulated by the PTHrP receptor antagonist, PTHrP(7–34) (Carvalho et al., 2015). Thus, PTHrP inhibits fluid uptake via the regulation of AQP1b and HCO_3^- secretion (Carvalho et al., 2015). In parallel with the adjustments to intestinal physiology that are made upon FW-entry, drinking must also be attenuated to avoid excessive hydration. In this regard, PTHrP further down regulates processes suited to marine environments by inhibiting drinking (Guerreiro et al., 2001). In addition to reducing intestinal Na^+, Cl^-, and fluid absorption (Morley, Chadwick, & El Tounsy, 1981; Utida et al., 1972), prolactin, like PTHrP, also inhibits intestinal HCO_3^- secretion (Ferlazzo et al., 2012), and thus an inhibitory action of prolactin on intestinal AQPs should be investigated. In eel, the intestinal hormone guanylin is secreted from goblet cells to affect ion and water metabolism (Takei & Yuge, 2007). For example, guanylin inhibited water absorption and short-circuit current in intestinal sacs prepared from SW-acclimated Japanese eel (Ando, Wong, & Takei, 2014; Yuge, Inoue, Hyodo, & Takei, 2003). Investigations into the potential for guanylin to repress AQP expression/function are warranted in light of guanylin inhibiting solute-linked water transport via $Na^+/K^+/2Cl^-$-cotransporter 2β (Ando et al., 2014).

4.2 AQPs in the kidney and urinary bladder

In order to support systemic water balance, the functional control of renal AQPs in fishes should align with the excretion of water in FW environments (via dilute urine) and/or the conservation of water in SW environments (Engelund & Madsen, 2011; Madsen et al., 2015). While renal AQP/*aqp* expression is highly responsive to salinity (Engelund & Madsen, 2011; Madsen et al., 2015), only a few studies have investigated hormonal control of renal AQPs (Deane et al., 2011; Martinez et al., 2005a). An 8-day cortisol infusion inhibited *aqp1aa* and *aqp1ab* mRNA levels in yellow European eels (Martinez et al., 2005a). Given that renal *aqp1aa* and *aqp1ab* levels are diminished when yellow eels are acclimated to SW versus FW (Martinez et al., 2005a), this finding is consistent with cortisol promoting phenotypes associated with SW acclimation. While precise roles for AQP1s in teleost nephrons are not defined, their apparent down regulation following SW exposure and cortisol treatment indicates they underlie water secretion by FW-acclimated fish (Engelund & Madsen, 2011). The finding that cortisol affected *aqp1* mRNA levels in eel was particularly noteworthy given the scant information on how cortisol affects renal activities in teleosts (Takei & McCormick, 2013).

In mammals, vasopressin plays an essential role in preventing dehydration by regulating AQP2 in renal collecting ducts (Valenti & Tamma, 2016). Interestingly, two studies have correlated *aqp* expression with AVT signaling in fish kidney. In black porgy, *aqp1* and *AVT receptor* mRNA levels are positively correlated in salinities ranging from FW to SW (An et al., 2008). In aestivating African lungfish (*Protopterus annectens*), the kidney supports the retention of body water in desiccated environments. When compared with lungfish inhabiting FW, aestivating lungfish showed concurrent elevations in *AVT* and *aqp0* mRNA expression in the hypothalamus and kidney, respectively (Konno, Hyodo, Yamaguchi, Matsuda, & Uchiyama, 2010). Importantly, Konno et al. (2010) demonstrated that lungfish *aqp0* encodes a functional water channel. Thus, regulatory systems linking vasopressin/AVT-family hormones with AQPs to support survival under terrestrial conditions are probably not restricted to tetrapods, and seemingly also operate in lungfishes.

The urinary bladder is an osmoregulatory tissue in teleosts exhibiting permeabilities to Na^+, Cl^-, and water generally above those found in tetrapod bladders (Bently, 1987). In euryhaline mudsucker (*Gillichthys mirabilis*) and flounder (*Platichthys* sp.), Na^+ and water are transported across urinary bladder epithelium at higher rates when fish are acclimated to SW versus FW (Hirano, Johnson, & Bern, 1971; Hirano et al., 1973). The reabsorption of water by the bladder reduces the amount of SW that must be ingested (and the associated salt load) to maintain water balance (Howe & Gutknecht, 1978).

There is no available information on whether AQPs are expressed in fish urinary bladder; thus, it stands entirely unresolved whether AQPs underlie water movement across this tissue. Cortisol and prolactin may activate and repress AQP expression/function, respectively, given their effects on Na^+ and water reabsorption at the tissue level (Doneen, 1976; Hirano et al., 1971). It will be important to investigate the hormonal regulation of AQPs in appropriate models because not all euryhaline species exhibit salinity-dependent capacities for water reabsorption by the urinary bladder (Hirano et al., 1973).

4.3 AQPs in the gill

The extensive surface area (up to 90% of the total body surface area) of teleost gills enables exchanges with the external environment that are necessary for hydromineral and acid-base balance (Evans et al., 2005). These exchanges occur via a suite of functionally distinct ionocytes (mitochondrion-rich cells) harbored in the gill that are specialized for Na^+, Cl^-, K^+, and Ca^{2+} transport (Dymowska, Hwang, & Goss, 2012; Evans et al., 2005). The coordination of ionocyte differentiation and proliferation, along with gene and protein expression within ionocytes, underlies the functional plasticity of branchial epithelium in euryhaline species (Dymowska et al., 2012; Kültz et al., 2007; McCormick, 2001).

Branchial AQP/*aqp* levels are correlated with environmental salinity and rapidly change following fluctuations in salinity (Cutler & Cramb, 2002; Lema, Carvalho, Egelston, Kelly, & McCormick, 2018; Lignot et al., 2002; Madsen et al., 2014; Moorman, Lerner, Grau, & Seale, 2015; Tipsmark, Sorensen, & Madsen, 2010; Tse, Au, & Wong, 2006; Whitehead, 2010). This is especially true for the aquaglyceroporin, AQP3 (Cutler, Martinez, & Cramb, 2007; Madsen et al., 2015). For example, in eel and Mozambique tilapia (*Oreochromis mossambicus*), *aqp3* levels were markedly reduced following transfer from FW to SW, whereas transfer from SW to FW induced *aqp3* levels (Breves et al., 2016; Cutler & Cramb, 2002; Lignot et al., 2002; Watanabe, Kaneko, & Aida, 2005). These patterns of *aqp3* regulation within euryhaline species imply that AQP3 has deleterious functions under SW conditions. Indeed, Whitehead et al. (2011) proposed that patterns of *aqp3* expression diverged by adaptive evolutionary mechanisms among mummichog populations distributed along salinity gradients. While additional *aqp* transcripts beyond *aqp3*, such as *aqp1a, -1b, -10*, and *-11*, also display salinity-dependent expression in salmon and medaka gill (Kim, Kim, Kim, Lee, & Nam, 2014; Tipsmark, Sorensen, & Madsen, 2010), the extent to which their expression is salinity-dependent in other euryhaline species is unknown.

A clear picture of how AQPs support adaptive functions in the gill remains elusive. Transepithelial water movement appears unlikely on the basis that branchial water exchange is disadvantageous for fish in both FW and SW environments (Cerdà & Finn, 2010). In rainbow wrasse (*Coris julis*) and eel, AQP3-immunoreactivity was observed in the extensive basolateral tubular system of ionocytes (Brunelli et al., 2010; Lignot et al., 2002). In turn, Cutler and Cramb (2002) and Tipsmark, Sorensen, and Madsen (2010) proposed that AQP3 enables efficient regulatory cell volume control upon salinity changes. Interestingly, mammalian sperm expressing mutated AQP3 fail to regulate cell volume when exposed to hyposmotic conditions (Chen & Duan, 2011). To tolerate variations in salinity, branchial cells in contact with the ambient environment must possess means for osmotic stress signaling (Kültz et al., 2007). AQP3 may permit cells to swell or shrink depending upon fluctuations in extracellular osmolality. Changes in cell volume can then couple with signal transduction pathways underlying the control of effectors of ion transport (Breves et al., 2016; Inokuchi et al., 2015; Watanabe et al., 2005). In a similar fashion, AQP3 plays a well-characterized role in mediating osmoreception in prolactin-secreting cells of the tilapia pituitary (Watanabe, Hirano, Grau, & Kaneko, 2009). It certainly cannot be discounted that AQP3 may mediate glycerol, urea, and/or ammonia transport by branchial cells (Cutler & Cramb, 2002; Lignot et al., 2002). While the functional role(s) for branchial AQPs remain unclear, this has not precluded investigations into whether hormones underlie their salinity-dependent expression patterns.

Cortisol was the first hormone shown to affect the branchial expression of an *aqp* transcript. FW-acclimated eels infused with cortisol showed a marked decrease in the expression of *aqp3* mRNA in the gill while intestinal *aqp3* levels were not affected (Cutler et al., 2007). Choi et al. (2013) subsequently reported that cortisol diminished branchial *aqp3* and -8 mRNAs in sockeye salmon (*Oncorhynchus nerka*) parr and smolts. These patterns suggest that elevated plasma cortisol following transfer from FW to SW is responsible, at least in part, for the concurrent attenuation of *aqp3* levels (Cutler & Cramb, 2002; Cutler et al., 2007; Mommsen et al., 1999). With cortisol established as a key controller of the SW phenotype (low AQP3), there remained no identified controller of the FW phenotype (elevated AQP3).

Plasma prolactin levels spike in euryhaline tilapia within several hours after transfer from SW to FW. Prolactin then plateaus at levels still elevated from those observed in SW-acclimated animals (Kajimura et al., 2004; Yada et al., 1994). The coincident increases in plasma prolactin and branchial

AQP3/*aqp3* that occur during FW acclimation motivated the application of hypophysectomy and hormone replacement (Nishioka, 1994) to assess whether prolactin signaling directs AQP3/*aqp3* expression. Hypophysectomy blocked the increase in *aqp3* expression that accompanies FW acclimation and reduced AQP3-immunoreactivity in ionocytes, goblet cells, and pavement cells (Breves et al., 2016). Replacement with prolactin rescued the inability of hypophysectomized animals to increase AQP3/*aqp3* upon exposure to FW. Prolactin similarly stimulated AQP3/*aqp3* expression in filaments co-incubated with prolactin *in vitro*. Interestingly, prolactin did not stimulate *aqp3* in SW-acclimated tilapia, while co-administration with cortisol blocked the effect of prolactin both *in vivo* and *in vitro*. Recall that cortisol and prolactin can be synergistic in their support of FW phenotypes (McCormick, 2001). This is decidedly not the case with respect to controlling AQP3. Thus, a complex interplay between prolactin, cortisol, and ambient salinity works to tightly-regulate branchial AQP3 expression.

With prolactin identified as a regulator of tilapia AQP3, it is interesting to consider the generality of this link. Prolactin signaling is essential for Mozambique tilapia to maintain osmotic balance in FW (Dharmamba & Maetz, 1972). By contrast, coho salmon (*Oncorhynchus kisutch*) and rainbow trout (*Salmo gairdneri*) maintain osmotic balance in FW following hypophysectomy (Björnsson & Hansson, 1983; Björnsson, Yamauchi, Nishioka, Deftos, & Bern, 1987). These contrasting dependencies upon the pituitary suggest that branchial phenotypes associated with FW are not regulated in the same fashion across teleost clades. Since there is consensus that AQP3/*aqp3* expression is generally increased under FW conditions (Cutler et al., 2007; Madsen et al., 2015), it will be interesting to learn whether salmonids exhibit a prolactin-AQP3 connection.

5. Endocrine control of AQPs during gamete production

Thus far this chapter has focused on the endocrine control of AQPs within tissues that mediate hydromineral balance of the organism; AQPs also underlie the production of teleost gametes. Eggs of marine teleosts encounter strong osmotic gradients upon their release into SW. AQP1 is indispensable to the adaptive physiologies that eggs leverage to minimize dehydration in hyperosmotic environments (Cerdà, Zapater, Chauvigné, & Finn, 2013; Chauvigné, Zapater, & Cerdà, 2011). The eggs of marine teleosts undergo hydration during final maturation. Hydration occurs because of the accumulation of osmolytes and lipids within the oocyte along with enhanced osmotic permeability of the plasma membrane (Fabra et al., 2006). Fabra, Raldúa, Power, Deen, and Cerda (2005) demonstrated that insertion of AQP1ab into the outer membrane is necessary for egg hydration in sea bream; Kagawa et al. (2011) subsequently implicated AQP1b in the hydration of eel eggs. In sea bream, transcription of the *aqp1ab* gene is stimulated by progesterone receptors bound by the progestin, 17α,20β-dihydroxy-4-pregnen-3-one (17,20β-P) (Zapater, Chauvigné, Tingaud-Sequeira, Finn, & Cerdà, 2013). Accordingly, the promoter of the *aqp1ab* gene contains two progestin responsive elements (Zapater et al., 2013). 17,20β-P is synthesized in granulosa cells under the control of follicle-stimulating hormone (FSH) during oocyte differentiation/early growth stage. Kagawa et al. (2011) further proposed that estradiol and IGF1, synthesized within ovarian follicles, may promote the regulation of AQP1 in post-translational fashions. By contrast, zebrafish (*Danio rerio*) eggs are not hydrated prior to their release into FW environments, and therefore, *aqp1b* is not highly expressed in the ovary (Tingaud-Sequeira et al., 2008). Interestingly, even though they are released into FW, catfish (*Heteropneustes fossilis*) eggs are hydrated (Singh & Joy, 2010) and ovarian *aqp1b* expression is elevated during the pre-spawning phase (Chaube et al., 2010). AVT stimulated *aqp1b* levels in incubated catfish ovarian follicles, leading Chaube et al. (2010) to propose a novel role for AVT in egg hydration.

In teleosts, AQPs support spermatogenesis, spermiation, sperm motility, and participate in controlling the composition of seminal fluid in the spermatic duct (Boj, Chauvigné, & Cerdà, 2015; Chauvigné et al., 2018; Chauvigné et al., 2011; Zilli et al., 2009). In sea bream testis, hormonal control of AQP/*aqp* expression varies depending upon the developmental stage of the tissue (Boj, Chauvigné, Zapater, & Cerdà, 2015). The enhanced expression of AQP1ab/*aqp1ab* during the resting phase is mediated by FSH and luteinizing hormone (LH) through the production of testosterone. Contrasting with its regulation in oocytes (Zapater et al., 2013), *aqp1ab* is not regulated by 17,20β-P in testis. During spermatogenesis and/or spermiation, AQP0a/*aqp0a*, -1aa/*-1aa*, -7/-7, *-8b*, *-9b*, and 10b/*-10b* are stimulated by FSH and/or LH through androgenic pathways (Boj, Chauvigné, Zapater, & Cerdà, 2015). Collectively, these patterns demonstrate that tight temporal regulation of multiple AQPs underlies fluid transport necessary for sperm production. It was recently revealed that nine AQPs are expressed in the spermatic duct of sea bream where they presumably support the further maturation and nourishment of sperm (Chauvigné et al., 2018). Whether hormones regulate AQPs in the spermatic duct is unknown; estrogens and progestins are candidates given their activities in mammalian spermatic duct (Chauvigné et al., 2018).

6. Concluding remarks

Basal vertebrates have proven to be powerful model systems within the discipline of comparative physiology. Agnathans, such as the sea lamprey (*Petromyzon marinus*), offer opportunities to consider how endocrine and osmoregulatory systems operate in basal vertebrates (Ferreira-Martins, Coimbra, Antunes, & Wilson, 2016; Kawauchi & Sower, 2006). The lamprey endocrine system reflects a condition that preceded the diversification of several hormone families and their receptors within the vertebrate lineage. For instance, lamprey possess only a single factor within the GH/prolactin/somatolactin-family of pituitary hormones. While this factor was designated GH in part because it stimulated hepatic *igf1* gene expression (Kawauchi et al., 2002), there is no information on whether it regulates osmoregulatory processes. Moreover, Close, Yun, McCormick, Wildbill, and Li (2010) proposed that 11-deoxycortisol is a mineralocorticoid (and a glucocorticoid) that promotes the SW tolerance of lamprey. By assessing the activities of these factors on lamprey AQPs (Finn et al., 2014), light can be shed on the extent to which hormones are linked with AQPs in extant members of basal vertebrate clades.

Another area for future study is to resolve how hormones underlie the expression of AQPs during early development. In mummichog embryos, AQP3 underlies water and solute transport processes that support the avoidance of dehydration and/or the acceleration of development following aerial exposure (Tingaud-Sequeira, Cerdà, Zapater, Chauvigné, & Otero, 2009). In zebrafish embryos, AQP1a1 is expressed in epidermal ionocytes where it regulates cell volume and/or water flux into the circulation (Kwong, Kumai, & Perry, 2013). Horng, Chao, Chen, Shih, and Lin (2015) reported that AQP1a1 facilitates water and CO_2 diffusion into a sub-type of ionocytes (HR cells) specialized for H^+ secretion. AQP1a1 knockdown decreased H^+ secretion by HR cells (Horng et al., 2015) as well as CO_2 and ammonia excretion by whole embryos (Talbot, Kwong, Gilmour, & Perry, 2015). Collectively, these studies demonstrate that AQPs confer critical physiological functions during early development. Embryonic endocrine systems are highly sensitive to osmoregulatory challenges and respond in fashions that promote homeostasis of the whole organism (Chou et al., 2011; Hoshijima & Hirose, 2007; Liu et al., 2006). With epidermal ionocytes already well-established targets of hormones (Guh et al., 2015), zebrafish are highly appropriate models to reveal how hormones regulate the activities of AQPs in a developmental context.

Acknowledgment

The author is currently supported by the National Science Foundation (IOS-1755131).

References

An, K., Kim, N., & Choi, C. (2008). Cloning and expression of aquaporin 1 and arginine vasotocin receptor mRNA from the black porgy, *Acanthopagrus schlegeli*: Effect of freshwater acclimation. *Fish Physiology and Biochemistry*, *34*(2), 185–194.

Ando, M., Wong, M. K., & Takei, Y. (2014). Mechanisms of guanylin action on water and ion absorption at different regions of seawater eel intestine. *American Journal of Physiology*, *307*(6), R653–R663.

Aoki, M., Kaneko, T., Katoh, F., Hasegawa, S., Tsutsui, N., & Aida, K. (2003). Intestinal water absorption through aquaporin 1 expressed in the apical membrane of mucosal epithelial cells in seawater-adapted Japanese eel. *Journal of Experimental Biology*, *206*(19), 3495–3505.

Bently, P. J. (1987). Actions of hormones on salt and water transport across cutaneous and urinary bladder epithelia. In K. T. Pang & M. P. Schreibman (Eds.), *Vertebrate endocrinology: Fundamentals and biomedical implications* (pp. 271–291). San Diego: Academic Press.

Björnsson, B. T. (1997). The biology of salmon growth hormone: From daylight to dominance. *Fish Physiology and Biochemistry*, *17*, 9–24.

Björnsson, B. T., & Hansson, T. (1983). Effects of hypophysectomy on the plasma ionic and osmotic balance in rainbow trout, *Salmo gairdneri*. *General and Comparative Endocrinology*, *49*, 240–247.

Björnsson, B. T., Yamauchi, K., Nishioka, R. S., Deftos, L. J., & Bern, H. A. (1987). Effects of hypophysectomy and subsequent hormonal replacement therapy on hormonal and osmoregulatory status of coho salmon, *Oncorhynchus kisutch*. *General and Comparative Endocrinology*, *68*, 421–430.

Boj, M., Chauvigné, F., & Cerdà, J. (2015). Coordinated action of aquaporins regulates sperm motility in a marine teleost. *Biology of Reproduction*, *93*(2), 40.

Boj, M., Chauvigné, F., Zapater, C., & Cerdà, J. (2015). Gonadotropin-activated androgen-dependent and independent pathways regulate aquaporin expression during teleost (*Sparus aurata*) spermatogenesis. *PLoS One*, *10*(11), e0142512.

Breves, J. P., Inokuchi, M., Yamaguchi, Y., Seale, A. P., Hunt, B. L., Watanabe, S., et al. (2016). Hormonal regulation of aquaporin 3: Opposing actions of prolactin and cortisol in tilapia gill. *Journal of Endocrinology*, *230*(3), 325–337.

Breves, J. P., McCormick, S. D., & Karlstrom, R. O. (2014). Prolactin and teleost ionocytes: New insights into cellular and molecular targets of prolactin in vertebrate epithelia. *General and Comparative Endocrinology*, *203*, 21–28.

Brunelli, E., Mauceri, A., Salvatore, F., Giannetto, A., Maisano, M., & Tripepi, S. (2010). Localization of aquaporin 1 and 3 in the gills of the rainbow wrasse *Coris julis*. *Acta Histochemica*, *112*(3), 251–258.

Cádiz, L., Román-Padilla, J., Gozdowska, M., Kulczykowska, E., Martínez-Rodríguez, G., Mancera, J. M., et al. (2015). Cortisol modulates vasotocinergic and isotocinergic pathways in the gilthead sea bream (*Sparus aurata*). *Journal of Experimental Biology*, *218*(2), 316–325.

Carvalho, E. S., Gregório, S. F., Canário, A. V., Power, D. M., & Fuentes, J. (2015). PTHrP regulates water absorption and aquaporin expression in the intestine of the marine sea bream (*Sparus aurata*, L.). *General and Comparative Endocrinology*, *213*, 24–31.
Cerdà, J., & Finn, R. N. (2010). Piscine aquaporins: An overview of recent advances. *Journal of Experimental Zoology A*, *313*(10), 623–650.
Cerdà, J., Zapater, C., Chauvigné, F., & Finn, R. (2013). Water homeostasis in the fish oocyte: New insights into the role and molecular regulation of a teleost-specific aquaporin. *Fish Physiology and Biochemistry*, *39*, 19–27.
Chaube, R., Chauvigné, F., Tingaud-Sequeira, A., Joy, K. P., Acharjee, A., Singh, V., et al. (2010). Molecular and functional characterization of catfish (*Heteropneustes fossilis*) aquaporin-1b: Changes in expression during ovarian development and hormone-induced follicular maturation. *General and Comparative Endocrinology*, *170*(1), 162–171.
Chauvigné, F., Parhi, J., Ducat, C., Ollé, J., Finn, R. N., & Cerdà, J. (2018). The cellular localization and redistribution of multiple aquaporin paralogs in the spermatic duct epithelium of a maturing marine teleost. *Journal of Anatomy*, *233*(2), 177–192.
Chauvigné, F., Zapater, C., & Cerdà, J. (2011). Role of aquaporins during teleost gametogenesis and early embryogenesis. *Frontiers in Physiology*, *2*, 66.
Chen, Q., & Duan, E. K. (2011). Aquaporins in sperm osmoadaptation: An emerging role for volume regulation. *Acta Pharmacologia Sinica*, *32*(6), 721–724.
Choi, C. Y., Choi, Y. J., Shin, H. S., Kim, N. N., Cho, S. H., Yamamoto, Y., et al. (2013). Expression of aquaporin-3 and -8 mRNAs in the parr and smolt stages of sockeye salmon, *Oncorhynchus nerka*: Effects of cortisol treatment and seawater acclimation. *Comparative Biochemistry and Physiology Part A*, *165*(2), 228–236.
Chou, M., Hung, J., Wu, L., Hwang, S., & Hwang, P. (2011). Isotocin controls ion regulation through regulating ionocyte progenitor differentiation and proliferation. *Cellular and Molecular Life Sciences*, *68*(16), 2797–2809.
Close, D. A., Yun, S. S., McCormick, S. D., Wildbill, A. J., & Li, W. (2010). 11-Deoxycortisol is a corticosteroid hormone in the lamprey. *Proceedings of the National Academy of Sciences of the United States of America*, *107*(31), 13942–13947.
Cornell, S. C., Portesi, D. M., Veillette, P. A., Sundell, K., & Specker, J. L. (1994). Cortisol stimulates intestinal fluid uptake in Atlantic salmon (*Salmo salar*) in the post-smolt stage. *Fish Physiology and Biochemistry*, *13*, 183–190.
Cutler, C. P., & Cramb, G. (2002). Branchial expression of an aquaporin 3 (AQP-3) homologue is downregulated in the European eel *Anguilla anguilla* following seawater acclimation. *Journal of Experimental Biology*, *205*(17), 2643–2651.
Cutler, C. P., Martinez, A., & Cramb, G. (2007). The role of aquaporin 3 in teleost fish. *Comparative Biochemistry and Physiology Part A*, *148*(1), 82–91.
Deane, E. E., Luk, J. C., & Woo, N. Y. (2011). Aquaporin 1a expression in gill, intestine, and kidney of the euryhaline silver sea bream. *Frontiers in Physiology*, *2*, 39.
Dharmamba, M., & Maetz, J. (1972). Effects of hypophysectomy and prolactin on the sodium balance of *Tilapia mossambica* in fresh water. *General and Comparative Endocrinology*, *19*(1), 175–183.
Doneen, B. A. (1976). Biological activities of mammalian and teleostean prolactins and growth hormones on mouse mammary gland and teleost urinary bladder. *General and Comparative Endocrinology*, *30*(1), 34–42.
Dymowska, A. K., Hwang, P. P., & Goss, G. G. (2012). Structure and function of ionocytes in the freshwater fish gill. *Respiratory Physiology & Neurobiology*, *184*(3), 282–292.
Engelund, M. B., Chauvigné, F., Christensen, B. M., Finn, R. N., Cerdà, J., & Madsen, S. S. (2013). Differential expression and novel permeability properties of three aquaporin 8 paralogs from seawater-challenged Atlantic salmon smolts. *Journal of Experimental Biology*, *216*(20), 3873–3885.

Engelund, M. B., & Madsen, S. S. (2011). The role of aquaporins in the kidney of euryhaline teleosts. *Frontiers in Physiology*, *2*, 51.

Evans, D. H., & Claiborne, J. B. (2008). Osmotic and ionic regulation in fishes. In D. H. Evans (Ed.), *Osmotic and ionic regulation: Cells and animals* (pp. 295–366). Boca Raton: CRC Press.

Evans, D. H., Piermarini, P. M., & Choe, K. P. (2005). The multifunctional fish gill: Dominant site of gas exchange, osmoregulation, acid-base regulation, and excretion of nitrogenous waste. *Physiological Reviews*, *85*(1), 97–177.

Fabra, M., Raldúa, D., Bozzo, M. G., Deen, P. M. T., Lubzens, E., & Cerdà, J. (2006). Yolk proteolysis and aquaporin-1o play essential roles to regulate fish oocyte hydration during meiosis resumption. *Developmental Biology*, *295*(1), 250–262.

Fabra, M., Raldúa, D., Power, D. M., Deen, P. M. T., & Cerda, J. (2005). Marine fish egg hydration is aquaporin-mediated. *Science*, *307*(5709), 545.

Ferlazzo, A., Carvalho, E. S. M., Gregorio, S. F., Power, D. M., Canario, A. V. M., Trischitta, F., et al. (2012). Prolactin regulates luminal bicarbonate secretion in the intestine of the sea bream (*Sparus aurata* L.). *Journal of Experimental Biology*, *215*, 3836–3844.

Ferreira-Martins, D., Coimbra, J., Antunes, C., & Wilson, J. M. (2016). Effects of salinity on upstream-migrating, spawning sea lamprey, *Petromyzon marinus*. *Conservation Physiology*, *4*, cov064, https://doi.org/10.1093/conphys/cov064.

Finn, R. N., & Cerdà, J. (2011). Aquaporin evolution in fishes. *Frontiers in Physiology*, *2*, 44.

Finn, R. N., Chauvigné, F., Hlidberg, J. B., Cutler, C. P., & Cerdà, J. (2014). The lineage-specific evolution of aquaporin gene clusters facilitated tetrapod terrestrial adaptation. *PLoS One*, *9*(11), e113686.

Fuentes, J., Brinca, L., Guerreiro, P. M., & Power, D. M. (2010). PRL and GH synthesis and release from the sea bream (*Sparus auratus* L.) pituitary gland *in vitro* in response to osmotic challenge. *General and Comparative Endocrinology*, *168*, 95–102.

Fukada, H., Ozaki, Y., Pierce, A. L., Adachi, S., Yamauchi, K., Hara, A., et al. (2004). Salmon growth hormone receptor: Molecular cloning, ligand specificity, and response to fasting. *General and Comparative Endocrinology*, *139*(1), 61–71.

Giffard-Mena, I., Boulo, V., Aujoulat, F., Fowden, H., Castille, R., Charmantier, G., et al. (2007). Aquaporin molecular characterization in the sea bass (*Dicentrarchus labrax*): The effect of salinity on AQP1 and AQP3 expression. *Comparative Biochemistry and Physiology Part A*, *148*, 430–444.

Grosell, M. (2014). Intestinal transport. In D. H. Evans, J. B. Claiborne, & S. Currie (Eds.), *The physiology of fishes* (pp. 175–203). Boca Raton: CRC Press.

Guerreiro, P. M., Fuentes, J., Power, D. M., Ingleton, P. M., Flik, G., & Canario, A. V. M. (2001). Parathyroid hormone-related protein: A calcium regulatory factor in sea bream (*Sparus aurata* L.) larvae. *American Journal of Physiology*, *281*(3), R855–R860.

Guerreiro, P. M., Renfro, J. L., Power, D. M., & Canario, A. V. M. (2007). The parathyroid hormone family of peptides: Structure, tissue distribution, regulation, and potential functional roles in calcium and phosphate balance in fish. *American Journal of Physiology*, *292*(2), R679–R696.

Guh, Y., Lin, C., & Hwang, P. (2015). Osmoregulation in zebrafish: Ion transport mechanisms and functional regulation. *EXCLI Journal*, *14*, 627–659.

Hillyard, S. D. (2015). Comparative and evolutionary physiology of water channels. In K. A. Hyndman & T. L. Pannabecker (Eds.), *Sodium and water homeostasis* (pp. 5–41). Berlin: Springer.

Hirano, T. (1986). The spectrum of prolactin action in teleosts. *Progress in Clinical and Biological Research*, *205*, 53–74.

Hirano, T., Johnson, D. W., & Bern, H. A. (1971). Control of water movement in flounder urinary bladder by prolactin. *Science*, *230*(5294), 469–470.

Hirano, T., Johnson, D. W., Bern, H. A., & Utida, S. (1973). Studies on water and ion movements in the isolated urinary bladder of selected freshwater, marine and euryhaline teleosts. *Comparative Biochemistry and Physiology Part A*, *45*(2), 529–540.
Hirano, T., & Mayer-Gostan, N. (1976). Eel esophagus as an osmoregulatory organ. *Proceedings of the National Academy of Sciences of the United States of America*, *73*(4), 1348–1350.
Hirano, T., & Utida, S. (1968). Effects of ACTH and cortisol on water movement in isolated intestine of the eel. *General and Comparative Endocrinology*, *11*, 373–380.
Hoar, W. S. (1988). The physiology of smolting salmonids. In W. S. Hoar & D. J. Randall (Eds.), *Fish physiology. Vol. XIB* (pp. 275–343). New York: Academic Press.
Horng, J. L., Chao, P. L., Chen, P. Y., Shih, T. H., & Lin, L. Y. (2015). Aquaporin 1 is involved in acid secretion by ionocytes of zebrafish embryos through facilitating CO_2 transport. *PLoS One*, *10*(8), e0136440.
Hoshijima, K., & Hirose, S. (2007). Expression of endocrine genes in zebrafish larvae in response to environmental salinity. *Journal of Endocrinology*, *193*, 481–491.
Howe, D., & Gutknecht, J. (1978). Role of urinary bladder in osmoregulation in marine teleost, *Opsanus tau*. *American Journal of Physiology*, *235*(1), R48–R54.
Inokuchi, M., Breves, J. P., Moriyama, S., Watanabe, S., Kaneko, T., Lerner, D. T., et al. (2015). Prolactin 177, prolactin 188, and extracellular osmolality independently regulate the gene expression of ion transport effectors in gill of Mozambique tilapia. *American Journal of Physiology*, *309*, R1251–R1263.
Jung, D., Adamo, M. A., Lehman, R. M., Barnaby, R., Jackson, C. E., Jackson, B. P., et al. (2015). A novel variant of aquaporin 3 is expressed in killifish (*Fundulus heteroclitus*) intestine. *Comparative Biochemistry and Physiology Part C*, *171*, 1–7.
Kagawa, H., Kishi, T., Gen, K., Kazeto, Y., Tosaka, R., Matsubara, H., et al. (2011). Expression and localization of aquaporin 1b during oocyte development in the Japanese eel (*Anguilla japonica*). *Reproductive Biology and Endocrinology*, *9*, 71.
Kajimura, S., Hirano, T., Moriyama, S., Vakkuri, O., Leppäluoto, J., & Grau, E. G. (2004). Changes in plasma concentrations of immunoreactive ouabain in the tilapia in response to changing salinity: Is ouabain a hormone in fish? *General and Comparative Endocrinology*, *135*(1), 90–99.
Kaneko, T., Watanabe, S., & Lee, K. M. (2008). Functional morphology of mitochondrion-rich cells in euryhaline and stenohaline teleosts. *Aqua-BioScience Monographs*, *1*, 1–62.
Kawauchi, H., & Sower, S. A. (2006). The dawn and evolution of hormones in the adenohypophysis. *General and Comparative Endocrinology*, *148*, 3–14.
Kawauchi, H., Suzuki, K., Yamazaki, T., Moriyama, S., Nozaki, M., Yamaguchi, K., et al. (2002). Identification of growth hormone in the sea lamprey, an extant representative of a group of the most ancient vertebrates. *Endocrinology*, *143*(12), 4916–4921.
Kim, D. S., Kim, B. S., Kim, Y. K., Lee, S. Y., & Nam, Y. K. (2014). Isolation and mRNA expression analysis of aquaporin isoforms in marine medaka *Oryzias dancena*, a euryhaline teleost. *Comparative Biochemistry and Physiology Part A*, *171*, 1–8.
Kim, Y., Watanabe, S., Kaneko, T., Huh, M., & Park, S. (2010). Expression of aquaporins 3, 8 and 10 in the intestines of freshwater- and seawater-acclimated Japanese eels *Anguilla japonica*. *Fisheries Science*, *76*(4), 695–702.
Konno, N., Hyodo, S., Yamaguchi, Y., Matsuda, K., & Uchiyama, M. (2010). Vasotocin/V2-type receptor/aquaporin axis exists in African lungfish kidney but is functional only in terrestrial condition. *Endocrinology*, *151*(3), 1089–1096.
Kültz, D. (2015). Physiological mechanisms used by fish to cope with salinity stress. *Journal of Experimental Biology*, *218*, 1907–1914.
Kültz, D., Fiol, D., Valkova, N., Gomez-Jimenez, S., Chan, S. Y., & Lee, J. (2007). Functional genomics and proteomics of the cellular osmotic stress response in 'non-model' organisms. *Journal of Experimental Biology*, *210*(9), 1593–1601.

Kwong, R. W. M., Kumai, Y., & Perry, S. F. (2013). The role of aquaporin and tight junction proteins in the regulation of water movement in larval zebrafish (*Danio rerio*). *PLoS One*, *8*(8), e70764.

Lema, S. C., Carvalho, P. G., Egelston, J. N., Kelly, J. T., & McCormick, S. D. (2018). Dynamics of gene expression responses for ion transport proteins and aquaporins in the gill of a euryhaline pupfish during freshwater and high-salinity acclimation. *Physiological and Biochemical Zoology*, *91*(6), 1148–1171.

Lignot, J., Cutler, C. P., Hazon, N., & Cramb, G. (2002). Immunolocalisation of aquaporin 3 in the gill and the gastrointestinal tract of the European eel *Anguilla anguilla* (L.). *Journal of Experimental Biology*, *205*(17), 2653–2663.

Liu, N., Liu, Q., Wawrowsky, K., Yang, Z., Lin, S., & Melmed, S. (2006). Prolactin receptor signaling mediates the osmotic response of embryonic zebrafish lactotrophs. *Molecular Endocrinology*, *20*(4), 871–880.

Madsen, S. S., Bujak, J., & Tipsmark, C. K. (2014). Aquaporin expression in the Japanese medaka (*Oryzias latipes*) in freshwater and seawater: Challenging the paradigm of intestinal water transport? *Journal of Experimental Biology*, *217*(17), 3108–3121.

Madsen, S. S., Engelund, M. B., & Cutler, C. P. (2015). Water transport and functional dynamics of aquaporins in osmoregulatory organs of fishes. *Biological Bulletin*, *229*(1), 70–92.

Madsen, S. S., Olesen, J. H., Bedal, K., Engelund, M. B., Velasco-Santamaria, Y. M., & Tipsmark, C. K. (2011). Functional characterization of water transport and cellular localization of three aquaporin paralogs in the salmonid intestine. *Frontiers in Physiology*, *2*, 56.

Mancera, J. M., & McCormick, S. D. (1998). Evidence for growth hormone/insulin-like growth factor I axis regulation of seawater acclimation in the euryhaline teleost *Fundulus heteroclitus*. *General and Comparative Endocrinology*, *111*(2), 103–112.

Martinez, A., Cutler, C. P., Wilson, G. D., Phillips, C., Hazon, N., & Cramb, G. (2005a). Cloning and expression of three aquaporin homologues from the European eel (*Anguilla anguilla*): Effects of seawater acclimation and cortisol treatment on renal expression. *Biology of the Cell*, *97*(8), 615–627.

Martinez, A. S., Cutler, C. P., Wilson, G. D., Phillips, C., Hazon, N., & Cramb, G. (2005b). Regulation of expression of two aquaporin homologs in the intestine of the European eel: Effects of seawater acclimation and cortisol treatment. *American Journal of Physiology*, *288*(6), R1733–R1743.

Martinez, A., Wilson, G., Phillips, C., Cutler, C., Hazon, N., & Cramb, G. (2005). Effect of cortisol on aquaporin expression in the esophagus of the European eel, *Anguilla anguilla*. *Annals of the New York Academy of Sciences*, *1040*(1), 395–398.

Martos-Sitcha, J. A., Campinho, M. A., Mancera, J. M., Martínez-Rodríguez, G., & Fuentes, J. (2015). Vasotocin and isotocin regulate aquaporin 1 function in the sea bream. *Journal of Experimental Biology*, *218*(5), 684–693.

Martos-Sitcha, J. A., Wunderink, Y. S., Gozdowska, M., Kulczykowska, E., Mancera, J. M., & Martínez-Rodríguez, G. (2013). Vasotocinergic and isotocinergic systems in the gilthead sea bream (*Sparus aurata*): An osmoregulatory story. *Comparative Biochemistry and Physiology Part A*, *166*(4), 571–581.

McCormick, S. D. (2001). Endocrine control of osmoregulation in teleost fish. *American Zoologist*, *41*, 781–794.

Mommsen, T. P., Vijayan, M. M., & Moon, T. W. (1999). Cortisol in teleosts: Dynamics, mechanisms of action, and metabolic regulation. *Reviews in Fish Biology and Fisheries*, *9*, 211–268.

Moorman, B. P., Lerner, D. T., Grau, E. G., & Seale, A. P. (2015). The effects of acute salinity challenges on osmoregulation in Mozambique tilapia reared in a tidally changing salinity. *Journal of Experimental Biology*, *218*, 731–739.

Morley, M., Chadwick, A., & El Tounsy, E. M. (1981). The effect of prolactin on water absorption by the intestine of the trout (*Salmo gairdneri*). *General and Comparative Endocrinology, 44*(1), 64–68.

Moyle, P. B., & Cech, J. J. (2004). *Fishes: An introduction to ichthyology* (5th ed.). San Francisco: Pearson, (chapter 17).

Nilsen, T. O., Ebbesson, L. O., Kiilerich, P., Björnsson, B. T., Madsen, S. S., McCormick, S. D., et al. (2008). Endocrine systems in juvenile anadromous and landlocked Atlantic salmon (*Salmo salar*): Seasonal development and seawater acclimation. *General and Comparative Endocrinology, 155*(3), 762–772.

Nishimura, H., & Fan, Z. (2003). Regulation of water movement across vertebrate renal tubules. *Comparative Biochemistry and Physiology Part A, 136*(3), 479–498.

Nishioka, R. S. (1994). Hypophysectomy of fish. In P. W. Hochachka & T. P. Mommsen (Eds.), *Biochemistry and molecular biology of fishes: Analytical techniques* (pp. 49–58). New York: Elsevier.

Pickford, G. E., & Phillips, J. G. (1959). Prolactin, a factor in promoting survival of hypophysectomized killifish in fresh water. *Science, 130*(3373), 454–455.

Raldúa, D., Otero, D., Fabra, M., & Cerdà, J. (2008). Differential localization and regulation of two aquaporin-1 homologs in the intestinal epithelia of the marine teleost *Sparus aurata. American Journal of Physiology, 294*(3), R993–1003.

Sakamoto, T., & McCormick, S. D. (2006). Prolactin and growth hormone in fish osmoregulation. *General and Comparative Endocrinology, 147*(1), 24–30.

Sakamoto, T., McCormick, S. D., & Hirano, T. (1993). Osmoregulatory actions of growth hormone and its mode of action in salmonids: A review. *Fish Physiology and Biochemistry, 11*, 155–164.

Schultz, E. T., & McCormick, S. D. (2013). Euryhalinity in an evolutionary context. In S. D. McCormick, A. P. Farrell, & C. J. Brauner (Eds.), *Euryhaline fishes* (pp. 477–529). Amsterdam: Academic Press.

Seale, A. P., Fiess, J. C., Hirano, T., Cooke, I. M., & Grau, E. G. (2006). Disparate release of prolactin and growth hormone from the tilapia pituitary in response to osmotic stimulation. *General and Comparative Endocrinology, 145*, 222–231.

Seale, A. P., Riley, L. G., Leedom, T. A., Kajimura, S., Dores, R. M., Hirano, T., et al. (2002). Effects of environmental osmolality on release of prolactin, growth hormone and ACTH from the tilapia pituitary. *General and Comparative Endocrinology, 128*(2), 91–101.

Seale, A. P., Watanabe, S., & Grau, E. G. (2012). Osmoreception: Perspectives on signal transduction and environmental modulation. *General and Comparative Endocrinology, 176*(3), 354–360.

Seidelin, M., Madsen, S. S., Byrialsen, A., & Kristiansen, K. (1999). Effects of insulin-like growth factor-I and cortisol on Na^+,K^+-ATPase expression in osmoregulatory tissues of brown trout (*Salmo trutta*). *General and Comparative Endocrinology, 113*(3), 331–342.

Singh, V., & Joy, K. P. (2010). An involvement of vasotocin in oocyte hydration in the catfish *Heteropneustes fossilis*: A comparison with effects of isotocin and hCG. *General and Comparative Endocrinology, 166*(3), 504–512.

Specker, J. L. (1982). Interrenal function and smoltification. *Aquaculture, 28*, 59–66.

Sundell, K., Jutfelt, F., Ágústsson, T., Olsen, R., Sandblom, E., Hansen, T., et al. (2003). Intestinal transport mechanisms and plasma cortisol levels during normal and out-of-season parr–smolt transformation of Atlantic salmon, *Salmo salar. Aquaculture, 222*(1), 265–285.

Sundell, K., & Sundh, H. (2012). Intestinal fluid absorption in anadromous salmonids: Importance of tight junctions and aquaporins. *Frontiers in Physiology, 3*, 388.

Takei, Y., Hiroi, J., Takahashi, H., & Sakamoto, T. (2014). Diverse mechanisms for body fluid regulation in teleost fishes. *American Journal of Physiology, 307*(7), R778–R792.

Takei, Y., & McCormick, S. D. (2013). Hormonal control of fish euryhalinity. In S. D. McCormick, A. P. Farrell, & C. J. Brauner (Eds.), *Euryhaline fishes* (pp. 69–123). Amsterdam: Academic Press.

Takei, Y., Wong, M. K., Pipil, S., Ozaki, H., Suzuki, Y., Iwasaki, W., et al. (2017). Molecular mechanisms underlying active desalination and low water permeability in the esophagus of eels acclimated to seawater. *American Journal of Physiology, 312*(2), R231–R244.

Takei, Y., & Yuge, S. (2007). The intestinal guanylin system and seawater adaptation in eels. *General and Comparative Endocrinology, 152*, 339–351.

Talbot, K., Kwong, R. W. M., Gilmour, K. M., & Perry, S. F. (2015). The water channel aquaporin-1a1 facilitates movement of CO_2 and ammonia in zebrafish (*Danio rerio*) larvae. *Journal of Experimental Biology, 218*, 3931–3940.

Tingaud-Sequeira, A., Calusinska, M., Finn, R. N., Chauvigné, F., Lozano, J., & Cerdà, J. (2010). The zebrafish genome encodes the largest vertebrate repertoire of functional aquaporins with dual paralogy and substrate specificities similar to mammals. *BMC Evolutionary Biology, 10*, 38.

Tingaud-Sequeira, A., Cerdà, J., Zapater, C., Chauvigné, F., & Otero, D. (2009). Adaptive plasticity of killifish (*Fundulus heteroclitus*) embryos: Dehydration-stimulated development and differential aquaporin-3 expression. *American Journal of Physiology, 296*(4), R1041–R1052.

Tingaud-Sequeira, A., Chauvigné, F., Fabra, M., Lozano, J., Raldúa, D., & Cerdà, J. (2008). Structural and functional divergence of two fish aquaporin-1 water channels following teleost-specific gene duplication. *BMC Evolutionary Biology, 8*, 259.

Tipsmark, C. K., Sorensen, K. J., Hulgard, K., & Madsen, S. S. (2010). Claudin-15 and -25b expression in the intestinal tract of Atlantic salmon in response to seawater acclimation, smoltification and hormone treatment. *Comparative Biochemistry and Physiology Part A, 155*(3), 361–370.

Tipsmark, C. K., Sorensen, K. J., & Madsen, S. S. (2010). Aquaporin expression dynamics in osmoregulatory tissues of Atlantic salmon during smoltification and seawater acclimation. *Journal of Experimental Biology, 213*(3), 368–379.

Tse, W. K. F., Au, D. W. T., & Wong, C. K. C. (2006). Characterization of ion channel and transporter mRNA expressions in isolated gill chloride and pavement cells of seawater acclimating eels. *Biochemical and Biophysical Research Communications, 346*(4), 1181–1190.

Utida, S., Hirano, T., Oide, H., Ando, M., Johnson, D. W., & Bern, H. A. (1972). Hormonal control of the intestine and urinary bladder in teleost osmoregulation. *General and Comparative Endocrinology, 3*(Suppl), 317–327.

Valenti, G., & Tamma, G. (2016). Vasopressin and the regulation of aquaporin-2 in health and disease. In G. Soveral, S. Nielsen, & A. Casini (Eds.), *Aquaporins in health and disease* (pp. 157–180). Boca Raton: CRC Press.

Veillette, P. A., Sundell, K., & Specker, J. L. (1995). Cortisol mediates the increase in intestinal fluid absorption in Atlantic salmon during parr-smolt transformation. *General and Comparative Endocrinology, 97*, 250–258.

Veillette, P. A., White, R. J., & Specker, J. L. (1993). Changes in intestinal fluid transport in Atlantic salmon (*Salmo salar* L) during parr-smolt transformation. *Fish Physiology and Biochemistry, 12*, 193–202.

Watanabe, S., Hirano, T., Grau, E. G., & Kaneko, T. (2009). Osmosensitivity of prolactin cells is enhanced by the water channel aquaporin-3 in a euryhaline Mozambique tilapia (*Oreochromis mossambicus*). *American Journal of Physiology, 296*(2), R446–R453.

Watanabe, S., Kaneko, T., & Aida, K. (2005). Aquaporin-3 expressed in the basolateral membrane of gill chloride cells in Mozambique tilapia *Oreochromis mossambicus* adapted to freshwater and seawater. *Journal of Experimental Biology, 208*, 2673–2682.

Whitehead, A. (2010). The evolutionary radiation of diverse osmotolerant physiologies in killifish (*Fundulus* sp.). *Evolution, 64*, 2070–2085.

Whitehead, A., Roach, J. L., Zhang, S., & Galvez, F. (2011). Genomic mechanisms of evolved physiological plasticity in killifish distributed along an environmental salinity gradient. *Proceedings of the National Academy of Sciences of the United States of America, 108*(15), 6193–6198.

Yada, T., Hirano, T., & Grau, E. G. (1994). Changes in plasma levels of the two prolactins and growth hormone during adaptation to different salinities in the euryhaline tilapia, *Oreochromis mossambicus*. *General and Comparative Endocrinology, 93*(2), 214–223.

Yuge, S., Inoue, K., Hyodo, S., & Takei, Y. (2003). A novel guanylin family (guanylin, uroguanylin, and renoguanylin) in eels: Possible osmoregulatory hormones in intestine and kidney. *Journal of Biological Chemistry, 278*(25), 22726–22733.

Zapater, C., Chauvigné, F., Tingaud-Sequeira, A., Finn, R. N., & Cerdà, J. (2013). Primary oocyte transcriptional activation of *aqp1ab* by the nuclear progestin receptor determines the pelagic egg phenotype of marine teleosts. *Developmental Biology, 377*(2), 345–362.

Zilli, L., Schiavone, R., Chauvigné, F., Cerdà, J., Storelli, C., & Vilella, S. (2009). Evidence for the involvement of aquaporins in sperm motility activation of the teleost gilthead sea bream (*Sparus aurata*). *Biology of Reproduction, 81*(5), 880–888.

CHAPTER THIRTEEN

Calcium sensing receptor exerts a negative regulatory action toward vasopressin-induced aquaporin-2 expression and trafficking in renal collecting duct

Marianna Ranieri, Annarita Di Mise, Grazia Tamma, Giovanna Valenti*
Department of Biosciences, Biotechnologies and Biopharmaceutics, University of Bari, Bari, Italy
*Corresponding author: e-mail address: giovanna.valenti@uniba.it

Contents

Abstract

Vasopressin (AVP) plays a major role in the regulation of water homeostasis by its antidiuretic action on the kidney, mediated by V2 receptors. An increase in plasma sodium concentration stimulates AVP release, which in turn promotes water reabsorption. Upon binding to the V2 receptors in the renal collecting duct, AVP induces the expression and apical membrane insertion of the aquaporin-2 (AQP2) water channels and subsequent water reabsorption. AVP regulates two independent mechanisms: the short-term regulation of AQP2 trafficking and long-term regulation of the total abundance of the AQP2 protein in the cells. On the other hand, several hormones, acting through specific receptors, have been reported to antagonize AVP-mediated water transport in kidney. In this respect, we previously described that high luminal Ca^{2+} in the renal collecting duct attenuates short-term AVP-induced AQP2 trafficking through activation of the Ca^{2+}-sensing receptor (CaSR). This effect is due to reduction of AVP-dependent cAMP

Vitamins and Hormones, Volume 112
ISSN 0083-6729
https://doi.org/10.1016/bs.vh.2019.08.008

generation and possibly hydrolysis. Moreover, CaSR signaling reduces AQP2 abundance both via AQP2-targeting miRNA-137 and the proteasomal degradation pathway.

This chapter summarizes recent data elucidating the molecular mechanisms underlying the physiological role of the CaSR-dependent regulation of AQP2 expression and trafficking.

1. Introduction

Regulation of water homeostasis is crucial for all terrestrial organisms. In most mammals, the maintenance of water balance is critically dependent on water intake, the sensation of thirst and the regulation of water excretion in the kidney, which is under the control of the antidiuretic hormone arginine vasopressin (AVP) (Knepper, Kwon, & Nielsen, 2015; Knepper, Nielsen, Chou, & DiGiovanni, 1994). AVP is secreted into the circulation by the posterior pituitary gland, in response to an increase in serum osmolality or a decrease in blood volume. In the kidney, AVP binds to the V2 vasopressin receptor (V2R), which belongs to the superfamily of G protein-coupled receptors (GPCRs) (Birnbaumer, 2000). V2Rs expressed in the basolateral membranes of collecting duct principal cells, in the last portion of the nephron, are functionally coupled to Gs protein causing an increase in intracellular cAMP. The cAMP/PKA signal transduction cascade results in multiple phosphorylating events in the C-terminus of aquaporin-2 (AQP2) representing important signals for triggering AQP2 trafficking toward the apical plasma membrane leading to an increase in luminal water permeability (Hoffert, Pisitkun, Wang, Shen, & Knepper, 2006; Jung & Kwon, 2016; Nedvetsky et al., 2009; Nielsen et al., 1995, 2002; Valenti, Procino, Tamma, Carmosino, & Svelto 2005). In addition to phosphorylation, AQP2 undergoes different regulated posttranslational modifications, such as ubiquitination and glutathionylation, which are likely to be fundamental for controlling AQP2 cellular localization, stability and function (Moeller, Olesen, & Fenton, 2011; Tamma, Di Mise, et al., 2014). Several proteins participating in the control of cAMP-dependent AQP2 trafficking have been identified including SNAREs, annexin-2, hsc70, A kinase anchoring proteins (AKAPs), small GTPases of the Rho family proteins controlling cytoskeletal dynamics (Centrone et al., 2017; Klussmann et al., 2001; McSorley et al., 2006; Procino, Barbieri, et al., 2008; Schrade et al., 2018; Tamma et al., 2003; Tamma, Procino, Mola, Svelto, & Valenti, 2008; Wang et al., 2010; Wu et al., 2018). In addition to regulate AQP2 trafficking, AVP also regulates the total amount of the water channel within the cell

(Hasler, Leroy, Martin, & Feraille, 2009). Alterations in the AQP2 abundance as well as defects in AVP signaling in the renal collecting duct can seriously compromise the maintenance of water balance in the body (Noda, Sohara, Ohta, & Sasaki, 2010). Loss-of-function mutations of both V2R and AQP2 are associated with congenital nephrogenic diabetes insipidus (NDI), a rare genetic disorder, which can be quite severe in infants, characterized by a failure to concentrate urine (despite normal or elevated levels of AVP), polyuria, polydipsia and hypernatremia (Bockenhauer & Bichet, 2017).

Hypercalcemia–induced hypercalciuria, can also causes acquired NDI, but the mechanisms underlying hypercalcemia-induced NDI are not well understood. Several studies point to a critical role of CaSR activation in reducing coupling efficiency between the V2R and adenylyl cyclase, reducing AQP2 transcription and trafficking (Bustamante et al., 2008; Procino, Mastrofrancesco, et al., 2008;Procino et al., 2004; Ranieri et al., 2015, 2018; Riccardi & Valenti, 2016). An opposite rare genetic disorder, the Nephrogenic Syndrome of Inappropriate Antidiuresis (NSIAD), has been identified, caused by gain-of-function mutations of the V2R and characterized by hyponatremia despite undetectable AVP levels (Erdélyi et al., 2015; Feldman et al., 2005). On the other hand, inappropriate antidiuresis is the most common cause of hyponatremia and it is characterized by inability to excrete a free water load, inappropriately concentrated urine, hypo-osmolality and natriuresis. The most common clinical entity is the idiopathic syndrome of inappropriate antidiuretic hormone secretion (SIADH), which is linked to hypersecretion of AVP (Cuesta & Thompson, 2016).

This chapter summarizes some of new understanding in the regulation of AQP2 trafficking and AQP2 protein abundance with a special focus on the negative feedback of CaSR signaling on AVP-induced AQP2 expression and trafficking in renal collecting duct.

2. Regulation of renal collecting duct water permeability by the vasopressin—aquaporin-2 system

The renal collecting duct water permeability is controlled by AVP through the regulation of the water channel AQP2 in two processes: short- and long-term regulation. The short-term regulation occurs over a period of minutes resulting in an increased trafficking of AQP2-bearing vesicles from an intracellular pool to the apical plasma membrane in response to AVP stimulus. The complex molecular signals controlling, at short-term,

AQP2 trafficking in response to AVP, include rapid changes of the activity of kinases, phosphatases, AKAPs, E3-ligases and regulation of cytoskeleton remodeling, all important steps for exocytosis and endocytosis of AQP2.

The long-term regulation, instead, occurs over a period of hours to days due to AVP regulation of cellular AQP2 total abundance.

2.1 Short-term regulation of AQP2 by vasopressin

The expression of AQP2 at the apical membrane is tightly regulated by AVP. Binding of AVP to its specific V2 receptor (V2R), localized to the basolateral membrane of collecting duct principal cells, increases the intracellular concentration of cAMP via adenylyl cyclase type III and VI, resulting in the activation of protein kinase A (PKA) (Hoffert, Chou, Fenton, & Knepper, 2005), although other kinases, such as Akt, Sgk, p38-MAPK, may play a role in response to AVP stimulation (Nedvetsky et al., 2010; Pisitkun et al., 2008; Rinschen et al., 2010). One target for PKA is AQP2 at serine-256, although AQP2 may be a substrate for other basophilic protein kinases such as Akt1 and protein kinase Cδ (Douglass et al., 2012). PKA-dependent phosphorylation of the water channel AQP2, at S256, has been considered essential to promote the translocation of AQP2 from an intracellular vesicles pool to the apical plasma membrane (Katsura, Gustafson, Ausiello, & Brown, 1997; Nielsen et al., 1995), although this concept has been recently questioned. Several studies clearly indicate that cAMP can also stimulate exocytosis independently of PKA action (Ozaki et al., 2000; Seino & Shibasaki, 2005). cAMP activates Epac (exchange protein activated by cAMP) and PKA similarly (Christensen et al., 2003). Epac is a cAMP downstream effector expressed in different tissues (Kawasaki et al., 1998) and functions as a guanine-nucleotide-exchange factor for Rap1 and Ras (de Rooij et al., 1998; Kawasaki et al., 1998). Stimulation of isolated renal tubules with a selective agonist of Epac, mimics AVP-induced intracellular calcium release and AQP2 targeting at the apical plasma membrane independently of PKA signaling activation (Yip, 2006). Recently, Cheung and coworkers (Cheung, Terlouw, Janssen, Brown, & Bouley, 2019) reported that Src kinase inhibition causes serine 256-independent AQP2 membrane trafficking and induces phosphorylation of serine 269 independently of serine 256 phosphorylation. In this line, previous study by Ando et al. (2016) demonstrated that Wnt5a induces accumulation of AQP2 to the apical membrane inducing phosphorylation of serine 269 independently of serine 256.Together these novel findings support the concept that AVP-mediated AQP2 trafficking may occur independently from cAMP (Lei et al., 2018).

Several studies have shown that cytoskeleton remodeling plays a key role in modulating AQP2 trafficking. Chemical disruption of microtubules inhibited the AVP-induced osmotic water permeability in the mammalian collecting duct (Brown & Stow, 1996; Sabolic, Katsura, Verbavatz, & Brown, 1995). Dynamics of actin cytoskeleton are also known to be involved in controlling the cellular distribution of AQP2 (Valenti et al., 2005).

Specifically, exposure of IMCD and renal CD8 cells to Clostridium toxin B, which inhibits proteins of Rho family (Rho, Rac and Cdc42) or Clostridium toxin C3, specifically inhibiting Rho, caused a partial depolymerization of actin filaments associated with an increase of the cell surface expression of AQP2 resulting in a significant increase of the osmotic water permeability, even in the absence of AVP stimulation (Klussmann et al., 2001; Tamma et al., 2001). Interestingly, immunoaffinity chromatography studies showed that the C-terminus of AQP2 strongly binds β- and γ-isoforms of actin (Noda, Horikawa, Katayama, & Sasaki, 2004).

Cell surface expression of AQP2 is the result of a balance between the exocytosis and the endocytosis. Dennis Brown group has convincingly showed that AVP exerts its action by stimulation of the exocytic pathway (Brown, Hasler, Nunes, Bouley, & Lu, 2008).

Docking and fusion of AQP2 vesicles to the apical membrane is a complex mechanism displaying several similarities with synaptic vesicles fusion.

Annexin-2 was found to be functionally involved in the fusion of highly purified AQP2 vesicles with plasma membrane (Tamma et al., 2008). Moreover, similarly to the crucial role played in synaptic vesicle fusion, SNARE proteins have been demonstrated to be essential for AQP2 vesicle fusion (Gouraud et al., 2002). Selective protein silencing of VAMP2, VAMP3 and SNAP23 inhibited AQP2 insertion at the apical plasma membrane (Procino, Barbieri, et al., 2008; Procino, Mastrofrancesco, et al., 2008). Moreover, knockout mice for VAMP8 develop hydronephrosis (Wang et al., 2010).

In a very recent study, we have provided evidence that the fusion of AQP2 vesicles to the plasma membrane requires Gi protein modulation of the potassium channel TASK-2 resulting in vesicle osmotic swelling, a mechanism that facilitates the fusion process (Centrone et al., 2018).

Conversely, vesicle-mediated internalization of plasma membrane, involves the coordinate activity of a dynamic complex of proteins including clathrin, dynamin, hsc70, endophilin, anphiphysin, synaptojanin, epsin, adaptor protein-2 (AP-2) and flotillin (Grant & Donaldson, 2009; Mettlen et al., 2009). The involvement of clathrin in AQP2 internalization is well described. AQP2 is accumulated in clathrin-enriched membrane domains during AVP stimulation and washout (Sun, Wu, & Wu, 2002).

AQP2 undergoes different regulated posttranslational modifications, such as phosphorylation, ubiquitination and glutathionylation, which are likely to be fundamental for controlling AQP2 cellular localization, stability and function (Kamsteeg et al., 2006; Medvar, Raghuram, Pisitkun, Sarkar, & Knepper, 2016; Moeller et al., 2011; Nedvetsky et al., 2010; Tamma, Ranieri, et al., 2014).

AQP2 phosphorylation. AVP-induced translocation of AQP2 vesicles to the plasma membrane requires PKA-dependent phosphorylation of AQP2 at S256. The crucial role of cAMP-dependent phosphorylation of AQP2 was first investigated by Kuwahara et al. (1995). Subsequently, phosphoproteomic studies revealed a more complex action of AVP on AQP2, showing that AVP stimulation increases S256, S264 and T269 while decreases S261 phosphorylation in the C-terminus (Hoffert et al., 2006). A time course study reveals that phosphorylation at S256 rapidly increases within few seconds followed by phosphorylation at S269, S264 and S261 dephosphorylation (Hoffert et al., 2008). Ser261 phosphorylation (followed by mono-ubiquitination at K270 and endocytosis) stabilizes AQP2 ubiquitination and intracellular localization (Tamma, Robben, Trimpert, Boone, & Deen, 2011). Short-term AVP stimulation (30 min) causes a relocalization of AQP2-pS264 from predominantly intracellular vesicles, to both the basolateral and apical plasma membranes. After 60 min, part of the AQP2-pS264 was observed in clathrin-coated vesicles, early endosomal compartments, and recycling compartments, but not lysosomes (Fenton et al., 2008). Quantitative mass spectrometry and immunoblotting analysis with phosphospecific antibodies showed that AVP increases the phosphorylation of AQP2 at Ser269 which is mainly located in the apical plasma membrane (Hoffert et al., 2008). This observation is confirmed analyzing the cellular localization of the mutant protein mimicking the constitutive phosphorylated AQP2 at S269 (AQP2-S269D), indicating that S269 phosphorylation might be a strong apical retention signal (Hoffert et al., 2008; Moeller, Knepper, & Fenton, 2009).

AQP2 ubiquitination. It has been reported that at short-term, the cAMP elevating agent forskolin, promoted K63-linked short chain ubiquitination of AQP2 at the apical plasma membrane (Kamsteeg et al., 2006). Interestingly, short chain ubiquitination increases during forskolin washout corresponding to a higher rate of AQP2 retrieval from the plasma membrane to cellular vesicles (Kamsteeg et al., 2006). Short chain ubiquitination of AQP2 has become recognized as a key signal for intracellular trafficking, endocytosis and degradation in lysosomes (Woelk et al., 2006). In situ mutagenesis experiments revealed that AQP2 is ubiquitinated, within the C-terminus, at lysine 270 (K270). On the other hand forskolin decreased AQP2 polyubiquitination resulting in a significant decrease of S261 phosphorylation (Nedvetsky et al., 2010).

AQP2 glutathionylation. We recently demonstrated for the first time that AQP2 is subjected to S-glutathionylation in native mammalian kidney and in renal cell culture and that this posttranslational modification is modulated by the oxidative stress (Tamma, Ranieri, et al., 2014). Topological analysis of AQP2 suggests that Cys75 and Cys79, on cytosolic B-loop, might be target of S-glutathionylation. Indeed, previous data have shown that AVP stimulation increased S-glutathionylation of different proteins in mpkccd cells, indicating the involvement of ROS in AVP activated signal transduction pathway (Sandoval et al., 2013). The physiological role of this posttranslational modification in AQP2 is, however, not clarified and additional studies will be needed to better elucidate the physiological relevance.

2.2 Aquaporin-2 protein abundance: Long-term regulation

The total cellular AQP2 abundance is the resulted balance of protein synthesis and protein removal via degradation or exosomal secretion. Prolonged exposure to AVP increases the total abundance of cellular AQP2 (Ecelbarger et al., 1997; Nielsen, DiGiovanni, Christensen, Knepper, & Harris, 1993).

AQP2 abundance depends on production of AQP2 mRNA or by direct regulation of translation. In the past few years it has become clear that alternative mechanisms controlling AQP2 abundance exist and AQP2 mRNA available for translation does not reflect the total AQP2 mRNA due to regulated sequestration of mRNAs and mRNA degradation via micro-RNA (miRNAs) (Alsaleh & Gottenberg, 2014).

As an example, it has been reported that expression of AQP2 is regulated by external osmolality and solute composition, via the tonicity-responsive enhancer binding protein (TonEBP), which plays a key role in protecting renal cells from hypertonic stress by stimulating transcription of specific genes (Storm, Klussmann, Geelhaar, Rosenthal, & Maric, 2003). In the mouse mpkCCD cell line, AVP increased endogenous AQP2 mRNA and protein levels (Hasler et al., 2002; Hasler, Vinciguerra, Vandewalle, Martin, & Feraille, 2005). Conversely, AQP2 mRNA as well as AQP2 protein abundance decreased in cells deficient for TonEBP (Hasler et al., 2006).

Urinary exosome excretion of AQP2. Urinary excretion of intact AQP2 was firstly described in 1995 by Kanno et al. The molecular mechanism responsible of AQP2 delivery into the urine is not fully clarified yet, although it is well accepted that it occurs via exosome secretion (Pisitkun, Shen, & Knepper, 2004). Under steady-state conditions, an increase in urinary AQP2 excretion could be due to an increased AQP2 synthesis or to a decrease in degradation. Both mechanisms depend on an increase in cellular abundance of AQP2, which could be reflective of AVP stimulation. AQP2 excretion with exosomes may represent a further mechanism of regulation of AQP2 abundance within the cell.

3. Aquaporin-2 and calcium-sensing receptor (CaSR) interplay

The kidney is a key organ for calcium homeostasis, and its ability to sense extracellular calcium levels in the urinary filtrate and the interstitial fluid is owing to the extracellular calcium-sensing receptor (CaSR), which is expressed in multiple sites along the nephron (Brown & MacLeod, 2001; Riccardi & Brown, 2010; Riccardi & Valenti, 2016; Ward & Riccardi, 2002).

The extracellular calcium-sensing receptor (CaSR) is a G-protein–coupled seven-transmembrane spanning receptor that resides on the cell surface as a dimer. It is highly expressed in the parathyroid glands, where it regulates the production and secretion of parathyroid hormone (PTH) in negative feedback fashion. It is also expressed in numerous other tissues, where it has diverse but less well defined functions (Riccardi & Kemp, 2012).

In the kidney, CaSR is expressed across the entire length of the nephron, with the highest expression on the basolateral membrane of thick ascending limb epithelial cells where it plays a crucial role in the regulation of divalent mineral cation transport by inhibiting calcium reabsorption in response to a stimulation by an increase in plasma calcium levels (Brown & MacLeod, 2001; Hebert, Brown, & Harris, 1997; Hofer & Brown, 2003; Pearce & Thakker, 1997; Ward & Riccardi, 2002).

In the proximal tubule, the CaSR is expressed apically where can directly and rapidly blunt the phosphaturic action of PTH, thereby modulating the inhibitory effect of PTH on Pi absorption (Ba, Brown, & Friedman, 2003).

In the collecting duct, CaSR is localized to the apical membrane of the principal cells and co-expressed with the aquaporin-2 (AQP2) water channel in intracellular vesicles (Brown, 1999; Riccardi & Brown, 2010; Sands et al., 1997). Several in vitro, in vivo, and human observational studies have suggested that CaSR signaling inhibits AVP-induced trafficking and expression of AQP2 (Bustamante et al., 2008; Earm et al., 1998; Procino et al., 2004; Procino et al., 2012; Ranieri et al., 2015, 2018; Renkema et al., 2009; Valenti et al., 2000, 2002).

A postulated mechanism for this process is that, during antidiuresis, the actions of AVP promoting water reabsorption from the lumen, causes an increase in urinary Ca2+ concentration secondary to urine concentration, which in turn activates the CaSR located on the apical membrane of the principal cells. The activation of the CaSR reduces the AVP-stimulated insertion of AQP2 into the plasma membrane and the rate of water reabsorption and consequently reduces the risk of Ca2+ supersaturation (Procino et al., 2004, 2012; Procino, Barbieri, et al., 2008; Procino, Mastrofrancesco, et al., 2008). Clinical evidence for an effect of luminal calcium on AQP2-mediated water reabsorption was provided for the first time in a previous study from our research group in enuretic children demonstrating that urinary AQP2 and calciuria correlate with the severity of enuresis in children (Valenti et al., 2000). Interestingly, hypercalciuric enuretic children receiving a low calcium diet to reduce hypercalciuria, had decreased overnight urine output (reduced nocturnal enuresis) paralleled by an increase in nighttime AQP2 excretion and osmolality (Valenti et al., 2002). More recently, we provided further evidence for the inverse relationship between calciuria and water metabolism in a bed rest study. Immobilization results in alterations of renal function, fluid redistribution and bone loss, which couples to a rise of urinary calcium excretion. Under these conditions it was observed that bed rest induced increase in blood hematocrit (reflecting water loss) which coincided to a reduction of urinary AQP2 likely paralleled by an increase in urinary calcium due to bone demineralization (Tamma, Di Mise, et al., 2014). These results further support the view that urinary calcium can modulate the AVP-dependent urine concentration through a down-regulation of AQP2 expression/trafficking.

Besides the effect on AQP2 trafficking, previous findings demonstrated that high external calcium reduces AQP2 expression both in the collecting duct cell line mpkCCD and in hypercalciuric rats (Bustamante et al., 2008; Sands, et al., 1998). In humans, vitamin D-elicited hypercalcemia/hypercalciuria is associated with polyuria. In rats dihydrotachysterol (DHT) induces AQP2 water channel down-regulation despite unaltered AQP2 mRNA expression, suggesting higher rate of AQP2 degradation attributed to activation of the calcium-sensitive protease calpain (Puliyanda, Ward, Baum, Hammond, & Harris, 2003). Altogether, these data support a direct effect of luminal calcium on AQP2 expression in collecting duct principal cells and point to a role of calcium in regulating both AQP2 trafficking and expression.

3.1 Role of CaSR in the regulation of AQP2 trafficking/abundance

Studies performed in cultured renal cells and microdissected collecting ducts have demonstrated that the inhibitory effect of CaSR signaling on AQP2 trafficking to the plasma membrane is associated to a significant reduction in cAMP-induced AQP2 phosphorylation at serine 256 and AQP2 trafficking, resulting in inhibition of osmotic water-permeability response (Ranieri et al., 2015).

Specifically, acute CaSR activation by selective agonists reduced AQP2 translocation to the plasma membrane in response to the cAMP elevating agent forskolin (FK). Moreover, in hypercalciuric patients, evaluation of renal response to dDAVP demonstrated a moderate urinary concentrating defect in these subjects paralleled by reduced urinary excretion of AQP2 (Procino et al., 2012). These findings support the hypothesis that the CaSR–AQP2 interplay represents an internal renal defense to mitigate the effect of hypercalciuria on the risk of calcium saturation during antidiuresis as initially suggested by Sands and coworkers (1997).

In our recent work we demonstrated that CaSR activation in renal cells modulates AQP2 trafficking/expression by alteration of its phosphorylation state (Ranieri et al., 2015). Interestingly, CaSR activation *per sé* induced a significant reduction of the basal AQP2-pS256 levels in HEK-293 cells thus having an opposite effect with respect to AVP action. These data were also confirmed in HEK-293 cells transfected with two gain-of-function variants of CaSR, the CaSR-N124K mutation and the CaSR-R990G polymorphism, used to mimic "tonic" activation of CaSR (Ranieri et al., 2013). Since, under basal conditions, the activation of CaSR by its positive allosteric modulator, NPS-R568, did not significantly alter the cAMP-dependent PKA activity, the hypothesis is that CaSR activation reduces basal

AQP2-pS256 levels as a consequence of the activation of the calcium-regulated protein phosphatase 2A (PP2A), known to dephosphorylate AQP2-pS256 (Tamma, Lasorsa, et al., 2014).

The physiological consequence of the negative feedback on cAMP-induced AQP2-pS256 phosphorylation and trafficking elicited by CaSR signaling is blunting the osmotic water-permeability response both in cells and in microdissected mouse collecting duct (Ranieri et al., 2015). Since a strong reduction in PKA activation in response to FK was observed upon CaSR activation, it is likely that the reduced cAMP accumulation is a consequence of inhibition of the calcium-sensitive adenylate cyclase (AC6) as demonstrated in HEK-293 cells (Ranieri et al., 2015).

The concept that high concentrations of urinary calcium counteract AVP action via the activation of the calcium-sensing receptor (CaSR) expressed in the luminal membrane of collecting duct cells, has been further validated in a mouse model double-knockout for Pendrin/NaCl cotransporter (Ranieri et al., 2018). These mice display significant calcium wasting and develop severe volume depletion, despite increased circulating AVP levels (Soleimani et al., 2012). The hypothesis that the CaSR-mediated impairment of AQP2 expression/trafficking underlies this AVP resistance in dKO mice found substantial confirmation in this study. Due to the severe hypercalciuria, a tonic activation of the luminal CaSR in the collecting duct is expected in this dKO mice model, and quite interestingly, those mice had a strong reduction in total AQP2 expression associated with a significantly higher expression ofAQP2-pS261 and ubiquitinated AQP2. In addition, in dKO mice, exposure of inner medulla kidney slices to the proteasome inhibitor MG132 increased total AQP2 by 50% indicating that the rate of AQP2 degradation via proteasome is significantly higher. Interestingly, the functional link between CaSR and AQP2 degradation was supported by the observation that the reduced total AQP2 and higher levels of AQP2-pS261 found in dKO mice are paralleled by higher levels of p38-MAPK, an enzyme activated by CaSR signaling and known to phosphorylate AQP2 at ser261 (Nedvetsky et al., 2010; Trepiccione et al., 2014). Of note, CaSR inhibition with the calcilytic NPS2143 reduced AQP2-pS261 levels in dKO, demonstrating that CaSR acts upstream of p38-MAPK and mediates the upregulation of AQP2-pS261.

Conversely, inhibition of p38-MAPK, caused a drastic decrease in AQP2-pS261, along with a nearly 5-fold increase in total AQP2. In addition, in dKO mice, p38-MAPK inhibition results in a drastic reduction in ubiquitinated AQP2 that is paralleled by a strong increase in total AQP2.

Previous work in a kidney proximal epithelial cell model, in airway smooth muscle cells, and in H-500 Leydig cancer cells also provided evidence that CaSR activation stimulates p38-MAPK phosphorylation (Maiti, Hait, & Beckman, 2008; Tfelt-Hansen & Brown, 2005; Yarova et al., 2015), demonstrating that p38-MAPK is a downstream effector of CaSR. In line with these data, specific CaSR inhibition caused a strong reduction in p38-MAPK paralleled by increased total AQP2 abundance and reduced p261-AQP2 and ubiquinated AQP2 levels pointing to CaSR as the upstream receptor that orchestrates the modulation of AQP2 expression in dKO mice (Ranieri et al., 2018).

The concept that CaSR expressed at the apical membrane of collecting duct principal cells could mediate effects of hypercalciuria in reducing AVP-elicited osmotic water permeability and urinary concentrating ability by activation of autophagic degradation of AQP2 has been recently suggested (Khositseth et al., 2017). Proteomic analysis of inner medullary collecting ducts isolated from parathyroid hormone–treated rats revealed increased autophagic degradation of a specific set of proteins including AQP2. Interestingly, PTH withdrawal was associated with termination of AQP2 autophagic degradation and normalization of total AQP2 levels and renal concentrating ability (Khositseth et al., 2017). Since CaSR expressed in parathyroid glands is the major player regulating the release of PTH these data further strengthen the link between CaSR signaling and regulation of AQP2 expression.

3.2 AQP2 regulation via microRNA pathway

MicroRNAs (miRNAs) are naturally occurring small non-coding RNAs that regulate posttranscriptional gene expression and are involved in several diseases. miRNAs may act as transcriptional or splicing regulators within the nucleus (Hwang, Wentzel, & Mendell, 2007), and be involved in genetic exchange with adjacent cells, through exosomes (Valadi et al., 2007). Approximately 60% of protein-coding genes are influenced by miRNAs (Friedman, Farh, Burge, & Bartel, 2009) that play crucial roles in several biological processes, including control of cell cycle and differentiation, proliferation and metabolism. As such, miRNA deregulation is being increasingly associated with several human pathologies.

Several studies highlight an emerging role of miRNAs in AQP regulation (reviewed in Gomes, da Silva, Rodrigues, Castro, & Soveral, 2018). Specifically, miRNAs have been identified as endogenous modulators of the expression of several AQPs (Chao et al., 2017; Chen, Shi, Liu, &

Sun, 2018; Huebert et al., 2011; Kim, Jung, Lee, & Kwon, 2015; Li, Shi, Gao, Ma, & Sun, 2018; Luo et al., 2018; Sepramaniam et al., 2010; Tang, Pei, Bai, & Wang, 2016; Wang et al., 2015; Xiong et al., 2018; Zheng et al., 2017). Regarding AQP2, two AQP2-targeting miRNAs, miR-32 and miR-137, were reported to decrease AQP2 expression in kidney collecting duct cells independently of AVP regulation (Kim et al., 2015). The authors demonstrated a significant decrease of AQP2 translation in mpkCCDc14 cells transfected with miR-32 or miR-137 providing novel insights into the regulation of AQP2 by RNA interference.

The concept that CaSR may regulate AQP2 expression also via miRNA, is an intriguing hypothesis. In a previous seminal work, Gong and coworkers demonstrated that in the thick ascending limb, high external calcium—via activation of CaSR—regulates the expression levels of 2 miRNAs, miR-9 and miR-374, which, in turn, reduce the expression of Claudin-14, a protein that blocks the paracellular cation channel, leading to decreases in cation permeation (Gong et al., 2012). These data indicate that the regulation of miRNA by CaSR signaling may occur on several layers within the kidney.

On the other hand, despite several studies that have demonstrated that transcriptional and posttranscriptional regulation of AQP2 play a crucial role in AQP2 expression levels within the cell, along with a profound impact on water homeostasis (Hasler et al., 2009; Hanne B Moeller et al., 2011), little is known about the role of miRNA in the regulation of AQP2 expression.

Of interest, we have recently correlated the AQP2-targeting miR-137 with the reduced expression of AQP2 in Pendrin/NaCl cotransporter dKO mice model, a mechanism found to be mediated by CaSR signaling (Ranieri et al., 2018). Therefore, in addition to the enhanced rate of AQP2 degradation, low AQP2 levels observed in dKO mice are also a result of reduced AQP2 mRNA translation via a novel miRNA pathway modulated by CaSR signaling.

Specifically, in dKO mice, miR-137 was found about 1.7-fold higher compared to WT mice, which was in line with the reduced translation of AQP2 mRNA. Worthy of attention, miR-137 transcript levels were increased by the calcimimetic NPS-R568 providing the first evidence that the CaSR signaling directly acts upstream of the miR-137-AQP2 axis. These findings represent the first demonstration that CaSR can regulate AQP2 expression via AQP2-targeting miRNA (see model in Fig. 1).

The discovery of miRNAs as endogenous modulators of AQPs offers a potential therapeutic approach for the regulation AQP-related disorders (Gomes et al., 2018).

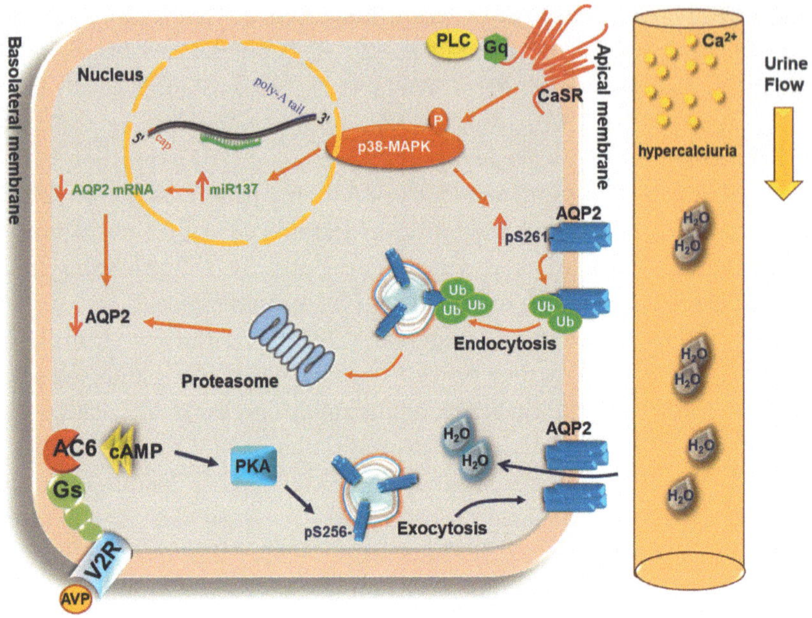

Fig. 1 Schematic model. The proposed model shows that high urinary calcium levels in the renal inner medulla activate CaSR, resulting in the activation and phosphorylation of p38-MAPK that, in turn, phosphorylates AQP2 at Ser261, causing AQP2 internalization, ubiquitination (Ub), and proteasomal degradation. In parallel, CaSR signaling promotes the synthesis of miRNA-137 via the activation of p38-MAPK, which results in reduced AQP2 mRNA translation. *Adapted from Ranieri, M., Zahedi, K., Tamma, G., Centrone, M., Di Mise, A., Soleimani, M., & Valenti, G. (2018). CaSR signaling down-regulates AQP2 expression via a novel microRNA pathway in pendrin and NaCl cotransporter knockout mice.* FASEB Journal 32, *2148–2159 doi:10.1096/fj.201700412RR.*

4. Conclusions

The identification of a novel physiological mechanism that, in the inner medulla, may link the activation of CaSR to the regulation of the AQP2 expression and trafficking, has opened new avenues for understanding AQP2 expression regulation. A bimodal down-regulation of AQP2 expression appears to be involved in this process: CaSR signaling (i) increases AQP2-pS261 and AQP2 ubiquitination resulting in AQP2 protein degradation, (ii) down-regulates AQP2 expression via AQP2-targeting miR-137 pathway. Both pathways contribute to reduce renal concentrating ability and may explain the AVP resistance observed in some pathological setting characterized by hypercalciuria expected to cause sustained stimulation of CaSR in the inner medulla.

Acknowledgments

The author acknowledges the support of Telethon (grant number GGP13227) and ASI (Italian Space Agency, grant number 2013-091-R.0). M.R. is a postdoctoral research fellow supported by "Intervento cofinanziato dal Fondo di Sviluppo e Coesione 2007-2013-APQ Ricerca Regione Puglia, Programma Regionale a Sostegno della Specializzazione Intelligente e della Sostenibilità Sociale ed Ambientale-FutureInResearch."

References

Alsaleh, G., & Gottenberg, J. E. (2014). Characterization of microRNAs and their targets. *Methods in Molecular Biology*, *1142*, 55–63. https://doi.org/10.1007/978-1-4939-0404-4_7.

Ando, F., Sohara, E., Morimoto, T., Yui, N., Nomura, N., Kikuchi, E., et al. (2016). Wnt5a induces renal AQP2 expression by activating calcineurin signalling pathway. *Nature Communications*, 7, 13636. https://doi.org/10.1038/ncomms13636.

Ba, J., Brown, D., & Friedman, P. A. (2003). Calcium-sensing receptor regulation of PTH-inhibitable proximal tubule phosphate transport. *American Journal of Physiology. Renal Physiology*, *285*(6), F1233–F1243. https://doi.org/10.1152/ajprenal.00249.2003.

Birnbaumer, M. (2000). Vasopressin receptors review. *Trends in Endocrinology and Metabolism*, *11*(10), 406–410. https://doi.org/10.1016/S1043-2760(00)00304-0.

Bockenhauer, D., & Bichet, D. (2017). Nephrogenic diabetes insipidus. *Current Opinion in Pediatrics*, *29*, 199–205.

Brown, E. M. (1999). Physiology and pathophysiology of the extracellular calcium-sensing receptor. *The American Journal of Medicine*, *106*(2), 238–253.

Brown, D., Hasler, U., Nunes, P., Bouley, R., & Lu, H. A. (2008). Phosphorylation events and the modulation of aquaporin 2 cell surface expression. *Current Opinion in Nephrology and Hypertension*, *17*(5), 491–498. https://doi.org/10.1097/MNH.0b013e3283094eb1.

Brown, E. M., & MacLeod, R. J. (2001). Extracellular calcium sensing and extracellular calcium signaling. *Physiological Reviews*, *81*(1), 239–297. https://doi.org/10.1152/physrev.2001.81.1.239.

Brown, D., & Stow, J. L. (1996). Protein trafficking and polarity in kidney epithelium: From cell biology to physiology. *Physiological Reviews*, *76*(1), 245–297. https://doi.org/10.1152/physrev.1996.76.1.245.

Bustamante, M., Hasler, U., Leroy, V., de Seigneux, S., Dimitrov, M., Mordasini, D., et al. (2008). Calcium-sensing receptor attenuates AVP-induced aquaporin-2 expression via a calmodulin-dependent mechanism. *Journal of the American Society of Nephrology: JASN*, *19*(1), 109–116. https://doi.org/10.1681/ASN.2007010092.

Centrone, M., De Santo, M. P., Nicotera, I., Labate, C., Ranieri, M., Di Mise, A., et al. (2018). Gi protein modulation of the potassium channel TASK-2 mediates vesicle osmotic swelling to facilitate the fusion of aquaporin-2 water channel containing vesicles. *Cell*, 7(12), 276. https://doi.org/10.3390/cells7120276.

Centrone, M., Ranieri, M., Di Mise, A., Berlingerio, S. P., Russo, A., Deen, P. M. T., et al. (2017). AQP2 abundance is regulated by the E3-ligase CHIP via HSP70. *Cellular Physiology and Biochemistry*, *44*(2), 515–531. https://doi.org/10.1159/000485088.

Chao, G., Wang, Y., Zhang, S., Yang, W., Ni, Z., & Zheng, X. (2017). MicroRNA-29a increased the intestinal membrane permeability of colonic epithelial cells in irritable bowel syndrome rats. *Oncotarget*, *8*(49), 85828–85837. https://doi.org/10.18632/oncotarget.20687.

Chen, G., Shi, Y., Liu, M., & Sun, J. (2018). circHIPK3 regulates cell proliferation and migration by sponging miR-124 and regulating AQP3 expression in hepatocellular carcinoma. *Cell Death & Disease*, *9*(2), 175. https://doi.org/10.1038/s41419-017-0204-3.

Cheung, P. W., Terlouw, A., Janssen, S. A., Brown, D., & Bouley, R. (2019). Inhibition of non-receptor tyrosine kinase Src induces phosphoserine 256-independent aquaporin-2 membrane accumulation. *The Journal of Physiology*, *597*(6), 1627–1642. https://doi.org/10.1113/jp277024.

Christensen, A. E., Selheim, F., de Rooij, J., Dremier, S., Schwede, F., Dao, K. K., et al. (2003). cAMP analog mapping of Epac1 and cAMP kinase. Discriminating analogs demonstrate that Epac and cAMP kinase act synergistically to promote PC-12 cell neurite extension. *The Journal of Biological Chemistry*, *278*(37), 35394–35402. https://doi.org/10.1074/jbc.M302179200.

Cuesta, M., & Thompson, C. J. (2016). The syndrome of inappropriate antidiuresis (SIAD). *Best Practice and Research: Clinical Endocrinology and Metabolism*, *30*(2), 175–187. https://doi.org/10.1016/j.beem.2016.02.009.

de Rooij, J., Zwartkruis, F. J., Verheijen, M. H., Cool, R. H., Nijman, S. M., Wittinghofer, A., et al. (1998). Epac is a Rap1 guanine-nucleotide-exchange factor directly activated by cyclic AMP. *Nature*, *396*(6710), 474–477. https://doi.org/10.1038/24884.

Douglass, J., Gunaratne, R., Bradford, D., Saeed, F., Hoffert, J. D., Steinbach, P. J., et al. (2012). Identifying protein kinase target preferences using mass spectrometry. *American Journal of Physiology. Cell Physiology*, *303*(7), C715–C727. https://doi.org/10.1152/ajpcell.00166.2012.

Earm, J. H., Christensen, B. M., Frokiaer, J., Marples, D., Han, J. S., Knepper, M. A., et al. (1998). Decreased aquaporin-2 expression and apical plasma membrane delivery in kidney collecting ducts of polyuric hypercalcemic rats. *Journal of the American Society of Nephrology*, *9*(12), 2181–2193.

Ecelbarger, C. A., Nielsen, S., Olson, B. R., Murase, T., Baker, E. A., Knepper, M. A., et al. (1997). Role of renal aquaporins in escape from vasopressin-induced antidiuresis in rat. *Journal of Clinical Investigation*, *99*(8), 1852–1863. https://doi.org/10.1172/JCI119352.

Erdélyi, L. S., Mann, W. A., Morris-Rosendahl, D. J., Groß, U., Nagel, M., Várnai, P., et al. (2015). Mutation in the V2 vasopressin receptor gene, AVPR2, causes nephrogenic syndrome of inappropriate diuresis. *Kidney International*, *88*(5), 1070–1078. https://doi.org/10.1038/ki.2015.181.

Feldman, B. J., Rosenthal, S., Vargas, G. A., Fenwick, R. G., Huang, E. A., Matsuda-Abedini, M., et al. (2005). Nephrogenic syndrome of inappropriate antidiuresis. *The New England Journal of Medicine*, *352*, 1884–1890.

Fenton, R. A., Moeller, H. B., Hoffert, J. D., Yu, M. J., Nielsen, S., & Knepper, M. A. (2008). Acute regulation of aquaporin-2 phosphorylation at Ser-264 by vasopressin. *Proceedings of the National Academy of Sciences of the United States of America*, *105*(8), 3134–3139. https://doi.org/10.1073/pnas.0712338105.

Friedman, R. C., Farh, K. K., Burge, C. B., & Bartel, D. P. (2009). Most mammalian mRNAs are conserved targets of microRNAs. *Genome Research*, *19*(1), 92–105. https://doi.org/10.1101/gr.082701.108.

Gomes, A., da Silva, I. V., Rodrigues, C. M. P., Castro, R. E., & Soveral, G. (2018). The emerging role of microRNAs in aquaporin regulation. *Frontiers in Chemistry*, *6*, 238. https://doi.org/10.3389/fchem.2018.00238.

Gong, Y., Renigunta, V., Himmerkus, N., Zhang, J., Renigunta, A., Bleich, M., et al. (2012). Claudin-14 regulates renal Ca(+)(+) transport in response to CaSR signalling via a novel microRNA pathway. *The EMBO Journal*, *31*(8), 1999–2012. https://doi.org/10.1038/emboj.2012.49.

Gouraud, S., Laera, A., Calamita, G., Carmosino, M., Procino, G., Rossetto, O., et al. (2002). Functional involvement of VAMP/synaptobrevin-2 in cAMP-stimulated aquaporin 2 translocation in renal collecting duct cells. *Journal of Cell Science*, *115* (Pt. 18), 3667–3674.

Grant, B. D., & Donaldson, J. G. (2009). Pathways and mechanisms of endocytic recycling. *Nature Reviews. Molecular Cell Biology*, *10*(9), 597–608. https://doi.org/10.1038/nrm2755.

Hasler, U., Jeon, U. S., Kim, J. A., Mordasini, D., Kwon, H. M., Feraille, E., et al. (2006). Tonicity-responsive enhancer binding protein is an essential regulator of aquaporin-2 expression in renal collecting duct principal cells. *Journal of the American Society of Nephrology*, *17*(6), 1521–1531. https://doi.org/10.1681/asn.2005121317.

Hasler, U., Leroy, V., Martin, P. Y., & Feraille, E. (2009). Aquaporin-2 abundance in the renal collecting duct: New insights from cultured cell models. *American Journal of Physiology. Renal Physiology*, *297*(1), F10–F18. https://doi.org/10.1152/ajprenal.00053.2009.

Hasler, U., Mordasini, D., Bens, M., Bianchi, M., Cluzeaud, F., Rousselot, M., et al. (2002). Long term regulation of aquaporin-2 expression in vasopressin-responsive renal collecting duct principal cells. *The Journal of Biological Chemistry*, *277*(12), 10379–10386. https://doi.org/10.1074/jbc.M111880200.

Hasler, U., Vinciguerra, M., Vandewalle, A., Martin, P. Y., & Feraille, E. (2005). Dual effects of hypertonicity on aquaporin-2 expression in cultured renal collecting duct principal cells. *Journal of the American Society of Nephrology*, *16*(6), 1571–1582. https://doi.org/10.1681/asn.2004110930.

Hebert, S. C., Brown, E. M., & Harris, H. W. (1997). Role of the Ca(2+)-sensing receptor in divalent mineral ion homeostasis. *The Journal of Experimental Biology*, *200*(Pt. 2), 295–302.

Hofer, A. M., & Brown, E. M. (2003). Extracellular calcium sensing and signalling. *Nature Reviews. Molecular Cell Biology*, *4*(7), 530–538. https://doi.org/10.1038/nrm1154.

Hoffert, J. D., Chou, C. L., Fenton, R. A., & Knepper, M. A. (2005). Calmodulin is required for vasopressin-stimulated increase in cyclic AMP production in inner medullary collecting duct. *The Journal of Biological Chemistry*, *280*(14), 13624–13630. https://doi.org/10.1074/jbc.M500040200.

Hoffert, J. D., Fenton, R. A., Moeller, H. B., Simons, B., Tchapyjnikov, D., McDill, B. W., et al. (2008). Vasopressin-stimulated increase in phosphorylation at Ser269 potentiates plasma membrane retention of aquaporin-2. *The Journal of Biological Chemistry*, *283*(36), 24617–24627. https://doi.org/10.1074/jbc.M803074200.

Hoffert, J. D., Pisitkun, T., Wang, G., Shen, R. F., & Knepper, M. A. (2006). Quantitative phosphoproteomics of vasopressin-sensitive renal cells: Regulation of aquaporin-2 phosphorylation at two sites. *Proceedings of the National Academy of Sciences of the United States of America*, *103*(18), 7159–7164. https://doi.org/10.1073/pnas.0600895103.

Huebert, R. C., Jagavelu, K., Hendrickson, H. I., Vasdev, M. M., Arab, J. P., Splinter, P. L., et al. (2011). Aquaporin-1 promotes angiogenesis, fibrosis, and portal hypertension through mechanisms dependent on osmotically sensitive microRNAs. *The American Journal of Pathology*, *179*(4), 1851–1860. https://doi.org/10.1016/j.ajpath.2011.06.045.

Hwang, H. W., Wentzel, E. A., & Mendell, J. T. (2007). A hexanucleotide element directs microRNA nuclear import. *Science*, *315*(5808), 97–100. https://doi.org/10.1126/science.1136235.

Jung, H. J., & Kwon, T.-H. (2016). Molecular mechanisms regulating aquaporin-2 in kidney collecting duct. *American Journal of Physiology. Renal Physiology*, *311*(6), F1318–F1328. https://doi.org/10.1152/ajprenal.00485.2016.

Kamsteeg, E. J., Hendriks, G., Boone, M., Konings, I. B., Oorschot, V., van der Sluijs, P., et al. (2006). Short-chain ubiquitination mediates the regulated endocytosis of the aquaporin-2 water channel. *Proceedings of the National Academy of Sciences of the United States of America*, *103*(48), 18344–18349. https://doi.org/10.1073/pnas.0604073103.

Kanno, K., Sasaki, S., Hirata, Y., Ishikawa, S., Fushimi, K., Nakanishi, S., et al. (1995). Urinary excretion of aquaporin-2 in patients with diabetes insipidus. *The New England Journal of Medicine*, *332*(23), 1540–1545. https://doi.org/10.1056/nejm199506083322303.

Katsura, T., Gustafson, C. E., Ausiello, D. A., & Brown, D. (1997). Protein kinase A phosphorylation is involved in regulated exocytosis of aquaporin-2 in transfected LLC-PK1 cells. *The American Journal of Physiology*, *272*(6 Pt. 2), F817–F822.

Kawasaki, H., Springett, G. M., Mochizuki, N., Toki, S., Nakaya, M., Matsuda, M., et al. (1998). A family of cAMP-binding proteins that directly activate Rap1. *Science*, *282*(5397), 2275–2279.

Khositseth, S., Charngkaew, K., Boonkrai, C., Somparn, P., Uawithya, P., Chomanee, N., et al. (2017). Hypercalcemia induces targeted autophagic degradation of aquaporin-2 at the onset of nephrogenic diabetes insipidus. *Kidney International*, *91*(5), 1070–1087. https://doi.org/10.1016/j.kint.2016.12.005.

Kim, J. E., Jung, H. J., Lee, Y. J., & Kwon, T. H. (2015). Vasopressin-regulated miRNAs and AQP2-targeting miRNAs in kidney collecting duct cells. *American Journal of Physiology. Renal Physiology*, *308*(7), F749–F764. https://doi.org/10.1152/ajprenal.00334.2014.

Klussmann, E., Tamma, G., Lorenz, D., Wiesner, B., Maric, K., Hofmann, F., et al. (2001). An inhibitory role of rho in the vasopressin-mediated translocation of aquaporin-2 into cell membranes of renal principal cells. *Journal of Biological Chemistry*, *276*(23), 20451–20457. https://doi.org/10.1074/jbc.M010270200.

Knepper, M. A., Kwon, T.-H., & Nielsen, S. (2015). Molecular physiology of water balance. *New England Journal of Medicine*, *372*(14), 1349–1358. https://doi.org/10.1056/NEJMra1404726.

Knepper, M. A., Nielsen, S., Chou, C. L., & DiGiovanni, S. R. (1994). Mechanism of vasopressin action in the renal collecting duct. *Seminars in Nephrology*, *14*(4), 302–321.

Kuwahara, M., Fushimi, K., Terada, Y., Bai, L., Marumo, F., & Sasaki, S. (1995). cAMP-dependent phosphorylation stimulates water permeability of aquaporin-collecting duct water channel protein expressed in *Xenopus* oocytes. *The Journal of Biological Chemistry*, *270*(18), 10384–10387.

Lei, L., Huang, M., Su, L., Xie, D., Mamuya, F. A., Ham, O., et al. (2018). Manganese promotes intracellular accumulation of AQP2 via modulating F-actin polymerization and reduces urinary concentration in mice. *American Journal of Physiology. Renal Physiology*, *314*(2), F306–f316. https://doi.org/10.1152/ajprenal.00391.2017.

Li, H., Shi, H., Gao, M., Ma, N., & Sun, R. (2018). Long non-coding RNA CASC2 improved acute lung injury by regulating miR-144-3p/AQP1 axis to reduce lung epithelial cell apoptosis. *Cell & Bioscience*, *8*, 15. https://doi.org/10.1186/s13578-018-0205-7.

Luo, L., Yang, R., Zhao, S., Chen, Y., Hong, S., Wang, K., et al. (2018). Decreased miR-320 expression is associated with breast cancer progression, cell migration, and invasiveness via targeting aquaporin 1. *Acta Biochimica et Biophysica Sinica Shanghai*, *50*(5), 473–480. https://doi.org/10.1093/abbs/gmy023.

Maiti, A., Hait, N. C., & Beckman, M. J. (2008). Extracellular calcium-sensing receptor activation induces vitamin D receptor levels in proximal kidney HK-2G cells by a mechanism that requires phosphorylation of p38alpha MAPK. *The Journal of Biological Chemistry*, *283*(1), 175–183. https://doi.org/10.1074/jbc.M707269200.

McSorley, T., Stefan, E., Henn, V., Wiesner, B., Baillie, G. S., Houslay, M. D., et al. (2006). Spatial organisation of AKAP18 and PDE4 isoforms in renal collecting duct principal cells. *European Journal of Cell Biology*, *85*(7), 673–678. https://doi.org/10.1016/j.ejcb.2006.01.005.

Medvar, B., Raghuram, V., Pisitkun, T., Sarkar, A., & Knepper, M. A. (2016). Comprehensive database of human E3 ubiquitin ligases: Application to aquaporin-2 regulation. *Physiological Genomics*, *48*(7), 502–512. https://doi.org/10.1152/physiolgenomics. 00031.2016.

Mettlen, M., Stoeber, M., Loerke, D., Antonescu, C. N., Danuser, G., & Schmid, S. L. (2009). Endocytic accessory proteins are functionally distinguished by their differential effects on the maturation of clathrin-coated pits. *Molecular Biology of the Cell*, *20*(14), 3251–3260. https://doi.org/10.1091/mbc.E09-03-0256.

Moeller, H. B., Knepper, M. A., & Fenton, R. A. (2009). Serine 269 phosphorylated aquaporin-2 is targeted to the apical membrane of collecting duct principal cells. *Kidney International*, *75*(3), 295–303. https://doi.org/10.1038/ki.2008.505.

Moeller, H. B., Olesen, E. T. B., & Fenton, R. A. (2011). Regulation of the water channel aquaporin-2 by posttranslational modification. *American Journal of Physiology. Renal Physiology*, *300*(5), F1062–F1073. https://doi.org/10.1152/ajprenal.00721.2010.

Nedvetsky, P. I., Tabor, V., Tamma, G., Beulshausen, S., Skroblin, P., Kirschner, A., et al. (2010). Reciprocal regulation of aquaporin-2 abundance and degradation by protein kinase A and p38-MAP kinase. *Journal of the American Society of Nephrology: JASN*, *21*(10), 1645–1656. https://doi.org/10.1681/ASN.2009111190.

Nedvetsky, P. I., Tamma, G., Beulshausen, S., Valenti, G., Rosenthal, W., & Klussmann, E. (2009). Regulation of aquaporin-2 trafficking. *Handbook of Experimental Pharmacology*, *190*, 133–157. https://doi.org/10.1007/978-3-540-79885-9_6.

Nielsen, S., Chou, C. L., Marples, D., Christensen, E. I., Kishore, B. K., & Knepper, M. A. (1995). Vasopressin increases water permeability of kidney collecting duct by inducing translocation of aquaporin-CD water channels to plasma membrane. *Proceedings of the National Academy of Sciences of the United States of America*, *92*(4), 1013–1017.

Nielsen, S., DiGiovanni, S. R., Christensen, E. I., Knepper, M. A., & Harris, H. W. (1993). Cellular and subcellular immunolocalization of vasopressin-regulated water channel in rat kidney. *Proceedings of the National Academy of Sciences of the United States of America*, *90*(24), 11663–11667.

Nielsen, S., Frokiaer, J., Marples, D., Kwon, T. H., Agre, P., & Knepper, M. A. (2002). Aquaporins in the kidney: From molecules to medicine. *Physiological Reviews*, *82*(1), 205–244. https://doi.org/10.1152/physrev.00024.2001.

Noda, Y., Horikawa, S., Katayama, Y., & Sasaki, S. (2004). Water channel aquaporin-2 directly binds to actin. *Biochemical and Biophysical Research Communications*, *322*(3), 740–745. https://doi.org/10.1016/j.bbrc.2004.07.195.

Noda, Y., Sohara, E., Ohta, E., & Sasaki, S. (2010). Aquaporins in kidney pathophysiology. *Nature Reviews. Nephrology*, *6*(3), 168–178. https://doi.org/10.1038/nrneph.2009.231.

Ozaki, N., Shibasaki, T., Kashima, Y., Miki, T., Takahashi, K., Ueno, H., et al. (2000). cAMP-GEFII is a direct target of cAMP in regulated exocytosis. *Nature Cell Biology*, *2*(11), 805–811. https://doi.org/10.1038/35041046.

Pearce, S. H., & Thakker, R. V. (1997). The calcium-sensing receptor: Insights into extracellular calcium homeostasis in health and disease. *The Journal of Endocrinology*, *154*(3), 371–378.

Pisitkun, T., Jacob, V., Schleicher, S. M., Chou, C. L., Yu, M. J., & Knepper, M. A. (2008). Akt and ERK1/2 pathways are components of the vasopressin signaling network in rat native IMCD. *American Journal of Physiology. Renal Physiology*, *295*(4), F1030–F1043. https://doi.org/10.1152/ajprenal.90339.2008.

Pisitkun, T., Shen, R. F., & Knepper, M. A. (2004). Identification and proteomic profiling of exosomes in human urine. *Proceedings of the National Academy of Sciences of the United States of America*, *101*(36), 13368–13373. https://doi.org/10.1073/pnas.0403453101.

Procino, G., Barbieri, C., Tamma, G., De Benedictis, L., Pessin, J. E., Svelto, M., et al. (2008). AQP2 exocytosis in the renal collecting duct—Involvement of SNARE isoforms and the regulatory role of Munc18b. *Journal of Cell Science*, *121*(Pt. 12), 2097–2106. https://doi.org/10.1242/jcs.022210.

Procino, G., Carmosino, M., Tamma, G., Gouraud, S., Laera, A., Riccardi, D., et al. (2004). Extracellular calcium antagonizes forskolin-induced aquaporin 2 trafficking in collecting duct cells. *Kidney International*, *66*(6), 2245–2255. https://doi.org/10.1111/j.1523-1755.2004.66036.x.

Procino, G., Mastrofrancesco, L., Mira, A., Tamma, G., Carmosino, M., Emma, F., et al. (2008). Aquaporin 2 and apical calcium-sensing receptor: New players in polyuric disorders associated with hypercalciuria. *Seminars in Nephrology*, *28*(3), 297–305. https://doi.org/10.1016/j.semnephrol.2008.03.007.

Procino, G., Mastrofrancesco, L., Tamma, G., Lasorsa, D. R., Ranieri, M., Stringini, G., et al. (2012). Calcium-sensing receptor and aquaporin 2 interplay in hypercalciuria-associated renal concentrating defect in humans. An in vivo and in vitro study. *PLoS One*, 7(3), e33145. https://doi.org/10.1371/journal.pone.0033145.

Puliyanda, D. P., Ward, D. T., Baum, M. A., Hammond, T. G., & Harris, H. W., Jr. (2003). Calpain-mediated AQP2 proteolysis in inner medullary collecting duct. *Biochemical and Biophysical Research Communications*, *303*(1), 52–58.

Ranieri, M., Tamma, G., Di Mise, A., Russo, A., Centrone, M., Svelto, M., et al. (2015). Negative feedback from CaSR signaling to aquaporin-2 sensitizes vasopressin to extracellular Ca2. *Journal of Cell Science*, *128*(13), 2350–2360. https://doi.org/10.1242/jcs.168096.

Ranieri, M., Tamma, G., Di Mise, A., Vezzoli, G., Soldati, L., Svelto, M., et al. (2013). Excessive signal transduction of gain-of-function variants of the calcium-sensing receptor (CaSR) are associated with increased ER to cytosol calcium gradient. *PLoS One*, *8*(11), e79113. https://doi.org/10.1371/journal.pone.0079113.

Ranieri, M., Zahedi, K., Tamma, G., Centrone, M., Di Mise, A., Soleimani, M., et al. (2018). CaSR signaling down-regulates AQP2 expression via a novel microRNA pathway in pendrin and NaCl cotransporter knockout mice. *FASEB Journal*, *32*(4), 2148–2159. https://doi.org/10.1096/fj.201700412RR.

Renkema, K. Y., Lee, K., Topala, C. N., Goossens, M., Houillier, P., Bindels, R. J., et al. (2009). TRPV5 gene polymorphisms in renal hypercalciuria. *Nephrology, Dialysis, Transplantation*, *24*(6), 1919–1924. https://doi.org/10.1093/ndt/gfn735.

Riccardi, D., & Brown, E. M. (2010). Physiology and pathophysiology of the calcium-sensing receptor in the kidney. *American Journal of Physiology. Renal Physiology*, *298*(3), F485–F499. https://doi.org/10.1152/ajprenal.00608.2009.

Riccardi, D., & Kemp, P. J. (2012). The calcium-sensing receptor beyond extracellular calcium homeostasis: Conception, development, adult physiology, and disease. *Annual Review of Physiology*, *74*, 271–297. https://doi.org/10.1146/annurev-physiol-020911-153318.

Riccardi, D., & Valenti, G. (2016). Localization and function of the renal calcium-sensing receptor. *Nature Reviews. Nephrology*, *12*(7), 414–425. https://doi.org/10.1038/nrneph.2016.59.

Rinschen, M. M., Yu, M. J., Wang, G., Boja, E. S., Hoffert, J. D., Pisitkun, T., et al. (2010). Quantitative phosphoproteomic analysis reveals vasopressin V2-receptor-dependent signaling pathways in renal collecting duct cells. *Proceedings of the National Academy of Sciences of the United States of America*, *107*(8), 3882–3887. https://doi.org/10.1073/pnas.0910646107.

Sabolic, I., Katsura, T., Verbavatz, J. M., & Brown, D. (1995). The AQP2 water channel: Effect of vasopressin treatment, microtubule disruption, and distribution in neonatal rats. *The Journal of Membrane Biology*, *143*(3), 165–175.

Sandoval, P. C., Slentz, D. H., Pisitkun, T., Saeed, F., Hoffert, J. D., & Knepper, M. A. (2013). Proteome-wide measurement of protein half-lives and translation rates in vasopressin-sensitive collecting duct cells. *Journal of the American Society of Nephrology*, *24*(11), 1793–1805. https://doi.org/10.1681/ASN.2013030279.

Sands, J. M., Flores, F. X., Kato, A., Baum, M. A., Brown, E. M., Ward, D. T., et al. (1998). Vasopressin-elicited water and urea permeabilities are altered in IMCD in hypercalcemic rats. *The American Journal of Physiology*, *274*(5), F978–F985.

Sands, J. M., Naruse, M., Baum, M., Jo, I., Hebert, S. C., Brown, E. M., et al. (1997). Apical extracellular calcium/polyvalent cation-sensing receptor regulates vasopressin-elicited water permeability in rat kidney inner medullary collecting duct. *The Journal of Clinical Investigation*, *99*(6), 1399–1405. https://doi.org/10.1172/jci119299.

Schrade, K., Tröger, J., Eldahshan, A., Zühlke, K., Abdul Azeez, K. R., Elkins, J. M., et al. (2018). An AKAP-Lbc-RhoA interaction inhibitor promotes the translocation of

aquaporin-2 to the plasma membrane of renal collecting duct principal cells. *PLoS One*, *13*(1), e0191423. https://doi.org/10.1371/journal.pone.0191423.
Seino, S., & Shibasaki, T. (2005). PKA-dependent and PKA-independent pathways for cAMP-regulated exocytosis. *Physiological Reviews*, *85*(4), 1303–1342. https://doi.org/10.1152/physrev.00001.2005.
Sepramaniam, S., Armugam, A., Lim, K. Y., Karolina, D. S., Swaminathan, P., Tan, J. R., et al. (2010). MicroRNA 320a functions as a novel endogenous modulator of aquaporins 1 and 4 as well as a potential therapeutic target in cerebral ischemia. *The Journal of Biological Chemistry*, *285*(38), 29223–29230. https://doi.org/10.1074/jbc.M110.144576.
Soleimani, M., Barone, S., Xu, J., Shull, G. E., Siddiqui, F., Zahedi, K., et al. (2012). Double knockout of pendrin and Na-Cl cotransporter (NCC) causes severe salt wasting, volume depletion, and renal failure. *Proceedings of the National Academy of Sciences of the United States of America*, *109*(33), 13368–13373. https://doi.org/10.1073/pnas.1202671109.
Storm, R., Klussmann, E., Geelhaar, A., Rosenthal, W., & Maric, K. (2003). Osmolality and solute composition are strong regulators of AQP2 expression in renal principal cells. *American Journal of Physiology. Renal Physiology*, *284*(1), F189–F198. https://doi.org/10.1152/ajprenal.00245.2002.
Sun, J. Y., Wu, X. S., & Wu, L. G. (2002). Single and multiple vesicle fusion induce different rates of endocytosis at a central synapse. *Nature*, *417*(6888), 555–559. https://doi.org/10.1038/417555a.
Tamma, G., Di Mise, A., Ranieri, M., Svelto, M., Pisot, R., Bilancio, G., et al. (2014). A decrease in aquaporin 2 excretion is associated with bed rest induced high calciuria. *Journal of Translational Medicine*, *12*, 133. https://doi.org/10.1186/1479-5876-12-133.
Tamma, G., Klussmann, E., Maric, K., Aktories, K., Svelto, M., Rosenthal, W., et al. (2001). Rho inhibits cAMP-induced translocation of aquaporin-2 into the apical membrane of renal cells. *American Journal of Physiology. Renal Physiology*, *281*(6), F1092–F1101. https://doi.org/10.1152/ajprenal.0091.2001.
Tamma, G., Klussmann, E., Procino, G., Svelto, M., Rosenthal, W., & Valenti, G. (2003). cAMP-induced AQP2 translocation is associated with RhoA inhibition through RhoA phosphorylation and interaction with RhoGDI. *Journal of Cell Science*, *116*(Pt. 8), 1519–1525.
Tamma, G., Lasorsa, D., Trimpert, C., Ranieri, M., Di Mise, A., Mola, M. G., et al. (2014). A protein kinase A-independent pathway controlling aquaporin 2 trafficking as a possible cause for the syndrome of inappropriate antidiuresis associated with polycystic kidney disease 1 haploinsufficiency. *Journal of the American Society of Nephrology*, *25*(10), 2241–2253. https://doi.org/10.1681/asn.2013111234.
Tamma, G., Procino, G., Mola, M. G., Svelto, M., & Valenti, G. (2008). Functional involvement of Annexin-2 in cAMP induced AQP2 trafficking. *Pflügers Archiv—European Journal of Physiology*, *456*(4), 729–736. https://doi.org/10.1007/s00424-008-0453-1.
Tamma, G., Ranieri, M., Di Mise, A., Centrone, M., Sveito, M., & Valenti, G. (2014). Glutathionylation of the aquaporin-2 water channel: A novel post-translational modification modulated by the oxidative stress. *Journal of Biological Chemistry*, *289*(40), 27807–27813. https://doi.org/10.1074/jbc.M114.586024.
Tamma, G., Robben, J. H., Trimpert, C., Boone, M., & Deen, P. M. (2011). Regulation of AQP2 localization by S256 and S261 phosphorylation and ubiquitination. *American Journal of Physiology. Cell Physiology*, *300*(3), C636–C646. https://doi.org/10.1152/ajpcell.00433.2009.
Tang, R., Pei, L., Bai, T., & Wang, J. (2016). Down-regulation of microRNA-126-5p contributes to overexpression of VEGFA in lipopolysaccharide-induced acute lung injury. *Biotechnology Letters*, *38*(8), 1277–1284. https://doi.org/10.1007/s10529-016-2107-2.
Tfelt-Hansen, J., & Brown, E. M. (2005). The calcium-sensing receptor in normal physiology and pathophysiology: A review. *Critical Reviews in Clinical Laboratory Sciences*, *42*(1), 35–70.

Trepiccione, F., Pisitkun, T., Hoffert, J. D., Poulsen, S. B., Capasso, G., Nielsen, S., et al. (2014). Early targets of lithium in rat kidney inner medullary collecting duct include p38 and ERK1/2. *Kidney International, 86*(4), 757–767. https://doi.org/10.1038/ki.2014.107.

Valadi, H., Ekstrom, K., Bossios, A., Sjostrand, M., Lee, J. J., & Lotvall, J. O. (2007). Exosome-mediated transfer of mRNAs and microRNAs is a novel mechanism of genetic exchange between cells. *Nature Cell Biology, 9*(6), 654–659. https://doi.org/10.1038/ncb1596.

Valenti, G., Laera, A., Gouraud, S., Pace, G., Aceto, G., Penza, R., et al. (2002). Low-calcium diet in hypercalciuric enuretic children restores AQP2 excretion and improves clinical symptoms. *American Journal of Physiology. Renal Physiology, 283*(5), F895–F903. https://doi.org/10.1152/ajprenal.00354.2001.

Valenti, G., Laera, A., Pace, G., Aceto, G., Lospalluti, M. L., Penza, R., et al. (2000). Urinary aquaporin 2 and calciuria correlate with the severity of enuresis in children. *Journal of the American Society of Nephrology, 11*(10), 1873–1881.

Valenti, G., Procino, G., Tamma, G., Carmosino, M., & Svelto, M. (2005). Minireview: Aquaporin 2 trafficking. *Endocrinology, 146*, 5063–5070.

Wang, Y., Huang, J., Ma, Y., Tang, G., Liu, Y., Chen, X., et al. (2015). MicroRNA-29b is a therapeutic target in cerebral ischemia associated with aquaporin 4. *Journal of Cerebral Blood Flow and Metabolism, 35*(12), 1977–1984. https://doi.org/10.1038/jcbfm.2015.156.

Wang, C. C., Ng, C. P., Shi, H., Liew, H. C., Guo, K., Zeng, Q., et al. (2010). A role for VAMP8/endobrevin in surface deployment of the water channel aquaporin 2. *Molecular and Cellular Biology, 30*(1), 333–343. https://doi.org/10.1128/mcb.00814-09.

Ward, D. T., & Riccardi, D. (2002). Renal physiology of the extracellular calcium-sensing receptor. *Pflügers Archiv, 445*(2), 169–176. https://doi.org/10.1007/s00424-002-0914-x.

Woelk, T., Oldrini, B., Maspero, E., Confalonieri, S., Cavallaro, E., Di Fiore, P. P., et al. (2006). Molecular mechanisms of coupled monoubiquitination. *Nature Cell Biology, 8*(11), 1246–1254. https://doi.org/10.1038/ncb1484.

Wu, Q., Moeller, H. B., Stevens, D. A., Sanchez-Hodge, R., Childers, G., Kortenoeven, M. L. A., et al. (2018). CHIP regulates aquaporin-2 quality control and body water homeostasis. *Journal of the American Society of Nephrology, 29*(3), 936–948. https://doi.org/10.1681/ASN.2017050526.

Xiong, W., Ran, J., Jiang, R., Guo, P., Shi, X., Li, H., et al. (2018). miRNA-320a inhibits glioma cell invasion and migration by directly targeting aquaporin 4. *Oncology Reports, 39*(4), 1939–1947. https://doi.org/10.3892/or.2018.6274.

Yarova, P. L., Stewart, A. L., Sathish, V., Britt, R. D., Jr., Thompson, M. A., P Lowe, A. P., et al. (2015). Calcium-sensing receptor antagonists abrogate airway hyperresponsiveness and inflammation in allergic asthma. *Science Translational Medicine*, 7(284), 284ra260. https://doi.org/10.1126/scitranslmed.aaa0282.

Yip, K. P. (2006). Epac-mediated Ca(2+) mobilization and exocytosis in inner medullary collecting duct. *American Journal of Physiology. Renal Physiology, 291*(4), F882–F890. https://doi.org/10.1152/ajprenal.00411.2005.

Zheng, L., Cheng, W., Wang, X., Yang, Z., Zhou, X., & Pan, C. (2017). Overexpression of microRNA-145 ameliorates astrocyte injury by targeting aquaporin 4 in cerebral ischemic stroke. *BioMed Research International, 2017*, 9530951. https://doi.org/10.1155/2017/9530951.

CHAPTER FOURTEEN

Aquaporins and placenta

Antonio Pérez-Pérez[a,†], Teresa Vilariño-García[a,†], Valeria Dietrich[b], Pilar Guadix[c], José L. Dueñas[c], Cecilia L. Varone[d], Alicia E. Damiano[b], Víctor Sánchez-Margalet[a,*]
[a]Department of Clinical Biochemistry, Virgen Macarena University Hospital, Medical School, University of Seville, Seville, Spain
[b]Cátedra de Biología Celular y Molecular, Departamento de Ciencias Biológicas, Facultad de Farmacia y Bioquímica, Universidad de Buenos Aires, Buenos Aires, Argentina
[c]Obstetrics and Gynecology Department, Virgen Macarena University Hospital, Medical School, University of Seville, Seville, Spain
[d]Department of Biological Chemistry, Facultad de Ciencias Exactas y Naturales, Universidad de Buenos Aires, Buenos Aires, Argentina
[*]Corresponding author: e-mail address: margalet@us.es

Contents

Abstract

Water is the major component of cells and tissues. The fetal body consists of about 70–90% water and its fluid balance is dependent on the mother. In fact, abortion, premature birth, amniotic fluid volume abnormality, malformation and fetal growth restrictions might result when the homeostasis of the maternal-fetal fluid exchange is disrupted. Thus, maternal-fetal fluid balance is critical during pregnancy. In this sense, several mechanisms, including aquaporins (AQPs) have been reported to play important roles in maternal-fetal fluid balance. AQPs are small membrane proteins (about 30 kDa), present in different organs, that increase the permeability of water, as well as other small uncharged molecules to be transported across the bilayer cell membranes. Several aquaporins are expressed in placenta, and aquaporins play key roles in the placental function. Even though aquaporins have a proven crucial role in water homeostasis, the physiological and pathological importance of aquaporins as glycerol channels is not fully understood. This review focuses on advances in our knowledge of the roles of aquaporins in placental cells, particularly the roles of AQP3 and AQP9 in placental metabolism and points to the pathophysiological importance of glycerol channels in placenta, as well as the signal transduction pathways activated by them. Moreover, the regulation of aquaporins expression by different placental hormones, such as leptin and the mechanisms involved will be discussed.

[†] These authors should be considered as first author.

Vitamins and Hormones, Volume 112
ISSN 0083-6729
https://doi.org/10.1016/bs.vh.2019.10.003

1. Introduction

Aquaporins (AQPs) are a class of membrane water channels, with a wide range of distribution in different organs (Gonen & Walz, 2006) and whose primary function is to facilitate the passive transport of water across the plasma membrane of the cell in response to osmotic gradients (Carbrey & Agre, 2009). It is well known that AQPs not only increase cell membrane water permeability (5- to 50-fold) as compared with membranes in which water moves primarily through the lipid bilayer, but also facilitate the permeability to uncharged small molecules, such as glycerol (Fu et al., 2000), urea and ammonia (Jelen et al., 2012; Musa-Aziz, Chen, Pelletier, & Boron, 2009). In this sense, 13 AQPs isoforms have been identified in humans, many of which have a broad tissue distribution (Gonen & Walz, 2006) and each one has a unique cellular and subcellular distribution with little overlap between homologs. AQPs are subcategorized as "classical aquaporins," the most numerous (AQP0, 1, 2, 4, 5, 6 and 8), that are selective only for water, "aquaglyceroporins" (AQP3, 7, 9 and 10) which enable the transport of water, glycerol and urea, although AQP9 also facilitates the flux of some solutes such as monocarboxylates, purines and pyrimidines (Ishibashi, 2009; Tsukaguchi, Weremowicz, Morton, & Hediger, 1999), and finally, "super-aquaporins" (AQP11 and 12) which are localized in the cytoplasm and whose permeabilities have not yet been fully determined (Ishibashi, 2009). In addition to the transport of water and glycerol, AQPs could potentially transport gasses such as CO_2, ammonia and nitric oxide (NO), as well as larger polar solutes such as sugars, hydrogen peroxide and even some ions (Bienert et al., 2007; Fu et al., 2000; Hara-Chikuma et al., 2012; Herrera, Hong, & Garvin, 2006; Musa-Aziz et al., 2009; Sui, Han, Lee, Walian, & Jap, 2001; Tsukaguchi et al., 1999; Yool & Weinstein, 2002), although, some evidences are conflicting. On the other hand, it is not known whether the transport of molecules other than water and glycerol is biologically important.

AQPs are involved in a variety of physiological and cellular functions, such as peritoneal dialysis (Ni et al., 2006), pleural fluid transport (Song, Yang, Matthay, Ma, & Verkman, 2000), intraocular pressure and aqueous fluid production (Zhang, Vetrivel, & Verkman, 2002), corneal endothelium fluid transport (Kuang et al., 2004), brain edema and neuroexcitation. Moreover, many authors have reported the expression of AQPs (particularly AQP1) in the human placenta and fetal membranes suggesting that AQPs may contribute to amniotic fluid volume regulation (Beall et al., 2007). In this sense, the developing fetus requires large volumes of water, which are obtained from the maternal circulation through the placenta. Placental function and the maternal circulatory system are intrinsically dependent upon one another, and perturbations in placental function are likely to influence the maternal circulation and vice versa (Damiano, 2011). Of note that human placenta is hemomonochorial, in which the syncytiotrophoblast, a single layer of polarized epithelium separates the maternal and fetal circulation (Stulc, 1997). Formation of this interface relies on coordinated interactions among transcriptional, epigenetic, and environmental factors (Maltepe, Bakardjiev, & Fisher, 2010). The diversity of functions performed by the placenta is very broad, ranging from anchoring the embryo to preventing its rejection by the maternal immune system to enabling the transport of nutrients, gasses, water, ions, and waste products between mother and embryo (Maltepe et al., 2010). Therefore, placenta is a key organ for supporting fetal growth and as the water is transported to the fetus from the maternal circulation across the placenta and amnion, the placental water transport is essential for fetal water acquisition. However, the actual mechanisms of the regulation of placenta water transfer and intramembranous reabsorption are poorly understood. Both phenomena involve the flow of water across biological membranes, which is a function of membrane water channels (Beall et al., 2007). In this sense, it has been demonstrated the expression of AQPs throughout pregnancy in placenta and uterus. More specifically, of the 13 known mammalian AQP genes, (AQP1, AQP3, AQP8, AQP9, and AQP11), 5 of them are known to be expressed in the human amnion (Damiano, 2011; Mann, Ricke, Yang, Verkman, & Taylor, 2002; Mobasheri, Wray, & Marples, 2005; Wang, Chen, Beall, Zhou, & Ross, 2004; Wang, Kallichanda, Song, Ramirez, & Ross, 2001), and each AQP may have a different function with specific roles in transport. For instance, AQP1 and AQP3 likely play a dominant role in passive water movement across the amnion. In this line, gene expression of AQP1 and AQP3 and their protein levels are increased in the human amnion. Other observations suggest that AQP8 and AQP9 water channels are likely fundamental to the regulation of fetal water and solutes flow both in intramembranous absorption and in placental water transfer from mother to fetus (Damiano, 2011; Wang et al., 2004, 2001). However, the role of AQPs in regulating amniotic fluid volume is poorly understood and further studies are needed to compare the relative water conductance of individual AQP water channels in the amnion.

Placenta has the particularity of being a temporary organ in continuous change, both in its macroscopic characteristics and in relation to the expression of different molecules with hormonal or transport functions. In fact, dynamic changes in fetal fluid accumulation and fetal weight, as well as placental weight, take place during gestation (Bazer, Thatcher, Martinat-Botte, & Terqui, 1988; Wu, Bazer, Hu, Johnson, & Spencer, 2005). In this regard, placenta and its associated extraembryonic membranes express also AQP in a spatial and temporal manner with higher mRNA levels for AQP1, 3, 9, and 11, but lower mRNA levels for AQP4, 5, and 8, in chorionic villi in the early stages of human pregnancy (10–14 weeks of gestation) (Escobar et al., 2012).

In humans, with the progression of placentation, two pathways of cytotrophoblast differentiation lead to the formation of two distinct phenotypes. In the villus, cytotrophoblast cells undergo cellular fusion and differentiation to form syncytiotrophoblast, while the extravillous trophoblast proliferates and migrates into the decidua, remodeling the pregnant endometrium (Malassiné & Cronier, 2002). In this sense, the AQP1 mRNA is detected in placental blood vessels and in the syncytiotrophoblast (Stulc, 1997), while AQP3 in trophectoderm (Zhu et al., 2009). Moreover, both AQP1 and AQP3 have been identified in amnion epithelium and chorionic cytotrophoblast cells (Zhu et al., 2009). Wang et al. further demonstrated the presence of AQP3 protein in human placental syncytiotrophoblast and cytotrophoblast cells of the chorionic and amnion epithelium (Wang, Amidi, Beall, Gui, & Ross, 2006). Conversely, AQP4 expression decreased in the syncytiotiophoblast, but increased in endothelial cells and stroma of placental villi between the first and the third trimesters of gestation (De Falco et al., 2007). These data demonstrate a distinct expression profile of AQPs at different stages of gestation in women and suggest that the regional and temporal regulation of AQPs plays important roles in normal pregnancy, fetal growth, and homeostasis of amniotic fluid volume. Therefore, the correct placental morphogenesis underpins pregnancy success and offspring phenotype through tight regulation of transport of nutrients, gasses and waste between the mother and fetus. That is why, the control of cell proliferation and apoptosis are critical for a correct placental development, which is a process finely regulated (Genbacev & Miller, 2000). In this context, it is well known that the role of leptin as a multifunctional hormone that regulates not only body weight homeostasis but also implantation and placentation in pregnancy (Chehab, 2014; Friedman & Halaas, 1998). Leptin promotes growth, proliferation and cell survival of trophoblastic cells

(Pérez-Pérez et al., 2008) by activating MAPK, and PI3K signaling pathways. Intriguingly, these signal transduction pathways as well as pathways that involve cyclic adenosine monophosphate (cAMP), and protein kinase C (PKC), are also known to regulate AQPs expression in animals in a cell-specific manner (Zhang, Ding, Shen, Wu, & Zhu, 2012). Moreover, expression of different AQPs can be regulated by distinct MAPK signal transduction pathways within the same cell. More specifically, the cAMP-PKA-dependent pathway plays a role in regulating expression of AQP1, 3, 8, and 9 in the placenta and fetal membranes (Belkacemi, Beall, Magee, Pourtemour, & Ross, 2008; Zhang et al., 2012). Moreover, it has been reported the regulation of aquaglyceroporins (AQP3, 7, and 9) through the PI3K/Akt/mTOR pathway in placental cells (Rodríguez Catalán, Gómez-Ambrosi, García-Navarro, et al., 2011), suggesting that, even though the physiological and pathological relevance of these aquaglyceroporins (as glycerol channels) is not fully understood, the highly expression in placenta seem to indicate an important role in the homeostasis of metabolism.

There are also several pregnancy molecules and hormones, commonly increased in the course of pregnancy, that involved in leptin up-regulation in the placenta. For example, leptin expression has been shown to be up-regulated by different pregnancy hormones, such as chorionic gonadotrophin, and 17 beta-estradiol, and by second messengers, such as cyclic adenosine 5′-monophosphate, mediated also through MAPK and PI3K signaling pathways (Gambino, Maymo, et al., 2012; Gambino, Perez, et al., 2012; Maymo et al., 2010, 2011, 2012, 2009). Consistent with these observations, high expression levels of AQPs by the same pregnancy molecules and steroid hormones has been also observed in placenta. For example, a concentration-dependent effect of human chorionic gonadotrophin (hCG) is to increase the abundance of AQP9 protein in human placentae (Marino, Castro-Parodi, Dietrich, & Damiano, 2010). Moreover, AQP1 expression in trophoblast-like cells is up-regulated by both arginine vasopressin and cAMP agonists, suggesting that modulation of AQP1 expression by maternal hormones may contribute to fetal-placental-amnion water homeostasis during gestation (Belkacemi, Beall, Magee, Pourtemour, & Ross, 2008). Therefore, all together, suggests that modulation of AQPs expression by maternal hormones contributes to fetal-placental-amnion water homeostasis during gestation. In fact, insulin and leptin regulate expression of AQP3 and AQP9 through the PI3K/Akt/mTOR pathway in placental cells. Moreover, exposure to insulin and leptin might increase the abundance of several AQPs, such as AQP3, which might facilitate their migration via PI3K-dependent and independent mechanisms (Cao et al., 2013). In fact, recently, unexpected cellular roles of AQPs have been reported, including proliferation, apoptosis, and cell migration (Verkman, 2011). AQP-dependent cell migration, in which the polarization of AQPs to the leading edge of migrating cells facilitates the entry of water into extending lamellipodia in the direction of cell movement, is one mechanism that has been proposed to facilitate AQP1-dependent tumor angiogenesis and metastasis (Papadopoulos, Saadoun, & Verkman, 2008). Moreover, it has been shown that AQP facilitates the proliferation of some types of tumor cells through mechanisms that may involve altered signaling pathways (Di Giusto et al., 2012; Hara-Chikuma & Verkman, 2008; Jung, Park, Jeon, & Kwon, 2011; Zhang et al., 2010). This is important as trophoblastic cells express an invasive phenotype during implantation.

In placental villi, cell turnover is tightly regulated, via apoptotic cascade (Heazell et al., 2008). In normal pregnancy, apoptosis is an essential feature of placental development and it is well-established that trophoblast apoptosis increases with placental growth and advancing gestation (Sharp, Heazell, Crocker, & Mor, 2010). The intracellular K + depletion generate an osmotic gradient that drives water out of the cell, which, finally, activate the apoptotic caspases (Chen et al., 2008). Regulators of apoptosis are now considered to have a major role in maintaining the integrity of villous trophoblast (Heazell et al., 2008). In fact, it has been suggesting a role central of leptin as a paracrine/autocrine signal, which could be an endogenous component of the proliferation/apoptosis regulatory machinery of trophoblast cells via MAPK pathway (Pérez-Pérez et al., 2008, 2009). Intriguingly, AQPs, stimulated by leptin, might mediate the loss water and subsequent cell shrinkage in the apoptotic cells (Jablonski, Webb, McConnell, Riley, & Hughes, 2004). Particularly, it has been demonstrated that only AQP3 might be important in the regulation of villous trophoblast apoptosis and consequently, an abnormal expression of this protein could alter this tightly regulated process (Szpilbarg et al., 2016). Moreover, as mentioned above, in early pregnancy, a moderately hypoxic environment is crucial for appropriate embryonic development since normal proliferation and differentiation of trophoblastic cells may be driven by low-oxygen concentration in the decidua (Castro-Parodi et al., 2013; James, Stone, & Chamley, 2006). Oxygen enrichment of fetal blood is promoted by partial pressure differences in the feto-maternal circulation and, for example, when PCO2 is incremented or the pH is reduced, hemoglobin affinity for oxygen decreases producing acidosis. Thus, an adequate exchange of oxygen and carbon dioxide by placenta is crucial to manage acid-base balance within a narrow pH, so, acidic environment could impair the correct development of the fetus as well as the anatomy and function of the placenta (Avagliano et al., 2015; Bobrow & Soothill, 1999). For example, failures in pHi homeostasis could alter various critical cellular functions such as water movements and cell volume regulation. In this respect, aquaporins are also key regulators of cell volume and intracellular ions. Five AQPs are expressed in the human placenta, although their functional significance remains unclear (Damiano, 2011). Possibly, changes in pHi by modifying might alter the selectivity of placental AQPs to water. Therefore, syncytiotrophoblast acid/base balance is necessary to maintain optimal solute transport rates across the placenta. However, the actual mechanisms of the regulation of acid/base balance by AQPs are poorly understood.

The developing fetus requires large volumes of water, which are obtained from the maternal circulation through the placenta. Once in the conceptus, water is partitioned into several compartments, including the amniotic fluid. In this, intramembranous transport pathway, which moves the fluid across the fetal amnion and chorion into the fetal vasculature, is considered a particularly important pathway for regulating amniotic fluid volume homeostasis (Beall et al., 2007). Moreover, changes in the expression of the AQP8 and AQP9 have been shown in pregnancies complicated by idiopathic polyhydramnios (large-for-gestational-age fetuses and macrosomia). In this sense, gestational diabetes mellitus (GDM) is a frequent pregnancy complication associated with increased incidence of polyhydramnios (Sohaey, Nyberg, Sickler, & Williams, 1994). Thus, GDM is a health problem that increases the risk of death during the perinatal period due to fetal macrosomia which is, in addition, accompanied by a larger size and weight of the placenta to support the increased needs of the macrosomic fetus. Therefore, in those subjects that develop polyhydramnios, altered *trans*-amnion water transport due to impared AQP gene expression in GDM could be the cause of the increased amniotic fluid volume (Dashe, Nathan, McIntire, & Leveno, 2000; Santolaya-Forgas, Mehta, & Castracane, 2006). For example, in vitro studies showed an up-regulation in various transport systems, such as amino acids (Vaughan, Rosario, Powell, & Jansson, 2017), fatty acids (Segura et al., 2017), and glucose (Stanirowski, Szukiewicz, Pazura-Turowska, Sawicki, & Cendrowski, 2018) in GDM placenta. Thus, alteration of placenta function may be the reason for abnormal fetal growth (Osmond, Nolan, King, Brennecke, & Gude, 2000) observed in this pathology of pregnancy (Jansson et al., 2006). In this context, it could be necessary to know the modulation factors, such as fetal or maternal hormones (Oliveira, Carnes, França, Hermo, & Hess, 2005; Pastor-Soler et al., 2002) or other endocrine factors (Vilariño-García et al., 2016), by which AQPs can be regulated. Intriguingly, women with GDM have increased plasma leptin levels (Lepercq et al., 1998) and insulin levels (Sagawa et al., 2002). Our group has recently described that insulin induces leptin expression in trophoblastic cells (Pérez-Pérez et al., 2016) and moreover, increased insulin and leptin levels may modulate the expression of AQPS, such as AQP9, as it was previously described in placenta (in syncytiotrophoblast; Castro Parodi et al., 2011), as well as adipocytes and hepatocytes (Rodríguez, Catalán, Gómez-Ambrosi, & Frühbeck, 2011). Since, AQP9 trophoblast was overexpressed in placentas from women with GDM, the leptin/leptin receptor system could play an important role, since the system is activated in the placenta from GDM by acting on the MAPK and PI3K pathways (Pérez-Pérez et al., 2013, 2015). Therefore, it could be speculate that leptin could mediate the increased expression of AQP9 (or others aquaglyceroporins), which might increase the transport of glycerol to the fetus and thus to contribute to attend the increased energy intake requirements in the macrosomic fetus in GDM. However, the functional role of AQPs in placenta remains to be elucidated and its expression and regulation in abnormal pregnancy may point to new potential therapeutic targets for the pathology of pregnancy.

Moreover, abnormal placentation and altered expression of a several trophoblast transporters are also associated to other pregnancy complication named preeclampsia. This condition is exclusively for human gestation and affects 7–10% of pregnancies worldwide (Giachini et al., 2017). Although its etiology remains unclear, it is well accepted that defects in placentation are also the main predisposing factors for preeclampsia (Hawfield & Freedman, 2009; Huppertz, 2008; Myatt, 2002). Briefly, an insufficient transformation of the maternal spiral arteries resulting in a decrease in blood supply to the fetoplacental unit and consequently, fluctuation in oxygen levels, increase of the oxidative stress and the apoptosis of the trophoblast, concluding in the maternal endothelial dysfunction. In this sense, villous syncytiotrophoblast is involved in the fetal-maternal oxygen and nutrient exchange and an altered expression of a variety of trophoblast transporters and channels are altered in preeclamptic placentas (Castro Parodi et al., 2011; Damiano, Zotta, & Ibarra, 2006; del Monaco et al., 2006; Dietrich, Szpilbarg, & Damiano, 2013; Szpilbarg & Damiano, 2017). More specifically, an abnormal expression of AQP3 and AQP9 were shown in these placentas, suggesting an altered transport of water, urea, and glycerol in placenta form preeclampsia.

The reduced expression of AQP3 in placentas from pregnancies complicated by preeclampsia is consistent with previous findings in explants exposed to oxygen changes (Szpilbarg et al., 2016), in which, AQP3 protein expression decreased after oxygen deprivation, and the subsequent reoxygenation failed to restore AQP3 to basal levels, possibly due to the oxidative damage of the plasma membrane of syncytiotrophoblast. Along with this idea, it has been speculated that the decreased expression of AQP3 might be an adaptive response of the placenta to reduce the trophoblast apoptosis, which is related to the clinical manifestations of preeclampsia. However, further studies are needed to elucidate this issue.

In conclusion, a number of AQPs have been found to be expressed in trophoblast and in fetal membranes with an important roles in facilitating water movement across the biological membranes of maternal- and fetal-placental tissues and organs, as well as between mother and fetus. These requirements are satisfied primarily via transplacental transfer from the maternal circulation, suggesting a need for increased placental water flux with advancing gestation. In this sense, AQP1, AQP3, AQP8 and AQP9 may be involved in regulation of the amniotic fluid volume. Moreover, it could be speculated that the maternal metabolic environment may generate stimuli within the placenta resulting in the altered production of aquaporins and inflammatory cytokines, such as leptin, whose expression is normal under control pregnancy. Thus, deregulation of leptin function in the placenta may be implicated in the pathogenesis of various disorders during pregnancy, such as GDM and preeclampsia (Bajoria, Sooranna, Ward, & Chatterjee, 2002; Sagawa et al., 2002) by alternating AQPs expression. In fact, while AQP3 and AQP9 are normally expressed in the normal trophoblastic cells, AQP9 is up-regulated in abnormal pregnancy. However, this could not be related to an increase in water flux and hypotheses concerning the function of placental AQPs are still speculative. There are many unanswered questions about the roles for AQPs in water transport by placenta, endometrium, and fetal membranes, as well as in relevant processes required for a normal placental development, such as cell migration and apoptosis. Moreover, although GDM and preeclampsia are complex and multisystemic disorders resulting from multiple simultaneous mechanisms, AQPs may be contributing as part of this network of alterations that give rise to the diverse clinical manifestations of these pathologies. Therefore, although the functional role of AQPs in placenta remains to be elucidated, AQPs expression and regulation in abnormal pregnancy might point to new potential therapeutic targets for the pathology of pregnancy.

References

Avagliano, L., Locatelli, A., Danti, L., Felis, S., Mecacci, F., & Pietro, B. G. (2015). Placental histology in clinically unexpected severe fetal acidemia at term. *Early Human Development, 91*, 339–343. Available at http://www.ncbi.nlm.nih.gov/pubmed/25875757. Accessed 5 April 2019.

Bajoria, R., Sooranna, S. R., Ward, B. S., & Chatterjee, R. (2002). Prospective function of placental leptin at maternal–fetal interface. *Placenta, 23*, 103–115. Available at http://linkinghub.elsevier.com/retrieve/pii/S0143400401907696. Accessed 5 April 2019.

Bazer, F. W., Thatcher, W. W., Martinat-Botte, F., & Terqui, M. (1988). Conceptus development in large white and prolific Chinese Meishan pigs. *Journal of Reproduction and Fertility, 84*, 37–42. Available at http://www.ncbi.nlm.nih.gov/pubmed/3184056. Accessed 5 April 2019.

Beall, M. H., Wang, S., Yang, B., Chaudhri, N., Amidi, F., & Ross, M. G. (2007). Placental and membrane aquaporin water channels: Correlation with amniotic fluid volume and composition. *Placenta, 28*, 421–428. Available at https://linkinghub.elsevier.com/retrieve/pii/S0143400406001597. Accessed 4 April 2019.

Belkacemi, L., Beall, M. H., Magee, T. R., Pourtemour, M., & Ross, M. G. (2008). AQP1 gene expression is upregulated by arginine vasopressin and cyclic AMP agonists in trophoblast cells. *Life Sciences, 82*(25–26), 1272–1280. https://doi.org/10.1016/j.lfs.2008.04.014.

Bienert, G. P., Møller, A. L. B., Kristiansen, K. A., Schulz, A., Møller, I. M., Schjoerring, J. K., et al. (2007). Specific aquaporins facilitate the diffusion of hydrogen peroxide across membranes. *The Journal of Biological Chemistry, 282*, 1183–1192. Available at http://www.jbc.org/lookup/doi/10.1074/jbc.M603761200. Accessed 4 April 2019.

Bobrow, C. S., & Soothill, P. W. (1999). Causes and consequences of fetal acidosis. *Archives of Disease in Childhood Fetal and Neonatal Edition, 80*, F246–F249. Available at http://www.ncbi.nlm.nih.gov/pubmed/10212094. Accessed 5 April 2019.

Cao, X.-C., Zhang, W.-R., Cao, W.-F., Liu, B.-W., Zhang, F., Zhao, H.-M., et al. (2013). Aquaporin3 is required for FGF-2-induced migration of human breast cancers. St-Pierre Y (ed) *PLoS One, 8* e56735 Available at https://dx.plos.org/10.1371/journal.pone.0056735. Accessed 5 April 2019.

Carbrey, J. M., & Agre, P. (2009). Discovery of the aquaporins and development of the field. In *Handbook of experimental pharmacology* (pp. 3–28). Available at http://link.springer.com/10.1007/978-3-540-79885-9_1. Accessed 4 April 2019.

Castro Parodi, M., Farina, M., Dietrich, V., Abán, C., Szpilbarg, N., Zotta, E., et al. (2011). Evidence for insulin-mediated control of AQP9 expression in human placenta. *Placenta, 32*, 1050–1056. Available at https://linkinghub.elsevier.com/retrieve/pii/S0143400411004826. Accessed 5 April 2019.

Castro-Parodi, M., Szpilbarg, N., Dietrich, V., Sordelli, M., Reca, A., Abán, C., et al. (2013). Oxygen tension modulates AQP9 expression in human placenta. *Placenta, 34*, 690–698. Available at https://linkinghub.elsevier.com/retrieve/pii/S0143400413002142. Accessed 5 April 2019.

Chehab, F. F. (2014). 20 years of leptin: Leptin and reproduction: Past milestones, present undertakings, and future endeavors. *The Journal of Endocrinology, 223*, T37–T48. Available at https://joe.bioscientifica.com/view/journals/joe/223/1/T37.xml. Accessed 5 April 2019.

Chen, J. M., Sepramaniam, S., Armugam, A., Shyan Choy, M., Manikandan, J., Melendez, A. J., et al. (2008). Water and ion channels: Crucial in the initiation and progression of apoptosis in central nervous system? *Current Neuropharmacology, 6*, 102–116. https://doi.org/10.2174/157015908784533879.

Damiano, A. E. (2011). Review: Water channel proteins in the human placenta and fetal membranes. *Placenta*, *32*(Suppl. 2), S207–S211.

Damiano, A. E., Zotta, E., & Ibarra, C. (2006). Functional and molecular expression of AQP9 channel and UT-A transporter in normal and preeclamptic human placentas. *Placenta*, *27*, 1073–1081. Available at http://www.ncbi.nlm.nih.gov/pubmed/16480766. Accessed 5 April 2019.

Dashe, J. S., Nathan, L., McIntire, D. D., & Leveno, K. J. (2000). Correlation between amniotic fluid glucose concentration and amniotic fluid volume in pregnancy complicated by diabetes. *American Journal of Obstetrics and Gynecology*, *182*, 901–904. Available at http://www.ncbi.nlm.nih.gov/pubmed/10764470. Accessed 5 April 2019.

De Falco, M., Cobellis, L., Torella, M., Acone, G., Varano, L., Sellitti, A., et al. (2007). Down-regulation of aquaporin 4 in human placenta throughout pregnancy. *In Vivo*, *21*, 813–817. Available at http://www.ncbi.nlm.nih.gov/pubmed/18019416. Accessed 5 April 2019.

del Monaco, S., Assef, Y., Damiano, A., Zotta, E., Ibarra, C., & Kotsias, B. A. (2006). Characterization of the epithelial sodium channel in human pre-eclampsia syncytiotrophoblast. *Medicina (B Aires)*, *66*, 31–35. Available at http://www.ncbi.nlm.nih.gov/pubmed/16555725. Accessed 5 April 2019.

Di Giusto, G., Flamenco, P., Rivarola, V., Fernández, J., Melamud, L., Ford, P., et al. (2012). Aquaporin 2-increased renal cell proliferation is associated with cell volume regulation. *Journal of Cellular Biochemistry*, *113*, 3721–3729. Available at http://doi.wiley.com/10.1002/jcb.24246. Accessed 5 April 2019.

Dietrich, V., Szpilbarg, N., & Damiano, A. E. (2013). Reduced expression of Na(+)/H(+) exchanger isoform 3 (NHE-3) in preeclamptic placentas. *Placenta*, *34*, 828–830. Available at http://www.ncbi.nlm.nih.gov/pubmed/23810111. Accessed 5 April 2019.

Escobar, J., Gormaz, M., Arduini, A., Gosens, K., Martinez, A., Perales, A., et al. (2012). Expression of aquaporins early in human pregnancy. *Early Human Development*, *88*, 589–594. Available at https://linkinghub.elsevier.com/retrieve/pii/S037837821200031X. Accessed 5 April 2019.

Friedman, J. M., & Halaas, J. L. (1998). Leptin and the regulation of body weight in mammals. *Nature*, *395*, 763–770. Available at http://www.ncbi.nlm.nih.gov/pubmed/9796811. Accessed 5 April 2019.

Fu, D., Libson, A., Miercke, L. J., Weitzman, C., Nollert, P., Krucinski, J., et al. (2000). Structure of a glycerol-conducting channel and the basis for its selectivity. *Science*, *290*, 481–486. Available at http://www.ncbi.nlm.nih.gov/pubmed/11039922. Accessed 4 April 2019.

Gambino, Y. P., Maymo, J. L., Perez, P. A., Calvo, J. C., Sanchez-Margalet, V., & Varone, C. L. (2012). Elsevier Trophoblast Research Award lecture: Molecular mechanisms underlying estrogen functions in trophoblastic cells—focus on leptin expression. *Placenta*, *33*(Suppl.), S63–S70.

Gambino, Y. P., Perez, P. A., Duenas, J. L., Calvo, J. C., Sanchez-Margalet, V., & Varone, C. L. (2012). Regulation of leptin expression by 17beta-estradiol in human placental cells involves membrane associated estrogen receptor alpha. *Biochimica et Biophysica Acta*, *1823*, 900–910.

Genbacev, O., & Miller, R. K. (2000). Post-implantation differentiation and proliferation of cytotrophoblast cells: In vitro models—A review. *Placenta*, **21**(Suppl. A), S45–S49. Available at http://www.ncbi.nlm.nih.gov/pubmed/10831121. Accessed 5 April 2019.

Giachini, F. R., Galaviz-Hernandez, C., Damiano, A. E., Viana, M., Cadavid, A., Asturizaga, P., et al. (2017). Vascular dysfunction in mother and offspring during preeclampsia: Contributions from Latin-American countries. *Current Hypertension Reports*, *19*, 83. Available at http://www.ncbi.nlm.nih.gov/pubmed/28986756. Accessed 5 April 2019.

Gonen, T., & Walz, T. (2006). The structure of aquaporins. *Quarterly Reviews of Biophysics, 39*, 361–396. Available at: http://www.journals.cambridge.org/abstract_S0033583506004458. Accessed 4 April 2019.
Hara-Chikuma, M., Chikuma, S., Sugiyama, Y., Kabashima, K., Verkman, A. S., Inoue, S., et al. (2012). Chemokine-dependent T cell migration requires aquaporin-3-mediated hydrogen peroxide uptake. *The Journal of Experimental Medicine, 209*, 1743–1752. Available at http://www.jem.org/lookup/doi/10.1084/jem.20112398. Accessed 4 April 2019.
Hara-Chikuma, M., & Verkman, A. S. (2008). Prevention of skin tumorigenesis and impairment of epidermal cell proliferation by targeted aquaporin-3 gene disruption. *Molecular and Cellular Biology, 28*, 326–332. Available at http://www.ncbi.nlm.nih.gov/pubmed/17967887. Accessed 5 April 2019.
Hawfield, A., & Freedman, B. I. (2009). Pre-eclampsia: The pivotal role of the placenta in its pathophysiology and markers for early detection. *Therapeutic Advances in Cardiovascular Disease, 3*, 65–73. Available at http://journals.sagepub.com/doi/10.1177/1753944708097114. Accessed 5 April 2019.
Heazell, A. E. P., Lacey, H. A., Jones, C. J. P., Huppertz, B., Baker, P. N., & Crocker, I. P. (2008). Effects of oxygen on cell turnover and expression of regulators of apoptosis in human placental trophoblast. *Placenta, 29*, 175–186. Available at https://linkinghub.elsevier.com/retrieve/pii/S0143400407002652. Accessed 5 April 2019.
Herrera, M., Hong, N. J., & Garvin, J. L. (2006). Aquaporin-1 transports NO across cell membranes. *Hypertension, 48*, 157–164. Available at https://www.ahajournals.org/doi/10.1161/01.HYP.0000223652.29338.77. Accessed 4 April 2019.
Huppertz, B. (2008). Placental origins of preeclampsia. *Hypertension, 51*, 970–975. Available at http://www.ncbi.nlm.nih.gov/pubmed/18259009. Accessed 5 April 2019.
Ishibashi, K. (2009). New members of mammalian aquaporins: AQP10-AQP12. In *Handbook of experimental pharmacology* (pp. 251–262). Available at http://link.springer.com/10.1007/978-3-540-79885-9_13. Accessed 4 April 2019.
Jablonski, E. M., Webb, A. N., McConnell, N. A., Riley, M. C., & Hughes, F. M. (2004). Plasma membrane aquaporin activity can affect the rate of apoptosis but is inhibited after apoptotic volume decrease. *The American Journal of Physiology, 286*, C975–C985. Available at http://www.ncbi.nlm.nih.gov/pubmed/14644770. Accessed 5 April 2019.
James, J. L., Stone, P. R., & Chamley, L. W. (2006). The effects of oxygen concentration and gestational age on extravillous trophoblast outgrowth in a human first trimester villous explant model. *Human Reproduction, 21*, 2699–2705. Available at https://academic.oup.com/humrep/article-lookup/doi/10.1093/humrep/del212. Accessed 5 April 2019.
Jansson, N., Pettersson, J., Haafiz, A., Ericsson, A., Palmberg, I., Tranberg, M., et al. (2006). Down-regulation of placental transport of amino acids precedes the development of intrauterine growth restriction in rats fed a low protein diet. *The Journal of Physiology, 576*, 935–946. Available at http://www.jphysiol.org/cgi/doi/10.1113/jphysiol.2006.116509. Accessed 5 April 2019.
Jelen, S., Gena, P., Lebeck, J., Rojek, A., Praetorius, J., Frøkiaer, J., et al. (2012). Aquaporin-9 and urea transporter-A gene deletions affect urea transmembrane passage in murine hepatocytes. *American Journal of Physiology. Gastrointestinal and Liver Physiology, 303*, G1279–G1287. Available at http://www.physiology.org/doi/10.1152/ajpgi.00153.2012. Accessed 4 April 2019.
Jung, H. J., Park, J.-Y., Jeon, H.-S., & Kwon, T.-H. (2011). Aquaporin-5: A marker protein for proliferation and migration of human breast cancer cells. De Wever O (ed) *PLoS One, 6*, e28492. Available at https://dx.plos.org/10.1371/journal.pone.0028492. Accessed 5 April 2019.

Kuang, K., Yiming, M., Wen, Q., Li, Y., Ma, L., Iserovich, P., et al. (2004). Fluid transport across cultured layers of corneal endothelium from aquaporin-1 null mice. *Experimental Eye Research*, *78*, 791–798. Available at https://linkinghub.elsevier.com/retrieve/pii/S0014483503003919. Accessed 4 April 2019.

Lepercq, J., Cauzac, M., Lahlou, N., Timsit, J., Girard, J., Auwerx, J., et al. (1998). Over-expression of placental leptin in diabetic pregnancy: A critical role for insulin. *Diabetes*, *47*, 847–850. Available at http://www.ncbi.nlm.nih.gov/pubmed/9588462. Accessed 5 April 2019.

Malassiné, A., & Cronier, L. (2002). Hormones and human trophoblast differentiation: A review. *Endocrine*, *19*, 3–11. Available at http://link.springer.com/10.1385/ENDO:19:1:3. Accessed 5 April 2019.

Maltepe, E., Bakardjiev, A. I., & Fisher, S. J. (2010). The placenta: Transcriptional, epigenetic, and physiological integration during development. *The Journal of Clinical Investigation*, *120*, 1016–1025. Available at http://www.jci.org/articles/view/41211. Accessed 5 April 2019.

Mann, S. E., Ricke, E. A., Yang, B. A., Verkman, A. S., & Taylor, R. N. (2002). Expression and localization of aquaporin 1 and 3 in human fetal membranes. *American Journal of Obstetrics and Gynecology*, *187*, 902–907. Available at http://www.ncbi.nlm.nih.gov/pubmed/12388974. Accessed 5 April 2019.

Marino, G. I., Castro-Parodi, M., Dietrich, V., & Damiano, A. E. (2010). High levels of human chorionic gonadotropin (hCG) correlate with increased aquaporin-9 (AQP9) expression in explants from human preeclamptic placenta. *Reproductive Sciences*, *17*(5), 444–453. https://doi.org/10.1177/1933719110361385.

Maymo, J. L., Perez, P. A., Duenas, J. L., Calvo, J. C., Sanchez-Margalet, V., & Varone, C. L. (2010). Regulation of placental leptin expression by cyclic adenosine 5′-monophosphate involves cross talk between protein kinase A and mitogen-activated protein kinase signaling pathways. *Endocrinology*, *151*, 3738–3751.

Maymo, J. L., Perez, A. P., Gambino, Y., Calvo, J. C., Sanchez-Margalet, V., & Varones, C. L. (2011). Review: Leptin gene expression in the placenta—regulation of a key hormone in trophoblast proliferation and survival. *Placenta*, *32*(Suppl. 2), S146–S153.

Maymo, J. L., Perez, P. A., Maskin, B., Duenas, J. L., Calvo, J. C., Sanchez, M., et al. (2012). The alternative Epac/cAMP pathway and the MAPK pathway mediate hCG induction of leptin in placental cells. *PLoS One*, *7*, e46216.

Maymo, J. L., Perez, P. A., Sanchez-Margalet, V., Duenas, J. L., Calvo, J. C., & Varone, C. L. (2009). Up-regulation of placental leptin by human chorionic gonadotropin. *Endocrinology*, *150*, 304–313.

Mobasheri, A., Wray, S., & Marples, D. (2005). Distribution of AQP2 and AQP3 water channels in human tissue microarrays. *Journal of Molecular Histology*, *36*, 1–14. Available at http://link.springer.com/10.1007/s10735-004-2633-4. Accessed 5 April 2019.

Musa-Aziz, R., Chen, L.-M., Pelletier, M. F., & Boron, W. F. (2009). Relative CO2/NH3 selectivities of AQP1, AQP4, AQP5, AmtB, and RhAG. *Proceedings of the National Academy of Sciences of the United States of America*, *106*, 5406–5411. Available at: http://www.pnas.org/cgi/doi/10.1073/pnas.0813231106. Accessed 4 April 2019.

Myatt, L. (2002). Role of placenta in preeclampsia. *Endocrine*, *19*, 103–112. Available at http://www.ncbi.nlm.nih.gov/pubmed/12583607. Accessed 5 April 2019.

Ni, J., Verbavatz, J.-M., Rippe, A., Boisdé, I., Moulin, P., Rippe, B., et al. (2006). Aquaporin-1 plays an essential role in water permeability and ultrafiltration during peritoneal dialysis. *Kidney International*, *69*, 1518–1525. Available at http://www.ncbi.nlm.nih.gov/pubmed/16508653. Accessed 4 April 2019.

Oliveira, C. A., Carnes, K., França, L. R., Hermo, L., & Hess, R. A. (2005). Aquaporin-1 and -9 are differentially regulated by oestrogen in the efferent ductule epithelium and initial segment of the epididymis. *Biology of the Cell, 97*, 385–395. Available at http://www.ncbi.nlm.nih.gov/pubmed/15850448. Accessed 5 April 2019.

Osmond, D. T., Nolan, C. J., King, R. G., Brennecke, S. P., & Gude, N. M. (2000). Effects of gestational diabetes on human placental glucose uptake, transfer, and utilisation. *Diabetologia, 43*, 576–582. Available at http://link.springer.com/10.1007/s001250051346. Accessed 5 April 2019.

Papadopoulos, M. C., Saadoun, S., & Verkman, A. S. (2008). Aquaporins and cell migration. *Pflügers Archiv—European Journal of Physiology, 456*, 693–700. Available at http://link.springer.com/10.1007/s00424-007-0357-5. Accessed 5 April 2019.

Pastor-Soler, N., Isnard-Bagnis, C., Herak-Kramberger, C., Sabolic, I., Van Hoek, A., Brown, D., et al. (2002). Expression of aquaporin 9 in the adult rat epididymal epithelium is modulated by androgens. *Biology of Reproduction, 66*, 1716–1722. Available at http://www.ncbi.nlm.nih.gov/pubmed/12021052. Accessed 5 April 2019.

Pérez-Pérez, A., Guadix, P., Maymó, J., Dueñas, J. L., Varone, C., Fernández-Sánchez, M., et al. (2016). Insulin and leptin signaling in placenta from gestational diabetic subjects. *Hormone and Metabolic Research, 48*, 62–69. Available at http://www.thieme-connect.de/DOI/DOI?10.1055/s-0035-1559722. Accessed 5 April 2019.

Pérez-Pérez, A., Maymó, J., Dueñas, J. L., Goberna, R., Calvo, J. C., Varone, C., et al. (2008). Leptin prevents apoptosis of trophoblastic cells by activation of MAPK pathway. *Archives of Biochemistry and Biophysics, 477*, 390–395. Available at http://www.ncbi.nlm.nih.gov/pubmed/18619412. Accessed 5 April 2019.

Pérez-Pérez, A., Maymó, J., Gambino, Y., Dueñas, J. L., Goberna, R., Varone, C., et al. (2009). Leptin stimulates protein synthesis-activating translation machinery in human trophoblastic cells1. *Biology of Reproduction, 81*, 826–832. Available at http://www.ncbi.nlm.nih.gov/pubmed/19553602. Accessed 5 April 2019.

Pérez-Pérez, A., Maymó, J. L., Gambino, Y. P., Guadix, P., Dueñas, J. L., Varone, C. L., et al. (2013). Activated translation signaling in placenta from pregnant women with gestational diabetes mellitus: Possible role of leptin. *Hormone and Metabolic Research, 45*, 436–442. Available at http://www.thieme-connect.de/DOI/DOI?10.1055/s-0032-1333276. Accessed 5 April 2019.

Pérez-Pérez, A., Sánchez-Jiménez, F., Maymó, J., Dueñas, J. L., Varone, C., & Sánchez-Margalet, V. (2015). Role of leptin in female reproduction. *Clinical Chemistry and Laboratory Medicine, 53*, 15–28. Available at http://www.ncbi.nlm.nih.gov/pubmed/25014521. Accessed 5 April 2019.

Rodríguez, A., Catalán, V., Gómez-Ambrosi, J., & Frühbeck, G. (2011). Aquaglyceroporins serve as metabolic gateways in adiposity and insulin resistance control. *Cell Cycle, 10*, 1548–1556. Available at http://www.tandfonline.com/doi/abs/10.4161/cc.10.10.15672. Accessed 5 April 2019.

Rodríguez, A., Catalán, V., Gómez-Ambrosi, J., García-Navarro, S., Rotellar, F., Valentí, V., et al. (2011). Insulin- and leptin-mediated control of aquaglyceroporins in human adipocytes and hepatocytes is mediated via the PI3K/Akt/mTOR signaling cascade. *The Journal of Clinical Endocrinology and Metabolism, 96*, E586–E597. Available at https://academic.oup.com/jcem/article-lookup/doi/10.1210/jc.2010-1408. Accessed 5 April 2019.

Sagawa, N., Yura, S., Itoh, H., Mise, H., Kakui, K., Korita, D., et al. (2002). Role of leptin in pregnancy—A review. *Placenta, 23*, S80–S86. Available at http://www.ncbi.nlm.nih.gov/pubmed/11978063. Accessed 5 April 2019.

Santolaya-Forgas, J., Mehta, S., & Castracane, V. A. (2006). Study to determine if acute Maternal and fetal hyperglycemia/Insulinemia induces leptin production during pregnancy. *Hormone and Metabolic Research*, *38*, 598–602. Available at http://www.ncbi.nlm.nih.gov/pubmed/16981143. Accessed 5 April 2019.

Segura, M. T., Demmelmair, H., Krauss-Etschmann, S., Nathan, P., Dehmel, S., Padilla, M. C., et al. (2017). Gestational diabetes alter placental lipid transporters and fatty acid composition. *Placenta*, *57*, 144–151. Available at http://www.ncbi.nlm.nih.gov/pubmed/28864004. Accessed 5 April 2019.

Sharp, A. N., Heazell, A. E. P., Crocker, I. P., & Mor, G. (2010). Placental apoptosis in health and disease. *American Journal of Reproductive Immunology*, *64*, 159–169. Available at http://doi.wiley.com/10.1111/j.1600-0897.2010.00837.x. Accessed 5 April 2019.

Sohaey, R., Nyberg, D. A., Sickler, G. K., & Williams, M. A. (1994). Idiopathic polyhydramnios: Association with fetal macrosomia. *Radiology*, *190*, 393–396. Available at http://www.ncbi.nlm.nih.gov/pubmed/8284386. Accessed 5 April 2019.

Song, Y., Yang, B., Matthay, M. A., Ma, T., & Verkman, A. S. (2000). Role of aquaporin water channels in pleural fluid dynamics. *American Journal of Physiology. Cell Physiology*, *279*, C1744–C1750. Available at http://www.physiology.org/doi/10.1152/ajpcell.2000.279.6.C1744. Accessed 4 April 2019.

Stanirowski, P. J., Szukiewicz, D., Pazura-Turowska, M., Sawicki, W., & Cendrowski, K. (2018). Placental expression of glucose transporter proteins in pregnancies complicated by gestational and pregestational diabetes mellitus. *Canadian Journal of Diabetes*, *42*, 209–217. Available at http://www.ncbi.nlm.nih.gov/pubmed/28583471. Accessed 5 April 2019.

Stulc, J. (1997). Placental transfer of inorganic ions and water. *Physiological Reviews*, *77*, 805–836. Available at http://www.physiology.org/doi/10.1152/physrev.1997.77.3.805. Accessed 5 April 2019.

Sui, H., Han, B. G., Lee, J. K., Walian, P., & Jap, B. K. (2001). Structural basis of water-specific transport through the AQP1 water channel. *Nature*, *414*, 872–878. Available at http://www.nature.com/articles/414872a. Accessed 4 April 2019.

Szpilbarg, N., Castro-Parodi, M., Reppetti, J., Repetto, M., Maskin, B., Martinez, N., et al. (2016). Placental programmed cell death: Insights into the role of aquaporins. *Molecular Human Reproduction*, *22*, 46–56. Available at https://academic.oup.com/molehr/article-lookup/doi/10.1093/molehr/gav063. Accessed 5 April 2019.

Szpilbarg, N., & Damiano, A. E. (2017). Expression of aquaporin-3 (AQP3) in placentas from pregnancies complicated by preeclampsia. *Placenta*, *59*, 57–60. Available at https://linkinghub.elsevier.com/retrieve/pii/S0143400417311244. Accessed 5 April 2019.

Tsukaguchi, H., Weremowicz, S., Morton, C. C., & Hediger, M. A. (1999). Functional and molecular characterization of the human neutral solute channel aquaporin-9. *The American Journal of Physiology*, *277*, F685–F696. Available at http://www.physiology.org/doi/10.1152/ajprenal.1999.277.5.F685. Accessed 4 April 2019.

Vaughan, O. R., Rosario, F. J., Powell, T. L., & Jansson, T. (2017). Regulation of placental amino acid transport and fetal growth. In *Vol. 145. Progress in molecular biology and translational science* (pp. 217–251). Available at http://www.ncbi.nlm.nih.gov/pubmed/28110752. Accessed 5 April 2019.

Verkman, A. S. (2011). Aquaporins at a glance. *Journal of Cell Science*, *124*, 2107–2112. Available at http://www.ncbi.nlm.nih.gov/pubmed/21670197. Accessed 5 April 2019.

Vilariño-García, T., Pérez-Pérez, A., Dietrich, V., Fernández-Sánchez, M., Guadix, P., Dueñas, J., et al. (2016). Increased expression of aquaporin 9 in trophoblast from gestational diabetic patients. *Hormone and Metabolic Research*, *48*, 535–539. Available at http://www.ncbi.nlm.nih.gov/pubmed/27082037. Accessed 5 April 2019.

Wang, S., Amidi, F., Beall, M., Gui, L., & Ross, M. G. (2006). Aquaporin 3 expression in human fetal membranes and its up-regulation by cyclic adenosine monophosphate in amnion epithelial cell culture. *Journal of the Society for Gynecologic Investigation, 13*, 181–185. Available at http://journals.sagepub.com/doi/10.1016/j.jsgi.2006.02.002. Accessed 5 April 2019.

Wang, S., Chen, J., Beall, M., Zhou, W., & Ross, M. G. (2004). Expression of aquaporin 9 in human chorioamniotic membranes and placenta. *American Journal of Obstetrics and Gynecology, 191*, 2160–2167. Available at http://www.ncbi.nlm.nih.gov/pubmed/15592307. Accessed 5 April 2019.

Wang, S., Kallichanda, N., Song, W., Ramirez, B. A., & Ross, M. G. (2001). Expression of aquaporin-8 in human placenta and chorioamniotic membranes: Evidence of molecular mechanism for intramembranous amniotic fluid resorption. *American Journal of Obstetrics and Gynecology, 185*, 1226–1231. Available at http://www.ncbi.nlm.nih.gov/pubmed/11717661. Accessed 5 April 2019.

Wu, G., Bazer, F. W., Hu, J., Johnson, G. A., & Spencer, T. E. (2005). Polyamine synthesis from PROLINE in the developing porcine placenta1. *Biology of Reproduction, 72*, 842–850. Available at http://www.ncbi.nlm.nih.gov/pubmed/15576824. Accessed 5 April 2019.

Yool, A. J., & Weinstein, A. M. (2002). New roles for old holes: Ion channel function in aquaporin-1. *News in Physiological Sciences, 17*, 68–72. Available at http://www.ncbi.nlm.nih.gov/pubmed/11909995. Accessed 4 April 2019.

Zhang, Z., Chen, Z., Song, Y., Zhang, P., Hu, J., & Bai, C. (2010). Expression of aquaporin 5 increases proliferation and metastasis potential of lung cancer. *The Journal of Pathology, 221*, 210–220. Available at http://doi.wiley.com/10.1002/path.2702. Accessed 5 April 2019.

Zhang, Y., Ding, S., Shen, Q., Wu, J., & Zhu, X. (2012). The expression and regulation of aquaporins in placenta and fetal membranes. *Frontiers in Bioscience, 17*, 2371–2382. Available at http://www.ncbi.nlm.nih.gov/pubmed/22652785. Accessed 5 April 2019.

Zhang, D., Vetrivel, L., & Verkman, A. S. (2002). Aquaporin deletion in mice reduces intraocular pressure and aqueous fluid production. *The Journal of General Physiology, 119*, 561–569. Available at http://www.ncbi.nlm.nih.gov/pubmed/12034763. Accessed 4 April 2019.

Zhu, X. Q., Jiang, S. S., Zhu, X. J., Zou, S. W., Wang, Y. H., & Hu, Y. C. (2009). Expression of aquaporin 1 and aquaporin 3 in fetal membranes and placenta in human term pregnancies with oligohydramnios. *Placenta, 30*, 670–676. Available at https://linkinghub.elsevier.com/retrieve/pii/S0143400409001763. Accessed 5 April 2019.

CHAPTER FIFTEEN

Aquaporins during pregnancy

Alicia E. Damiano*

Laboratorio de Biología de la Reproducción, Instituto de Fisiología y Biofísica Bernardo Houssay (IFIBIO)-CONICET-Facultad de Medicina, Universidad de Buenos Aires, Buenos Aires, Argentina
Cátedra de Biología Celular y Molecular, Departamento de Ciencias Biológicas, Facultad de Farmacia y Bioquímica, Universidad de Buenos Aires, Buenos Aires, Argentina
*Corresponding author: e-mail address: adamiano@ffyb.uba.ar

Contents

Abstract

Aquaporins (AQPs) are water channels proteins that facilitate water flux across cell membranes in response to osmotic gradients. Despite of the differences in the mammalian placentas, the conserved combination of AQPs expressed in placental and fetal membranes throughout gestation suggests that these proteins may be important in the regulation of fetal water homeostasis. Thus, AQPs may regulate the amniotic fluid volume and participate in the trans-placental transfer of water. Apart from their classical roles, recent studies have revealed that placental AQPs may also cooperate in cellular processes such as the migration and the apoptosis of the trophoblasts. Aquaglyceroporins can also participate in the energy metabolism and in the urea elimination across the placenta.

Many factors including oxygen, hormones, acid-basis homeostasis, maternal dietary status, interaction with other transport proteins and osmotic stress are proposed to regulate their expression and function during gestation and alterations result in pathological pregnancies.

Vitamins and Hormones, Volume 112
ISSN 0083-6729
https://doi.org/10.1016/bs.vh.2019.08.009

1. From the blastocyst to the placenta: The expression of aquaporins

Shortly after fecundation takes place, the fertilized egg, called zygote, begins a process of rapid internal cellular division and forms a ball of cells called morula. Immediately, compactation goes on by increasing the cell-to-cell contact and activates the development of the trophectoderm. The trophectoderm is a specialized epithelium that acquires the capacity to initiate and regulate the subsequent events of cavitation to form the blastocyst. The blastocyst consists of an inner cell mass and a fluid-filled cavity surrounded by the trophectoderm.

During cavitation, it was proposed that water channel proteins, known as aquaporins (AQPs), may facilitate the trans-trophectodermal fluid accumulation in response to a relatively small osmotic gradient (Watson & Barcroft, 2001). In this regard, mRNAs encoding AQP1, AQP3, AQP5, AQP6, AQP7 and AQP9 were identified in murine embryos from one-cell stage up to the blastocyst stage (Barcroft, Offenberg, Thomsen, & Watson, 2003; Offenberg et al., 2000). It was observed that AQP3 mRNA expression increased at the morula-blastocyst transition while AQP8 mRNA was not detected in early cleavage stages but it was present in the morula and the blastocyst-stage embryos (Offenberg et al., 2000). In addition, Barcroft and co-workers reported that in mouse blastocysts, AQP3 and AQP8 were found in the basolateral membrane of the trophectoderm, while AQP3 was also observed in cell margins of all inner cell mass. In contrast, AQP9 was predominantly observed within the apical membrane domains of the trophectoderm (Barcroft et al., 2003). The localization of these AQPs in both, the apical and the basolateral membrane domains, may provide a route for trans-trophectoderm water fluxes during blastocyst formation. Furthermore, the authors demonstrated that these proteins mediate the trans-epithelial water movements during cavitation (Barcroft et al., 2003).

In preimplantation human embryos, mRNAs of AQPs 1, 2, 3, 4, 5, 7, 9, 11 and 12 were detected (Xiong et al., 2013). Nevertheless, only the expressions of AQP3 and AQP7 persist from zygote to blastocyst stages, suggesting that both proteins may be crucial for early embryonic development (Xiong et al., 2013).

AQP11 was also identified in all preimplantation stages in mouse (Offenberg & Thomsen, 2005; Park & Cheon, 2015). Its expression decreased until two-cell stage and increased at both four-cell stage and blastocyst stages. Regarding its localization, AQP11 was mainly found in cytoplasm in all stages of embryos in maternal tracts (Park & Cheon, 2015). Even though the role of AQP11 in blastocoel formation has not been studied, the authors suppose that this protein may not participate in the transcellular water movement during cavitation given that its cytoplasmatic location. Instead, they propose that AQP11 may be a critical intracellular molecule for the development of certain organs during early stage embryo development.

From the trophectoderm derives the trophoblast cells which mediate embryonic implantation, the process by which an embryo initiates and maintains contact with the maternal uterine epithelia or stroma. In humans, trophoblast cells differentiate into the invasive extravillous trophoblast (EVT) or the villous trophoblast (VT) cells (Castellucci, Scheper, Scheffen, Celona, & Kaufmann, 1990; Golos, Giakoumopoulos, & Gerami-Naini, 2013; Pfeffer & Pearton, 2012). EVT cells proliferate, migrate and invade into the maternal decidua and the inner myometrium, remodeling the maternal spiral arteries to ensure the appropriate supply of blood to the fetal-placental unit. On the other hand, VT cells coat the chorionic villi, the main functional units of the placenta. The chorionic villi contain a system of fetal capillaries (blood vessels) to allow maximum contact area with the maternal blood (also known as the intervillous space). Thus, VT cells mediate the transfer of nutrients and waste products between the mother and the fetus (Castellucci et al., 1990).

The trophectoderm cells also give rise to the fetal membranes: the amnion, the chorion, the yolk sac and the allantois. The amnion limits the amniotic cavity which contains the amniotic fluid. It provides a protective environment for the developing embryo. The chorion is formed by extraembryonic mesoderm and two layers of trophoblast, which give rise to the fetal face of the placenta. In other mammals, the yolk sac and the allantois are the primary most important mean of nutritional, oxygen and waste product exchange. However, in humans, part of the yolk sac is eventually incorporated into the embryo gut tube and the allantois is appended to the urinary bladder.

Many reports described the expression of AQPs in the placentas and fetal membranes from diverse mammalian species with anatomically and histologically different placentas (Table 1).

It is important to remark that mammalian placentas can be classified according to the number of tissue layers separating maternal and fetal blood, and by the pattern of distribution of the chorionic villi on the surface of the placenta facing the maternal endometrium (Fig. 1) (Carter & Enders, 2004; Grosser, 1909). Therefore, some differences in the transport properties of the placenta are clearly related to differences in structure

Table 1 Expression and localization of AQPs in placenta and fetal membranes from diverse mammalian species.

Species	Type of placenta	AQP1	AQP2	AQP3	AQP4	AQP5	AQP8	AQP9	AQP11	References
Human	Hemomonochorial	*Yes*, vascular endothelial cells, amnion, chorion	*Yes*, fetal membranes, trophoblasts, syncytio-trophoblasts	*Yes*, syncytio-trophoblasts, amnion, chorion	*Yes*, vascular endothelial cells, cyto-trophoblasts, syncytio-trophoblasts, stroma of placental villi	–	*Yes*, syncytio-trophoblasts, amnion, chorion	*Yes*, cyto-trophoblasts, syncytio-trophoblasts, amnion, chorion	*Yes*, amnion, chorion	Damiano et al. (2001), Wang et al. (2001, 2004), Mann et al. (2002), De Falco et al. (2007), Escobar et al. (2012), Prat et al. (2012), Saadoun et al. (2013), and Zhao et al. (2018)
Macaques	Hemomonochorial	*Yes*, amnion	–	*Yes*, amnion	–	–	*Yes*, amnion	*Yes*, amnion	*Yes*, amnion	Cheung et al. (2018)
Mouse	Hemotrichorial	Yes, vascular endothelial cells, placental labyrinth, amnion	–	Yes, labyrinth trophoblasts, amnion	–	–	Yes, labyrinth trophoblasts, fetal membranes	Yes, labyrinth trophoblasts, fetal membranes	–	Ma et al. (1997), Beall et al. (2007), and Kobayashi and Yasui (2010)
Rat	Hemotrichorial	Yes, placental basal and labyrinth zones	–	–	–	–	Yes, placental basal and labyrinth zones	Yes, placental basal and labyrinth zones	–	Belkacemi et al. (2011)
Dog	Endoteliochorial	Yes, vascular endothelial cells, amnion, allantochorion, allantois, yolk sac	–	Yes, labyrinth, amnion, allantois, yolk sac	–	Yes, amnion, allantochorion	Yes, amnion, allantois sac	Yes, amnion, allantois sac	–	Aralla et al. (2012)

Giraffe	Epitheliochorial	Yes, trophoblasts	–	Yes, trophoblasts	–	–	–	–	–	Wooding et al. (2015)
Sheep	Epitheliochorial	Yes, vascular endothelial cells	–	Yes, trophoblasts, chorion, amnion	–	–	Yes, trophoblasts, amnion	Yes, amnion, allantois	Yes, chorion, amnion	Johnston et al. (2000), Liu et al. (2004), and Cheung et al. (2016)
Pig	Epitheliochorial	Yes, localization not determined	–	Yes, localization not determined	–	Yes, localization not determined	–	Yes, localization not determined	–	Zhu et al. (2015)

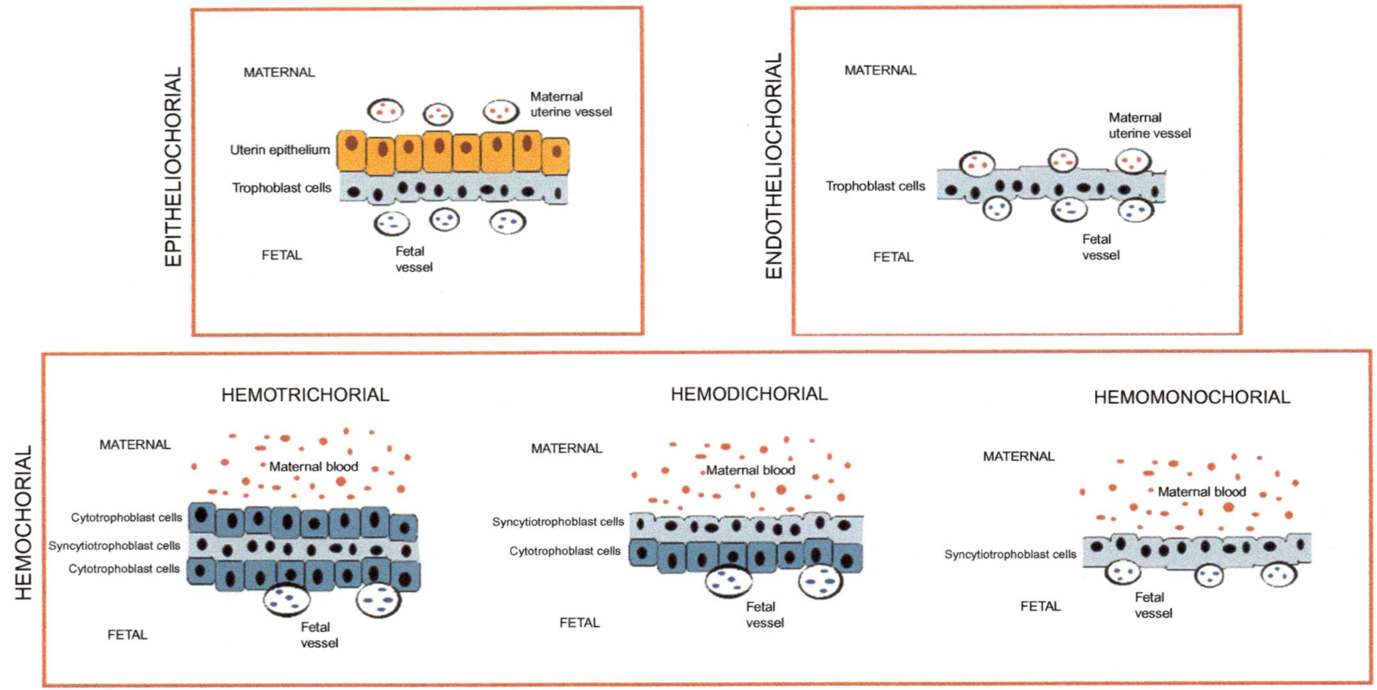

Fig. 1 Schematic drawing of the main types of placental barriers: epitheliochorial placenta (the chorionic villi are next to the lining of the uterus but does not invade or erode the lining, e.g., sheep, giraffe, pig); endotheliochorial placenta (the chorionic villi are in contact with the endothelium of maternal blood vessels, e.g., dog) and hemochorial placenta (the chorionic villi are in direct contact with maternal blood). Hemochorial placentas are classified in trihemochorial (e.g., rat, mouse), dihemochorial (e.g., rabbit) and hemomonochorial (e.g., human, primates, guinea pig) placentas.

(Dilworth & Sibley, 2013). In this regard, the human placenta can be considered hemomonochorial in which only one layer of syncytiotrophoblast is the barrier between the fetal and maternal circulations.

At early stages of human pregnancy, Escobar and co-workers reported that mRNA of AQPs 1, 3, 4, 5, 8, 9 and 11 are expressed in chorionic villi (Escobar et al., 2012). In this work, a high expression of AQP3 and AQP9 mRNAs was described. Both AQPs belong to the aquaglyceroporin family which can permeate urea and glycerol, and in the case of AQP9 it also allows the passage of monocarboxylates, purines and pyrimidines (Tsukaguchi et al., 1998). The authors speculated that these proteins could have a key role in the fetal energy metabolism throughout gestation (Escobar et al., 2012).

De Falco and co-workers have also described the expression of AQP4 mRNA and protein in chorionic villi samples from first trimester, and its downregulation throughout pregnancy (De Falco et al., 2007). In addition, Saadoun et al. have reported changes from moderate and intense levels in the second trimester to weak or undetectable in the third trimester (Saadoun et al., 2013).

Concerning the expression of AQPs in the human trophoblast at term, AQP3 and AQP9 were the first AQPs found in the apical membranes of the human villous syncytiotrophoblast (Damiano, Zotta, Goldstein, Reisin, & Ibarra, 2001). AQP8 was also localized in the trophoblast but the cell polarity of this protein has not yet been determined (Wang, Kallichanda, Song, Ramirez, & Ross, 2001). Regarding AQP4, De Falco et al. reported that expression of AQP4 was weak in the syncytiotrophoblast at term (De Falco et al., 2007) in contrast with an increased expression in endothelial cells and stroma of placental villi. On the other hand, Saadoun and co-workers have reported that this protein was almost undetectable at term (Saadoun et al., 2013).

The expression of AQP2 in human placenta and fetal membranes is controversial. Although many authors have reported that AQP2 was not detectable (Escobar et al., 2012; Mobasheri, Wray, & Marples, 2005), Zhao and co-workers have observed expression of AQP2 in chorionic villi and fetal membranes. They have also found that AQP2 increased in placentas from pregnancies complicated by preeclampsia (Zhao, Lin, & Lai, 2018).

In the same way, the expression of AQPs in human fetal membranes was widely studied. Prat and co-workers found the expression of AQP1, 3, 8, 9, and 11 mRNAs and proteins in human amnion and chorion throughout human gestation (Prat et al., 2012). These authors also reported that AQP1, AQP3 and AQP8 are highly expressed during the first trimester, while

AQP9 and AQP11 are increased in the second trimester (Prat et al., 2012). It is possible that the time-specific expression pattern of these AQPs may be related to the changes in human amniotic fluid (AF) composition and volume during gestation.

In human fetal membranes at term, Mann and co-workers reported the expression of AQP1 and AQP3 and they also localized AQP1 in the capillary endothelium of placental vessels (Mann, Ricke, Yang, Verkman, & Taylor, 2002). In addition, Wang and co-workers described the expression of AQP8 and AQP9 in chorioamniotic membranes (Wang, Chen, Beall, Zhou, & Ross, 2004; Wang et al., 2001). Recently, AQP11 expression was also detected in amniotic membranes at term (Prat et al., 2012).

Human amnion can be separated into two different regions, the reflected amnion that covers the membranous chorion and the placental amnion which overlays the placenta. Bednar and co-workers have found regional differences in the expression of all five AQPs expressed in the amnion (Bednar, Beardall, Brace, & Cheung, 2015). They also observed higher AQP mRNA levels and lower AQP protein levels in the placental compared to reflected amnion.

Likewise, studies in animal models showed that AQPs display a dynamic temporal expression pattern along gestation (Beall, Wang, et al., 2007; Cheung, Anderson, & Brace, 2016; Johnston et al., 2000).

Regarding epitheliochorial placentas, in sheep AQP3 and AQP8 were found in the trophoblast cells (Johnston et al., 2000; Liu, Koukoulas, Ross, Wang, & Wintour, 2004). It was also reported differences in the expression of these genes during sheep pregnancy, being AQP3 the highly expressed (Liu et al., 2004). In the ovine amnion, similarly to human, AQP1, AQP3, AQP8, AQP9 and AQP11 were detected (Cheung et al., 2016; Wang, Chen, Huang, & Ross, 2005). However, AQP8 mRNA levels were the highest among the five AQPs while AQP9 mRNA levels were least expressed (Cheung et al., 2016), while in humans, AQP8 was least expressed and AQP9 mRNA was expressed at high levels (Bednar et al., 2015). Contrary to human, amniotic membranes in the sheep are vascularized and AQP1 localized only in the blood vessels (Johnston et al., 2000). Additionally, in the giraffe epitheliochorial placenta, AQP1 and AQP3 were found in trophoblast cytoplasm (Wooding et al., 2015). Furthermore, in the porcine placenta, Zhu and co-workers found that the mRNAs of AQP1, 3, 4, 5, 6, 7, 8, 9, and 11 are expressed while only AQP1, 3, 5 and 9 proteins were identified (Zhu et al., 2015).

In endotheliochorial placentas, Aralla et al. reported that AQP1, 3, 5, 8 and 9 are expressed in canine fetal membranes and placenta with distinct expression patterns thought gestation. AQP1 was found in endothelium, allantochorion, amnion, allantois and yolk sac. AQP3 was present in the placental labyrinth, amnion, allantois and yolk sac. AQP5 was found in the amniocytes and in the columnar cells of allantochorion while AQP8 and AQP9 were identified in the amniotic and allantois sacs (Aralla, Mobasheri, Groppetti, Cremonesi, & Arrighi, 2012).

As regards hemochorial placentas, Ma and co-workers reported a strongly expression of AQP8 mRNA in mice placentas at gestation days 8 and 18 (Ma, Yang, & Verkman, 1997). In addition, Beall et al., identified the expression of AQP1, AQP3, AQP8 and AQP9 mRNA and proteins in mouse placenta and fetal membranes, and found that AQP1 was localized in the vascular endothelium while AQP3 was present in the trophoblast (Beall, Wang, et al., 2007; Kobayashi & Yasui, 2010). They also described changes in AQPs with advance of gestation, and a correlation with the AF volume. AQP1, 8 and 9 were also found in rat placentas (Belkacemi et al., 2011). In Japanese macaques, AQP1, AQP3, AQP8, AQP9 and AQP11 are expressed in all regions of the amnion. The expression profiles of the five AQPs showed that AQP1 and AQP3 mRNA levels were the highest while AQP8 the lowest among the AQPs expressed in the amnion. In the reflected placental amnion, it was also observed a high expression of AQP1 and a low expression of AQP8 (Cheung, Roberts, Frias, & Brace, 2018).

2. Classical roles of AQPs throughout gestation

The conserved combination of AQPs expressed in the placental and fetal membranes throughout gestation supports the hypothesis that these proteins may be important in the regulation of fetal water homeostasis.

Water moves across the fetal amnion and chorion into the fetal blood vessels (intramembranous transport), and into the chorionic villi across maternal barrier which in human is the syncytiotrophoblast (trans-placental transfer).

Intramembranous transport pathways regulate the AF volume. AF volume progressively increases throughout pregnancy accordingly with fetal growth (Beall, van den Wijngaard, van Gemert, & Ross, 2007; Brace, 1997; Gillibrand, 1969; Mann, Nijland, & Ross, 1996). Consequently, by 28 weeks, AF reaches a volume of about 800 mL, while near term it decreases to about 400 mL (Mann et al., 1996). AF initially originates from maternal plasma, but by the second half of pregnancy, fetal urination and swallowing contribute significantly to the volume of AF, thus its regulation is determined by a balance between its production and resorption.

In normal pregnancy, water moves down its gradient from the hypotonic AF (255 mOsm/kg) into the fetal circulation (280 mOsm/kg) (Liu, Zheng, & Wintour, 2008). Lloyd and co-workers have demonstrated that trans-epithelial permeability coefficient was too high to be explained by simple diffusion of water across the amnion, but it was consistent with the presence of water channels (Lloyd, Garlid, Reba, & Seeds, 1969).

In addition, emerging evidence shows that an altered expression of AQPs may be associated with AF volume abnormalities such as polyhydramnios (excessive AF volume) or oligohydramnios (reduced AF volume) (Harman, 2008; Mann, Dvorak, Gilbert, & Taylor, 2006; Zhu et al., 2010, 2009). These obstetrical pathologies are related to fetal malformations and increased perinatal morbi-mortality (Harman, 2008). Several studies explored the relationship between the expression of AQPs in fetal membranes and AF volume disorders. An increase in AQP1 expression was reported in the amnion in pregnancies complicated by idiopathic polyhydramnios (Mann et al., 2006). Furthermore, the expressions of AQP8 and AQP9 were significantly increased, but their expressions in the trophoblast were significantly decreased (Zhu et al., 2010). These differences suggest that some factors may be inducing changes in the expression of both AQPs that may enhance the intramembranous absorption and decrease the maternal-to-fetal water flow to maintain AF homeostasis.

On the other hand, a decrease in AQP1 levels was found in the amnion in pregnancies complicated by oligohydramnios. However, no significant changes in the chorion and in the placenta were observed (Zhu et al., 2009). In addition, it was detected that AQP3 decreased in the amnion and in the chorion in this pathology while its expression increased in the placenta (Zhu et al., 2009). Regarding AQP8 and AQP9 in oligohydramnios, Jiang and co-workers have reported that both proteins were significantly decreased in amnion while their expressions in the placenta were significantly increased. In addition, they found that AQP9 expression was also significantly decreased in the chorion, while AQP8 was unchanged (Jiang et al., 2012).

Although the expression of AQPs has been demonstrated in human fetal membranes, to date, no study has examined their functional regulation and significance in the AF volume during human pregnancy. However, some AQP-knockout mice were developed to explore the role of AQPs in the regulation of AF volume (Mann, Ricke, Torres, & Taylor, 2005; Seo, Lim, Kim, & Bae, 2018; Sha, Xiong, Liu, Zheng, & Ma, 2011). Thus, in AQP1-knockout mice it was observed that the AF volume increased while its osmolarity was reduced (Mann et al., 2005). Although, it was proposed that AQP1 null mice could be a good model to study polyhydramnios, in humans, this alteration is not related to a decrease of AQP1 expression (Mann et al., 2006). In addition, Luo and co-workers have recently demonstrated that the lack of AQP1 resulted in the downregulation of AQP9 and the upregulation of AQP8 in fetal membranes (Luo et al., 2018). Furthermore, Sha et al. showed in AQP8-knockout mice an increase in the amount of AF (Sha et al., 2011). Take into consideration that AQP8 and AQP1 are highly selective for the passage of water, these findings suggested a compensatory mechanism between both AQPs in the regulation of AF volume.

Interestingly, recently it was demonstrated that AQP3-knockout mice neither developed olygohydramnios nor polyhydramnios (Seo et al., 2018). In fact, it was shown that the deficiency of AQP3 in placenta and fetal membranes is associated with reduced metabolite concentrations in the amniotic fluid and impaired fetal growth proposing a role for AQP3 in providing nutrients (Seo et al., 2018).

On the other hand, trans-placental water transfer depends on the pressure that drives it and the resistance to flow in the barrier. In hemochorial placenta, physiological evidence shows that both a transcellular and a paracellular pathway are available for water transfer (Edwards, Jones, Sibley, & Nelson, 1993; Jansson, Powell, & Illsley, 1999). Previous investigations using isolated membrane vesicles have suggested that the movement of water across the human syncytiotrophoblast occurs by a lipid diffusion pathway (Jansson & Illsley, 1993). However, the placenta behaves like a very low permeability membrane, and diffusional water transfer across the placenta is inadequate to support the fetal water requirements throughout gestation (Stulc, 1997). In subsequent studies, the expressions of AQP3 and AQP9 were described in the apical membrane of the human syncytiotrophoblast (Damiano et al., 2001) and functional uptake experiments using placental explants confirmed that these proteins may facilitate the transcellular water transport across the placenta (Damiano, Zotta, & Ibarra, 2006). A possible interpretation for the discrepancy in the functional experiments is that, in those performed in isolated vesicles, some of the components of the cytoskeleton which may be important for the regulation and functionality of AQPs may be excluded (Noda & Sasaki, 2008).

Currently, it is well-known that water moves across the human hemomonochorial placenta through both the paracellular and transcellular routes, and its transfer may be facilitated by AQPs (Damiano, 2011). These AQPs could participate not only in the water transport between the mother and the fetus, but also in the rapid movement of solutes across cell membranes, with minimal osmotic perturbation (Damiano et al., 2001). However, the role of human trophoblast AQPs exclusively in the facilitation of transcellular water transport across the placenta is still debated.

In this regard, several reports showed an altered expression of AQP3, AQP4 AQP9 in placentas from pregnancies complicated by preeclampsia (Damiano et al., 2006; Szpilbarg & Damiano, 2017; Szpilbarg, Seyahian, et al., 2018). AQP3 and AQP4 decreased and AQP9 increased in these placentas, whereas the water uptake mediated by these AQPs was reduced. However, there is no evidence that the water transfer between the mother and the fetus is abnormal in preeclampsia.

Moreover, in diabetic placentas AQP9 expression was increased with no impact on the feto-maternal water flux (Vilariño-García et al., 2016). These results suggest that these proteins may be involved in other cellular processes.

Regarding AQP11, it was described by Escobar and co-workers in human first trimester chorionic villi and by Prat and co-workers in human fetal membranes (Escobar et al., 2012; Prat et al., 2012). Unlike the rest of AQPs detected along gestation, AQP11 is expressed at the membrane of intracellular organelles such as the endoplasmic reticulum (Gorelick, Praetorius, Tsunenari, Nielsen, & Agre, 2006; Morishita et al., 2005; Rojek et al., 2013; Yakata, Tani, & Fujiyoshi, 2011). Currently, the ability of AQP11 to function as a water channel is not clear due to its unusual location in the cell (Gorelick et al., 2006; Yakata et al., 2007, 2011). It was proposed that AQP11 may be regulating the permeability of the endoplasmic reticulum membrane to water, in order to keep an appropriate environment necessary for the correct protein translation and folding (Escobar et al., 2012; Prat et al., 2012). Thus, the enhanced expression of AQP11 observed during embryogenesis could be associated with the development of fetal organs (Martínez & Damiano, 2017; Park & Cheon, 2015).

3. Non-classical roles of AQPs throughout gestation

Apart from the classical functions of AQPs in transcellular water transport, recent studies have revealed that AQPs may cooperate in different cellular processes, such as apoptosis and cell migration. All these processes require transient changes in cell volume and activity of certain ion transporters.

During placental development, trophoblast cells differentiate along the VT or the EVT pathways (Castellucci et al., 1990; Golos et al., 2013;

Pfeffer & Pearton, 2012). EVT cells migrate away from the placenta and invade into the endometrium remodeling the maternal spiral arteries to provide an adequate blood supply to the growing fetus. These events resemble those of malignant cells to growth and invade. However, unlike tumors, the trophoblast migration and invasion are tightly controlled (Piechowski, 2016; Soundararajan & Jagannadha Rao, 2004).

Recently, in the EVT cell line Swan 71, isolated from 7 weeks chorionic villi, mRNA and protein of AQP3 and AQP9 were detected and it was demonstrated that only AQP3 participates in the migration of EVT cells (Reca, Szpilbarg, & Damiano, 2018). The silencing of AQP3 expression or the blocking of its function resulted in an impaired migration. Therefore, it was proposed that an abnormal expression of this protein during the first differentiation step of the trophoblast cell lineage may affect EVT cells and lead to a shallow trophoblast invasion and poor remodeling of the maternal spiral arteries, resulting in fetal growth restriction (Reca et al., 2018; Szpilbarg, Martínez, et al., 2018). In many cases, placentation takes place under fluctuations of oxygen tension, which are believed to increase reactive oxidative and nitrative species, being a potent stimulus for VT apoptosis. The consequences of this set of alterations may give rise to a combination of preeclampsia and growth restriction.

Nevertheless, so far, the expression and function of AQPs in early placentas that at the end of gestation will result in preeclampsia or fetal growth restriction is not known.

On the other hand, apoptosis is a physiological process required for the fusion of the mononuclear cytotrophoblast cells into the multinucleate syncytium to allow the continuous renewal of VT. Thus, VT apoptosis increases progressively throughout pregnancy (Huppertz, Kadyrov, & Kingdom, 2006; Sharp, Heazell, Crocker, & Mor, 2010; Smith & Baker, 1999).

The role of AQPs in the programmed cell death was first demonstrated in ovarian granulosa cells by Jablonski, Webb, and Hughes (2004). They established that AQPs may mediate the movement of water across the plasma membrane in dying cells. The proposed mechanism is that the loss of intracellular K^+ concentration generates an osmotic gradient that drives water out of the cell and forces the cell to shrink. This event is known as apoptotic volume decrease (AVD). After AVD inactivation of AQPs produces changes in the plasma membrane, which become significantly less permeable to water while the continuous efflux of ions K^+ decreases the ionic strength of the cytoplasm and activates the apoptotic caspases (Jablonski et al., 2004).

Recently, the participation of AQPs in the apoptosis of the human VT cells was explored. It was observed that the inhibition of these proteins, and in particular the blocking of AQP3, abrogates the apoptotic response (Szpilbarg et al., 2016). Thus, it was proposed that AQP3 may be critical in the regulation of villous trophoblast differentiation and turn over.

Since AQP3 and AQP9 pertain to the aquaglyceroporin family, it cannot be excluded a possible role in the diffusion of glycerol and urea (Méndez-Giménez, Rodríguez, Balaguer, & Frühbeck, 2014). Consequently, both AQPs may act in the energy metabolism through the transport of glycerol. Glycerol transport may modulate fat degradation and glucose metabolism. AQP9 also permeates other solutes such as lactate and monocarboxylates associated with energy provision and purines and pyrimidines related to DNA metabolism (Tsukaguchi et al., 1998).

In this context, it was recently reported that AQP9 mRNA and protein expressions are elevated in placentas from women with Gestational diabetes mellitus. These data could suggest that during this disorder the over-expression of AQP9, which correlates with higher leptin plasma levels, increments glycerol transport to the fetus and may help to cover the increase in energy needs that occur during this gestational metabolic disorder (Vilariño-García et al., 2016).

In contrast, urea is the main final product of protein metabolism. It is well-known that in the absence of a transport protein, urea has a low permeability across cell membranes. During fetal stage, urea is considered to be excreted transplacentally from the fetus into the maternal body. Although, it was previously reported the expression of a urea/H^+ exchanger and a urea transporter type A (UT-A), the responsible mechanisms for urea elimination across the human placenta are not well known (Damiano et al., 2006; Hisanaga et al., 1991). Therefore, it could not be discarded that the presence of aquaglyceroporins, AQP3 and AQP9, in the syncytiotrophoblast could also contribute to the excretion of urea across the placenta.

4. Regulation of placenta and fetal membranes AQPs

Little is known about the mechanisms that control and regulate the expression and function of AQPs in the placenta and fetal membranes. However, in the last years some reports explored the changes of placental AQPs in normal and pathological conditions and proposed the possible mechanisms involved in these changes.

4.1 Oxygen regulation

Hemochorial placentas develop initially in a relatively hypoxic environment and after setting the intervillous circulation, oxygen tension increases (Genbacev, Zhou, Ludlow, & Fisher, 1997; James, Stone, & Chamley, 2006). Therefore, this physiological switch in oxygen tension is a prerequisite for proper placental development. In this context, it is well-documented that oxygen controls trophoblast differentiation and the placenta adapts to these profound changes in oxygenation by modulation of the hypoxia inducible factor-1 (HIF-1) (Patel, Landers, Mortimer, & Richard, 2010). HIF-1 regulates the expression of many genes that enable cells to adapt to this stress condition (Liu, Shen, Zhao, & Chen, 2012). Disruptions in oxygen homeostasis are associated with human placental diseases such as preeclampsia (Hung & Burton, 2006).

It was found that in human placental explant cultures, AQP9 protein decreased abruptly when HIF-1α is expressed by deprivation of O_2 (Castro-Parodi et al., 2013). The same result was observed after $CoCl_2$ treatment. $CoCl_2$ is known to activate hypoxia-dependent pathways under normal O_2 levels by inhibiting prolyl-hydroxylase domain-containing enzymes, a family of enzymes that plays a key role in the oxygen-dependent degradation of HIF-1α and consequently stabilizing HIF-1 (Castro-Parodi et al., 2013). The *in-silico* analysis of the human AQP9 gene showed the presence of 14 putative hypoxia-response element (HRE) sites (5′-ACGTGC-3′), but none of them was in the promoter region. Although, in other HIF-1α-inducible genes functional HRE sites were localized outside the promoter region, it is possible that HREs in the promoter region should be required to induce an upregulation of AQP9 transcription. Therefore, it was proposed that HIF-1α may enhance the expression of some intermediate which impacts directly downregulating AQP9 expression (Castro-Parodi et al., 2013).

Regarding AQP3, hypoxia decreased AQP3 protein expression and changed its cellular distribution being mainly found in the cytosol of the syncytiotrophoblasts (Szpilbarg et al., 2016).

Recently, the effect of low oxygen tension on the expression of AQP4 was also studied. It was found that unlike AQP3 and AQP9, hypoxia and HIF-1α increase AQP4 mRNA and protein expression in human placental explants. *In-silico* analysis showed three putative binding sites for HIF-1α in the AQP4 promotor region (Szpilbarg, Seyahian, et al., 2018).

All these changes in AQPs expression, do not correlate with the functional experiments. Water uptake in explants exposed to hypoxia or treated with $CoCl_2$ showed a significantly decreased. However, water flux after hypoxia and not after $CoCl_2$ treatment was unresponsive to $HgCl_2$, a known blocker of AQPs, suggesting that water is not passing through AQPs (Castro-Parodi et al., 2013). A possible explanation for this discrepancy is that $CoCl_2$ only affects HIF-1α expression while the low-oxygen tension may also modify other factors such as intracellular pH (pHi) which could also be altering AQPs permeability.

4.2 Hormone regulation

The syncytiotrophoblasts represents the endocrine tissue of the placenta, secreting large amounts of protein hormones including human chorionic gonadotropin hormone (hCG) and leptin.

After implantation, hCG is the first trophoblast signal detected in the maternal blood. In normal placental explants treated with different concentrations of recombinant hCG, AQP9 expression increased significantly compared to non-treated explants. This effect on AQP9 expression was dependent on hCG concentrations (Marino, Castro-Parodi, Dietrich, & Damiano, 2010). Additionally, it was also reported that hCG exerts its stimulatory effect on AQP9 expression via cAMP (Marino et al., 2010). In addition, consistent with the protein increase, it was observed an increase of 1.6-fold in water transcellular flux (Marino et al., 2010). It was also found a correlation with the high levels of serum hCG observed in women with pregnancies complicated by preeclampsia (Basirat, Barat, & Hajiahmadi, 2006) and the levels of AQP9 protein expression in the placentas (Damiano et al., 2006; Marino et al., 2010).

Regarding leptin, it is a peptide hormone that regulates the growth and metabolism of the trophoblast cells. In explants culture, leptin produces an increase in AQP9 transcriptional and protein levels (Vilariño-García et al., 2018). Recently, increase leptin serum levels and placental AQP9 expression were related to Gestational Diabetes Mellitus (Vilariño-García et al., 2016). The typical macrosomia observed in newborn from diabetic pregnant women requires a greater availability of energy nutrients such as glycerol. In this sense, AQP9 could facilitate the transport of glycerol to the fetus, contributing to attend the increased energy intake requirements in the macrosomic fetus.

On the other hand, recent data in placental and fetal membranes provide evidence that AQPs have a tissue-selective response to insulin. In trophoblast tissue, insulin downregulates the levels of AQP9 in a concentration-dependent manner through a negative insulin-response elements (IRE) in its promoter gene (Castro-Parodi et al., 2011). In the case of placental AQP3, insulin has no effect on its expression (Damiano, 2011). Although AQP9 expression dramatically decreased after insulin treatment, water uptake mediated by AQPs did not change, suggesting that AQP9 is not exclusively involved in water transport between the mother and the fetus (Castro-Parodi et al., 2011). In this context, preeclampsia is associated to a marked hyperinsulinemia, insulin resistance and an increased in AQP9 (Castro-Parodi et al., 2011; Scioscia et al., 2006). It was reported that the systemic inflammation via the tumor necrosis factor α (TNF-α) present in preeclamptic women may impair insulin signaling by inducing serine phosphorylation of the insulin receptor substrate-1 (IRS-1) (Kanety, Feinstein, Papa, Hemi, & Karasik, 1995; Scioscia et al., 2006). Consistent with this, AQP9 expression was not modified in preeclamptic placental explants treated with insulin. Even more, in normal explants treated with TNF-α previously to insulin treatment, AQP9 expression was unaltered (Castro-Parodi et al., 2011). Therefore, a disturbed insulin signaling could explain the poor response of AQP9 to the insulin stimuli observed in preeclamptic placentas.

In fetal membranes, Bouvier and co-workers have recently reported that insulin represses significantly the transcriptional expression of AQP3 and AQP9 in the amnion, but not in the chorion. This decreased expression of both AQPs correlated with a reduction in the glycerol transport mediated by these proteins. They also described that the AQPs downregulation by insulin is blocked by a phosphatidylinositol 3-kinase inhibitor (Bouvier et al., 2015). Pregnancies complicated by Diabetes Mellitus type 2 and Gestational Diabetes Mellitus, are usually associated to polyhydramnios. Thus, the authors postulated that in these disorders, because of the high levels of insulin, aquaglyceroporins may be repressed in the amnion promoting the retention of amniotic fluid (Bouvier et al., 2015).

The effect of arginine vasopressin (AVP) on AQPs expression was also studied. Several data showed that AVP may control the placental water homeostasis (Bajoria, Ward, & Sooranna, 2004; Ervin, Ross, Leake, & Fisher, 1986; Ross, Cedars, Nijland, & Ogundipe, 1996). Belkacemi and co-workers have found that AVP regulates AQP1 in trophoblast-like cells. They observed that AVP enhanced the expression of AQP1 in JEG-3 cells that have features of VT cells and in HTR-8/SVneo which are EVT cells (Belkacemi, Beall, Magee, Pourtemour, & Ross, 2008). This effect on AQP1 gene expression is mediated by cAMP (Belkacemi et al., 2008).

These results propose that AVP may modulate AQP1 expression and facilitate water exchange across the placenta. The effect of AVP on AQP1 may also enhanced the migration of EVT cells during the remodulation of maternal spiral arteries. Moreover, high levels of AVP were found in oligohydramnios (Bajoria et al., 2004) and AQP1 detected in human fetal membranes responded to cAMP (Moore & Whitsett, 1982). Thus, AVP might have also a stimulatory effect on AQP1 expressed in fetal membranes and in the regulation of AF volume.

Although, in other tissues, there is some evidence that the expression of various AQPs can be regulated by steroid sex hormone, in placenta and fetal membranes it was not investigated yet.

4.3 pH regulation

Acid-base homeostasis is critical for a successful pregnancy and changes in the pH microenvironment may alter the expression and the activity of several enzymes and transporters. In this regard, AQPs are highly pH sensitive and structural pH sensors were identified in these proteins (Engel, Fujiyoshi, & Agre, 2000; Morishima et al., 2008; Németh-Cahalan, 2004; Yasui, 2009; Zeuthen, 1999).

The intracellular pH (pHi) of the syncytiotrophoblast may be modified within normal limits. It was reported that the isoforms 1, 2 and 3 of the Na^+/H^+ exchanger family and the isoforms 1 and 2 of the Cl^-/HCO_3^- anion exchangers are involved in the maintenance of the pHi of the syncytiotrophoblast (Grassl, 1989; Lacey, Nolan, Greenwood, Glazier, & Sibley, 2005; Sibley et al., 2002; Speake, Mynett, Glazier, Greenwood, & Sibley, 2005).

Recently, in experiments carried out in normal placental explants it was observed that the cytosolic acidification did not affect the transcellular water movement mediated by AQPs (Dietrich & Damiano, 2015). It is possible that the presence of functional NHEs can restore the physiological pH of the cells. However, the disturbances of pH and the blocking of NHEs significantly decreased water uptake and, in this case, water uptake was not sensitive to $HgCl_2$, suggesting that the permeability of the AQPs to water has changed. Thus, it was proposed that NHEs are necessary to maintain the pHi of the syncytiotrophoblast and the transcellular water transport mediated by AQPs (Dietrich & Damiano, 2015). Consequently, the changes in the pHi of the syncytiotrophoblast may alter the permeability of these AQPs to water.

In some pathological conditions, the expression of NHEs is abnormal and the acid-basis homeostasis is altered promoting the development of fetal acidosis. In this regard, in preeclamptic placentas the expression of NHE-1 and NHE-3 were reduced (Dietrich, Szpilbarg, & Damiano, 2013; Khan, al-Yatama, & Nandakumaran, 1999) while in diabetic placentas, it was observed a decrease in NHE-1 protein levels (Khan, Al-Awadi, Nandakumaran, Abul, & Al-Azemi, 2003). In addition, in placentas from pregnancies complicated by IUGR the activity and the expression of NHE-1, NHE-3 and NHE-3 were significantly reduced (Johansson, Glazier, Sibley, Jansson, & Powell, 2002). In this context, these alterations may also have a negative consequence on the permeability of placental AQPs.

4.4 CFTR regulation

Cystic fibrosis transmembrane conductance regulator (CFTR) is a cAMP-regulated chloride channel that may interact with various membrane proteins controlling their transport activity (Kunzelmann, 2001). In this regard, water transport mediated by AQPs may be regulated by CFTR (Cheung, Leung, Leung, & Wong, 2003). In human placenta, it was demonstrated that CFTR co-localizes with AQP9 in the apical membrane of syncytiotrophoblast (Castro-Parodi, Levi, Dietrich, Zotta, & Damiano, 2009). Water uptake experiments showed that the blocking of CFTR with diphenylamine-2-carboxylate (DPC), 4,4′-diisothiocyanatostilbene-2,2′-disulfonic acid (DIDs) (inhibitors of chloride channels) and glibenclamide (an open-channel blocker of CFTR) considerably reduced the water transport mediated by AQPs suggesting that CFTR is required to preserve the normal functionality of AQPs (Castro-Parodi et al., 2009). These findings highlight that both proteins work in a synergistic manner. The reduced protein expression of CFTR in preeclamptic placentas could also explain the reduced transcellular water transport mediated by AQPs observed in these placentas (Castro-Parodi et al., 2009).

4.5 cAMP regulation

cAMP (3′, 5′-cyclic adenosine monophosphate) is one of the most common second messengers in the endocrine signal transduction pathway. Cyclic AMP-dependent PKA is the typical immediate effecter of the cAMP signaling pathway.

It was reported that cAMP enhances mRNA expression of AQP1, 3, 8 and 9 in human amnion epithelial cells (Wang, Amidi, Beall, Gui, & Ross, 2006; Wang, Amidi, Yin, Beall, & Ross, 2007) but the response of each AQP to cAMP stimulation is different. In the case of AQP3, increase of intracellular cAMP level by SP-cAMP (a protein kinase A-PKA-activator) or forskolin (a potent activator of adenyl cyclase), results in rapid but temporary upregulation of AQP3 expression via a PKA dependent pathway. After the rapid increased of AQP3 expression at 2 h, its expression returned to normal level by 10 h of treatment, suggesting that AQP3 mRNA has a very short biological half-life (Wang et al., 2006). Regarding AQP8, forskolin treatment resulted in a rapid and sustained increase of the gene expression while the response of AQP1 and 9 genes were delayed and persistent. (Wang et al., 2007). AQP8 mRNA expression was also upregulated by cAMP in WISH cell line that was originally derived from human amnion epithelia (Wang, Chen, Au, & Ross, 2003). Additionally, it was also found a lack of effect of SP-cAMP on AQP1, 8, and 9 mRNA expressions, suggesting that cAMP upregulates these proteins via a PKA independent pathway (Wang et al., 2007).

4.6 Osmotic stress regulation

In many tissues, the expression of AQPs changes in response to osmotic stress. These changes in AQPs are of special interest in oligohydramnios and polyhydramnios.

AQP8 was found to be regulated by osmotic stress in amnion epithelial cells (Qi, Li, Zong, Hyer, & Huang, 2009). It was observed that a hypotonic media significantly enhanced AQP8 mRNA and protein expression while hypertonic media significantly decreased its expression. In addition, the cellular localization of the protein also changed. AQP8 was predominantly found on the plasma membrane in hypotonic media while mainly located in cytoplasm in hyperosmolar media (Qi et al., 2009). These findings correlated with the AQP8 decreased expression in oligohydramnios and its increased in polyhydramnios, proposing that this protein may participate in the balance of the AF (Jiang et al., 2012; Zhu et al., 2010).

4.7 Retinoic acid regulation

All-trans-retinoic acid, a vitamin A derivative, is needed to assure a correct embryonic and placental development (Rhinn & Dolle, 2012). Lack of vitamin A could lead to fetal malformation while an excess could produce teratogenesis (Marceau, Gallot, Lemery, & Sapin, 2007). In other tissues, retinoic acid is involved in the regulation of several target genes including AQPs (Al Tanoury, Piskunov, & Rochette-Egly, 2013; Bellemere, Von Stetten, & Oddos, 2008).

Recently, it was found that AQP3 gene was regulated by all-trans-retinoic acid in human amnion and epithelial amniotic cells (Prat et al., 2015). Thus, retinoic acid may stimulate AQP3 transcripts which results in an increase in AQP3 protein and an uptake of glycerol, suggesting an increased in AQP3 function. However, the significance of these results in the amniotic fluid homeostasis is still unknown.

5. Influence of the membrane lipid bilayer of the trophoblast cells on placental AQPs

It was well-established that transmembrane transport activities are greatly influenced by the fluidity of the lipid bilayer (Gross & Han, 2011).

In this regard, studies on human placental syncytiotrophoblast also demonstrated that an essential role is played by the membrane lipid bilayer fluidity in the modulation of membrane transport processes (Levi et al., 2016). In preeclamptic placentas, the apical membranes of syncytiotrophoblast are more rigid than normal ones. It was found that an increase in sphingomyelin, with no changes in the cholesterol amount, may reduce the number of caveolae and the expression of Caveolin-1 (Levi et al., 2016). Transcellular water flux was considerably reduced in these placentas and experimental data showed that AQPs seemed to be not functional (Damiano et al., 2006). Moreover, the analysis of the primary structure of AQPs protein revealed a putative caveolin-1-binding site which is required for AQPs functionality (Jablonski & Hughes, 2006). Consequently, it was speculated that an altered lipid composition may disrupt the ability of sphingomyelin and cholesterol to assemble into caveolae in the apical leaflet of the bilayer, affecting protein expression and cell signaling (Levi et al., 2016). This may produce a negative environment for AQPs insertion or function into the plasma membrane.

Furthermore, recently it found that was the lack of caveolae may also affect the ability of EVT cells to migrate during placentation in a mechanism in which AQP3 is required (Reca et al., 2018).

6. Influence of maternal dietary status on placental AQPs

Emerging evidence shows that maternal disorders of nutrition adversely affect the fetal and placental development (Lowensohn, Stadler, & Naze, 2016) and produce serious long-term health consequences (Contu & Hawkes, 2017; Roberts, Frias, & Grove, 2015).

It was described in sheep that obesity during gestation reduced amniotic and allantoic fluid volumes (Satterfield, Dunlap, Keisler, Bazer, & Wu, 2012). In pregnant macaques, high-fat-diet modified AQP1 and AQP8 expressions, but AF volume was not altered. In addition, a correlation between AQP11 expression and AF volume was observed.

On the other hand, it was found that maternal undernutrition may modify the expression of AQPs in rat placenta. AQP1 significantly decreased, while AQP8 and AQP9 significantly increased (Belkacemi et al., 2011).

Up to now, in humans the effect of maternal dietary status on the expression of placenta and fetal AQPs was not investigated.

7. Conclusions

Despite of the anatomically and histologically differences in the placental structure, a similar pattern of AQPs are expressed in the placental and fetal membranes of different mammals throughout pregnancy.

Although the functional importance of AQPs remains to be elucidated, emerging evidence suggests that these proteins may play an essential role not only in the regulation of fetal water homeostasis but also in the formation and development of the placenta.

Many factors including oxygen, hormones, acid-basis homeostasis, maternal dietary status, interaction with other transport proteins and osmotic stress are proposed to control their expression and function. In addition, alterations in these control mechanisms result in pathological gestations, suggesting that the regulation of these proteins could be a potential target for the treatment of abnormal pregnancies.

References

Al Tanoury, Z., Piskunov, A., & Rochette-Egly, C. (2013). Vitamin A and retinoid signaling: Genomic and nongenomic effects. *Journal of Lipid Research*, *54*(7), 1761–1775.

Aralla, M., Mobasheri, A., Groppetti, D., Cremonesi, F., & Arrighi, S. (2012). Expression of aquaporin water channels in canine fetal adnexa in respect to the regulation of amniotic fluid production and absorption. *Placenta*, *33*(6), 502–510.

Bajoria, R., Ward, S., & Sooranna, S. R. (2004). Influence of vasopressin in the pathogenesis of oligohydramnios-polyhydramnios in monochorionic twins. *European Journal of Obstetrics, Gynecology, and Reproductive Biology, 113*(1), 49–55.

Barcroft, L. C., Offenberg, H., Thomsen, P., & Watson, A. J. (2003). Aquaporin proteins in murine trophectoderm mediate transepithelial water movements during cavitation. *Developmental Biology, 256*(2), 342–354.

Basirat, Z., Barat, S., & Hajiahmadi, M. (2006). Serum beta human chorionic gonadotropin levels and preeclampsia. *Saudi Medical Journal, 27*(7), 1001–1004.

Beall, M. H., van den Wijngaard, J. P., van Gemert, M. J., & Ross, M. G. (2007). Regulation of amniotic fluid volume. *Placenta, 28*(8–9), 824–832.

Beall, M. H., Wang, S., Yang, B., Chaudhri, N., Amidi, F., & Ross, M. G. (2007). Placental and membrane aquaporin water channels: Correlation with amniotic fluid volume and composition. *Placenta, 28*(5–6), 421–428.

Bednar, A. D., Beardall, M. K., Brace, R. A., & Cheung, C. Y. (2015). Differential expression and regional distribution of aquaporins in amnion of normal and gestational diabetic pregnancies. *Physiological Reports, 3*(3), e12320.

Belkacemi, L., Beall, M. H., Magee, T. R., Pourtemour, M., & Ross, M. G. (2008). AQP1 gene expression is upregulated by arginine vasopressin and cyclic AMP agonists in trophoblast cells. *Life Sciences, 82*(25–26), 1272–1280.

Belkacemi, L., Desai, M., Beall, M. H., Liu, Q., Lin, J. T., Nelson, D. M., et al. (2011). Early compensatory adaptations in maternal undernourished pregnancies in rats: Role of the aquaporins. *Journal of Maternal-Fetal and Neonatal Medicine, 24*(5), 752–759.

Bellemere, G., Von Stetten, O., & Oddos, T. (2008). Retinoic acid increases aquaporin 3 expression in normal human skin. *Journal of Investigative Dermatology, 128*(3), 542e8.

Bouvier, D., Rouzaire, M., Marceau, G., Prat, C., Pereira, B., Lemarié, R., et al. (2015). Aquaporins and fetal membranes from diabetic parturient women: Expression abnormalities and regulation by insulin. *Journal of Clinical Endocrinology and Metabolism, 100*(10), E1270–E1279.

Brace, R. A. (1997). Physiology of amniotic fluid volume regulation. *Clinical Obstetrics and Gynecology, 40*(2), 280–289.

Carter, A. M., & Enders, A. C. (2004). Comparative aspects of trophoblast development and placentation. *Reproductive Biology and Endocrinology, 2*, 46.

Castellucci, M., Scheper, M., Scheffen, I., Celona, A., & Kaufmann, P. (1990). The development of the human placental villous tree. *Anatomy and Embryology (Berlin), 181*(2), 117–128.

Castro-Parodi, M., Farina, M., Dietrich, V., Szpilbarg, N., Aban, C., Zotta, E., et al. (2011). Evidence for insulin-mediated control of AQP9 expression in human placenta. *Placenta, 32*(12), 1050–1056.

Castro-Parodi, M., Levi, L., Dietrich, V., Zotta, E., & Damiano, A. E. (2009). CFTR may modulate AQP9 functionality in preeclamptic placentas. *Placenta, 30*(7), 642–648.

Castro-Parodi, M., Szpilbarg, N., Dietrich, V., Sordelli, M., Reca, A., Abán, C., et al. (2013). Oxygen tension modulates AQP9 expression in human placenta. *Placenta, 34*(8), 690–698.

Cheung, C. Y., Anderson, D. F., & Brace, R. A. (2016). Aquaporins in ovine amnion: Responses to altered amniotic fluid volumes and intramembranous absorption rates. *Physiological Reports, 4*(14), e12868.

Cheung, K. H., Leung, C. T., Leung, G. P. H., & Wong, P. Y. D. (2003). Synergistic effects of cystic fibrosis transmembrane conductance regulator and aquaporin-9 in the rat epididymis. *Biology of Reproduction, 68*(5), 1505–1510.

Cheung, C. Y., Roberts, V. H. J., Frias, A. E., & Brace, R. A. (2018). High-fat diet effects on amniotic fluid volume and amnion aquaporin expression in non-human primates. *Physiological Reports, 6*(14), e13792.

Contu, L., & Hawkes, C. A. (2017). A review of the impact of maternal obesity on the cognitive function and mental health of the offspring. *International Journal of Molecular Sciences, 18*(5), 1093.

Damiano, A. E. (2011). Review: Water channel proteins in the human placenta and fetal membranes. *Placenta, 32*(Suppl. 2), S207–S211.

Damiano, A. E., Zotta, E., Goldstein, J., Reisin, I., & Ibarra, C. (2001). Water channel proteins AQP3 and AQP9 are present in syncytiotrophoblast of human term placenta. *Placenta, 22*(8–9), 776–781.

Damiano, A. E., Zotta, E., & Ibarra, C. (2006). Functional and molecular expression of AQP9 channel and UT-A transporter in normal and preeclamptic human placentas. *Placenta, 27*(11–12), 1073–1081.

De Falco, M., Cobellis, L., Torella, M., Acone, G., Varano, L., Sellitti, A., et al. (2007). Down-regulation of aquaporin 4 in human placenta throughout pregnancy. *In Vivo, 21*(5), 813–817.

Dietrich, V., & Damiano, A. E. (2015). Activity of NA(+)/H(+) exchangers alters aquaporin-mediated water transport in human placenta. *Placenta, 36*(12), 1487–1489.

Dietrich, V., Szpilbarg, N., & Damiano, A. E. (2013). Reduced expression of Na^+/H^+ exchanger isoform 3 (NHE-3) in preeclamptic placentas. *Placenta, 34*(9), 828–830.

Dilworth, M. R., & Sibley, C. P. (2013). Review: Transport across the placenta of mice and women. *Placenta, 34*(Suppl), S34–S39.

Edwards, D., Jones, C. J., Sibley, C. P., & Nelson, D. M. (1993). Paracellular permeability pathways in the human placenta: A quantitative and morphological study of maternal fetal transfer of horseradish peroxidase. *Placenta, 14*(1), 63–73.

Engel, A., Fujiyoshi, Y., & Agre, P. (2000). The importance of aquaporin water channel protein structures. *EMBO Journal, 19*(5), 800–806.

Ervin, M. G., Ross, M. G., Leake, R. D., & Fisher, D. A. (1986). Fetal recirculation of amniotic fluid arginine vasopressin. *American Journal of Physiology, 250*(3 Pt. 1), E253–E258.

Escobar, J., Gormaz, M., Arduini, A., Gosens, K., Martinez, A., Perales, A., et al. (2012). Expression of aquaporins early in human pregnancy. *Early Human Development, 88*(8), 589–594.

Genbacev, O., Zhou, Y., Ludlow, J. W., & Fisher, S. J. (1997). Regulation of human placental development by oxygen tension. *Science, 277*(5332), 1669–1672.

Gillibrand, P. N. (1969). Changes in the electrolytes, urea and osmolality of the amniotic fluid with advancing pregnancy. *The Journal of Obstetrics and Gynaecology of the British Commonwealth, 76*(10), 898–905.

Golos, T. G., Giakoumopoulos, M., & Gerami-Naini, B. (2013). Review: Trophoblast differentiation from human embryonic stem cells. *Placenta, 34*(Suppl), S56–S61.

Gorelick, D. A., Praetorius, J., Tsunenari, T., Nielsen, S., & Agre, P. (2006). Aquaporin-11: A channel protein lacking apparent transport function expressed in brain. *BMC Biochemistry*, 7, 14.

Grassl, S. M. (1989). Cl/HCO3 exchange in human placental brush border membrane vesicles. *Journal of Biological Chemistry, 264*(19), 11103–11106.

Gross, R. W., & Han, X. (2011). Lipidomics at the interface of structure and function in systems biology. *Chemistry & Biology, 18*(3), 284–291.

Grosser, O. (1909). *Vergleichende Anatomie und Entwicklungs-geschichte der Eih&ute und der Placenta.* Vienna: Braumuller.

Harman, C. R. (2008). Amniotic fluid abnormalities. *Seminars in Perinatology, 32*(4), 288–294.

Hisanaga, H., Iioka, H., Moriyama, I., Nabuchi, K., Morimoto, K., & Ichijo, M. (1991). The mechanism of human placental urea transport: A study using placental brush border (microvillous) membrane vesicles. *Asia-Oceania Journal of Obstetrics and Gynaecology, 17*(1), 67–72.

Hung, T. H., & Burton, G. J. (2006). Hypoxia and reoxygenation: A possible mechanism for placental oxidative stress in preeclampsia. *Taiwanese Journal of Obstetrics & Gynecology, 45*(3), 189–200.

Huppertz, B., Kadyrov, M., & Kingdom, J. C. (2006). Apoptosis and its role in the trophoblast. *American Journal of Obstetrics and Gynecology, 195*(1), 29–39.
Jablonski, E. M., & Hughes, F. M., Jr. (2006). The potential role of caveolin-1 in inhibition of aquaporins during the AVD. *Biology of the Cell, 98*(1), 33–42.
Jablonski, E., Webb, A., & Hughes, F. M., Jr. (2004). Water movement during apoptosis: A role for aquaporins in the apoptotic volume decrease (AVD). *Advances in Experimental Medicine and Biology, 559*, 179–188.
James, J. L., Stone, P. R., & Chamley, L. W. (2006). The regulation of trophoblast differentiation by oxygen in the first trimester of pregnancy. *Human Reproduction Update, 12*(2), 137–144.
Jansson, T., & Illsley, N. P. (1993). Osmotic water permeabilities of human placental microvillous and basal membranes. *Journal of Membrane Biology, 132*(2), 147–155.
Jansson, T., Powell, T. L., & Illsley, N. P. (1999). Gestational development of water and nonelectrolyte permeability of human syncytiotrophoblast plasma membranes. *Placenta, 20*(2–3), 155–160.
Jiang, S. S., Zhu, X. J., Ding, S. D., Wang, J. J., Jiang, L. L., Jiang, W. X., et al. (2012). Expression and localization of aquaporins 8 and 9 in term placenta with oligohydramnios. *Reproductive Sciences, 19*(12), 1276–1284.
Johansson, M., Glazier, J. D., Sibley, C. P., Jansson, T., & Powell, T. L. (2002). Activity and protein expression of the Na+/H+ exchanger is reduced in syncytiotrophoblast microvillous plasma membranes isolated from preterm intrauterine growth restriction pregnancies. *Journal of Clinical Endocrinology and Metabolism, 87*(12), 5686–5694.
Johnston, H., Koukoulas, I., Jeyaseelan, K., Armugam, A., Earnest, L., Baird, R., et al. (2000). Ontogeny of aquaporins 1 and 3 in ovine placenta and fetal membranes. *Placenta, 21*(1), 88–99.
Kanety, H., Feinstein, R., Papa, M. Z., Hemi, R., & Karasik, A. (1995). Tumor necrosis factor alpha-induced phosphorylation of insulin receptor substrate-1 (IRS-1). Possible mechanism for suppression of insulin-stimulated tyrosine phosphorylation of IRS-1. *Journal of Biological Chemistry, 270*(40), 23780–23784.
Khan, I., Al-Awadi, F. M., Nandakumaran, M., Abul, H., & Al-Azemi, M. (2003). Suppression of Na+/H+ exchanger-1 in placentas of type 2 diabetic pregnant women: Possible functional implication. *Acta Diabetologica, 40*(1), 28–36.
Khan, I., al-Yatama, M., & Nandakumaran, M. (1999). Expression of the Na(+)-H+ exchanger isoform-1 and cyclooxygenases in human placentas: Their implications in preeclampsia. *Biochemistry and Molecular Biology International, 47*(4), 715–722.
Kobayashi, K., & Yasui, M. (2010). Cellular and subcellular localization of aquaporins 1, 3, 8, and 9 in amniotic membranes during pregnancy in mice. *Cell and Tissue Research, 342*(2), 307–316.
Kunzelmann, K. (2001). CFTR: Interacting with everything? *News in Physiological Sciences, 16*, 167–170.
Lacey, H. A., Nolan, T., Greenwood, S. L., Glazier, J. D., & Sibley, C. P. (2005). Gestational profile of Na+/H+ exchanger and Cl-/HCO3- anion exchanger mRNA expression in placenta using real-time QPCR. *Placenta, 26*(1), 93–98.
Levi, L., Castro-Parodi, M., Martínez, N., Piehl, L. L., Rubín De Celis, E., Herlax, V., et al. (2016). The unfavorable lipid environment reduced caveolin-1 expression in apical membranes from human preeclamptic placentas. *Biochimica et Biophysica Acta, 1858*(9), 2171–2180.
Liu, H., Koukoulas, I., Ross, M. C., Wang, S., & Wintour, E. M. (2004). Quantitative comparison of placental expression of three aquaporin genes. *Placenta, 25*(6), 475–478.
Liu, W., Shen, S. M., Zhao, X. Y., & Chen, G. Q. (2012). Targeted genes and interacting proteins of hypoxia inducible factor-1. *International Journal of Biochemistry and Molecular Biology, 3*(2), 165–178.

Liu, H., Zheng, Z., & Wintour, E. M. (2008). Aquaporins and fetal fluid balance. *Placenta, 29*(10), 840–847.
Lloyd, S. J., Garlid, K. D., Reba, R. C., & Seeds, A. E. (1969). Permeability of different layers of the human placenta to isotopic water. *Journal of Applied Physiology, 26*(3), 274–276.
Lowensohn, R. I., Stadler, D. D., & Naze, C. (2016). Current concepts of maternal nutrition. *Obstetrical & Gynecological Survey, 71*(7), 413–426.
Luo, H., Xie, A., Hua, Y., Wang, J., Liu, Y., & Zhu, X. (2018). Aquaporin 1 gene deletion affects the amniotic fluid volume and composition as well as the expression of other aquaporin water channels in placenta and fetal membranes. *Clinica Chimica Acta, 482*, 161–165.
Ma, T., Yang, B., & Verkman, A. S. (1997). Cloning of a novel water and urea-permeable aquaporin from mouse expressed strongly in colon, placenta, liver, and heart. *Biochemical and Biophysical Research Communications, 240*(2), 324–328.
Mann, S. E., Dvorak, N., Gilbert, H., & Taylor, R. N. (2006). Steady-state levels of aquaporin 1 mRNA expression are increased in idiopathic polyhydramnios. *American Journal of Obstetrics and Gynecology, 194*(3), 884–887.
Mann, S. E., Nijland, M. J., & Ross, M. G. (1996). Mathematic modeling of human amniotic fluid dynamics. *American Journal of Obstetrics and Gynecology, 175*(4 Pt. 1), 937–944.
Mann, S. E., Ricke, E. A., Torres, E. A., & Taylor, R. N. (2005). A novel model of polyhydramnios: Amniotic fluid volume is increased in aquaporin 1 knockout mice. *American Journal of Obstetrics and Gynecology, 192*(6), 2041–2044.
Mann, S. E., Ricke, E. A., Yang, B. A., Verkman, A. S., & Taylor, R. N. (2002). Expression and localization of aquaporin 1 and 3 in human fetal membranes. *American Journal of Obstetrics and Gynecology, 187*(4), 902–907.
Marceau, G., Gallot, D., Lemery, D., & Sapin, V. (2007). Metabolism of retinol during mammalian placental and embryonic development. *Vitamins and Hormones, 75*, 97–115.
Marino, G., Castro-Parodi, M., Dietrich, V., & Damiano, A. E. (2010). High levels of human chorionic gonadotropin (HCG) may increase aquaporin-9 (AQP9) expression in human preeclamptic placenta. *Reproductive Sciences, 17*(5), 444–453.
Martínez, N., & Damiano, A. E. (2017). Aquaporins in fetal development. *Advances in Experimental Medicine and Biology, 969*, 199–212.
Méndez-Giménez, L., Rodríguez, A., Balaguer, I., & Frühbeck, G. (2014). Role of aquaglyceroporins and caveolins in energy and metabolic homeostasis. *Molecular and Cellular Endocrinology, 397*(1–2), 78–92.
Mobasheri, A., Wray, S., & Marples, D. (2005). Distribution of AQP2 and AQP3 water channels in human tissue microarrays. *Journal of Molecular Histology, 36*(1–2), 1–14.
Moore, J. J., & Whitsett, J. A. (1982). The beta-adrenergic receptor in mammalian placenta: Species differences and ontogeny. *Placenta, 3*(3), 257–268.
Morishima, T., Aoyama, M., Iida, Y., Yamamoto, N., Hirate, H., Arima, H., et al. (2008). Lactic acid increases aquaporin 4 expression on the cell membrane of cultured rat astrocytes. *Neuroscience Research, 61*(1), 18–26.
Morishita, Y., Matsuzaki, T., Hara-chikuma, M., Andoo, A., Shimono, M., & Matsuki, A. (2005). Disruption of aquaporin-11 produces polycystic kidneys following vacuolization of the proximal tubule. *Molecular and Cellular Biology, 25*(17), 7770–7779.
Németh-Cahalan, K. (2004). Molecular basis of pH and Ca2+ regulation of aquaporin water permeability. *Journal of General Physiology, 123*(5), 573–580.
Noda, Y., & Sasaki, S. (2008). Actin-binding channels. *Progress in Brain Research, 170*, 551–557.
Offenberg, H., Barcroft, L. C., Caveney, A., Viuff, D., Thomsen, P. D., & Watson, A. J. (2000). mRNAs encoding aquaporins are present during murine preimplantation development. *Molecular Reproduction and Development, 57*(4), 323–330.
Offenberg, H., & Thomsen, P. D. (2005). Functional challenge affects aquaporin mRNA abundance in mouse blastocysts. *Molecular Reproduction and Development, 71*(4), 422–430.

Park, J. W., & Cheon, Y. P. (2015). Temporal aquaporin 11 expression and localization during preimplantation embryo development. *Development & Reproduction, 19*(1), 53–60.
Patel, J., Landers, K., Mortimer, R. H., & Richard, K. (2010). Regulation of hypoxia inducible factors (HIF) in hypoxia and normoxia during placental development. *Placenta, 31*(11), 951–957.
Pfeffer, P. L., & Pearton, D. (2012). Trophoblast development. *Reproduction, 143*(3), 231–246.
Piechowski, J. (2016). Trophoblastic-like transdifferentiation: A key to oncogenesis. *Critical Reviews in Oncology/Hematology, 101*, 1–11.
Prat, C., Blanchon, L., Borel, V., Gallot, D., Herbet, A., Bouvier, D., et al. (2012). Ontogeny of aquaporins in human fetal membranes. *Biology of Reproduction, 86*(2), 48.
Prat, C., Bouvier, D., Comptour, A., Marceau, G., Belville, C., Clairefond, G., et al. (2015). All-trans-retinoic acid regulates aquaporin-3 expression and related cellular membrane permeability in the human amniotic environment. *Placenta, 36*(8), 881–887.
Qi, H., Li, L., Zong, W., Hyer, B. J., & Huang, J. (2009). Expression of aquaporin 8 is diversely regulated by osmotic stress in amnion epithelial cells. *Journal of Obstetrics and Gynaecology Research, 35*(6), 1019–1025.
Reca, A., Szpilbarg, N., & Damiano, A. E. (2018). The blocking of aquaporin-3 (AQP3) impairs extravillous trophoblast cell migration. *Biochemical and Biophysical Research Communications, 499*(2), 227–232.
Rhinn, M., & Dolle, P. (2012). Retinoic acid signalling during development. *Development, 139*(5), 843–858.
Roberts, V. H., Frias, A. E., & Grove, K. L. (2015). Impact of maternal obesity on fetal programming of cardiovascular disease. *Physiology (Bethesda), 30*(3), 224–231.
Rojek, A., Fuchtbauer, E. M., Fuchtbauer, A., Jelen, S., Malmendal, A., & Fenton, R. A. (2013). Liver-specific aquaporin 11 knockout mice show rapid vacuolization of the rough endoplasmic reticulum in periportal hepatocytes after amino acid feeding. *American Journal of Physiology. Gastrointestinal and Liver Physiology, 304*(5), G501–G515.
Ross, M. G., Cedars, L., Nijland, M. J., & Ogundipe, A. (1996). Treatment of oligohydramnios with maternal 1-deamino-[8-D-arginine] vasopressin-induced plasma hypoosmolality. *American Journal of Obstetrics and Gynecology, 174*(5), 1608–1613.
Saadoun, S., Waters, P., Leite, M. I., Bennett, J. L., Vincent, A., & Papadopoulos, M. C. (2013). Neuromyelitis optica IgG causes placental inflammation and fetal death. *Journal of Immunology, 191*(6), 2999–3005.
Satterfield, C. M., Dunlap, K. A., Keisler, D. H., Bazer, F. W., & Wu, G. (2012). Arginine nutrition and fetal brown adipose tissue development in diet-induced obese sheep. *Amino Acids, 43*(4), 1593–1603.
Scioscia, M., Gumaa, K., Kunjara, S., Paine, M. A., Selvaggi, L. E., Rodeck, C. H., et al. (2006). Insulin resistance in human preeclamptic placenta is mediated by serine phosphorylation of insulin receptor substrate-1 and -2. *Journal of Clinical Endocrinology and Metabolism, 91*(2), 709–717.
Seo, M. J., Lim, J. H., Kim, D. H., & Bae, H. R. (2018). Loss of aquaporin-3 in placenta and fetal membranes induces growth restriction in mice. *Development & Reproduction, 22*(3), 263–273.
Sha, X. Y., Xiong, Z. F., Liu, H. S., Zheng, Z., & Ma, T. H. (2011). Pregnant phenotype in aquaporin 8-deficient mice. *Acta Pharmacologica Sinica, 32*(6), 840–844.
Sharp, A. N., Heazell, A. E., Crocker, I. P., & Mor, G. (2010). Placental apoptosis in health and disease. *American Journal of Reproductive Immunology, 64*(3), 159–169.
Sibley, C. P., Glazier, J. D., Greenwood, S. L., Lacey, H., Mynett, K., Speake, P., et al. (2002). Regulation of placental transfer: The Na(+)/H(+) exchanger e a review. *Placenta, 23*(Suppl. A), S39–S46.
Smith, S. C., & Baker, P. N. (1999). Placental apoptosis is increased in post-term pregnancies. *British Journal of Obstetrics and Gynaecology, 106*(8), 861–862.

Soundararajan, R., & Jagannadha Rao, A. (2004). Trophoblast 'pseudo-tumorigenesis': Significance and contributory factors. *Reproductive Biology and Endocrinology*, *2*, 15.
Speake, P. F., Mynett, K. J., Glazier, J. D., Greenwood, S. L., & Sibley, C. P. (2005). Activity and expression of Na+/H+ exchanger isoforms in the syncytiotrophoblast of the human placenta. *Pflügers Archiv—European Journal of Physiology*, *450*(2), 123–130.
Stulc, J. (1997). Placental transfer of inorganic ions and water. *Physiological Reviews*, *77*(3), 805–836.
Szpilbarg, N., Castro-Parodi, M., Reppetti, J., Repetto, M., Maskin, B., Martinez, N., et al. (2016). Placental programmed cell death: Insights into the role of aquaporins. *Molecular Human Reproduction*, *22*(1), 46–56.
Szpilbarg, N., & Damiano, A. E. (2017). Expression of aquaporin-3 (AQP3) in placentas from pregnancies complicated by preeclampsia. *Placenta*, *59*, 57–60.
Szpilbarg, N., Martínez, N., Di Paola, M., Reppetti, J., Medina, Y., Seyahian, A., et al. (2018). New insights into the role of placental aquaporins and the pathogenesis of preeclampsia. *Frontiers in Physiology*, *9*, 1507.
Szpilbarg, N., Seyahian, A., Di Paola, M., Castro-Parodi, M., Martínez, N., Farina, M., et al. (2018). Oxygen regulation of aquaporin-4 in human placenta. *Reproductive BioMedicine Online*, *37*(5), 601–612.
Tsukaguchi, H., Shayakul, C., Berger, U. V., Mackenzie, B., Devidas, S., Guggino, W. B., et al. (1998). Molecular characterization of a broad selectivity neutral solute channel. *Journal of Biological Chemistry*, *273*(38), 24737–24743.
Vilariño-García, T., Pérez-Pérez, A., Dietrich, V., Fernández-Sánchez, M., Guadix, P., Dueñas, J. L., et al. (2016). Increased expression of aquaporin 9 in trophoblast from gestational diabetic patients. *Hormone and Metabolic Research*, *48*(8), 535–539.
Vilariño-García, T., Pérez-Pérez, A., Dietrich, V., Guadix, P., Dueñas, J., Varone, C., et al. (2018). Leptin up-regulates aquaporin 9 expression in human placenta in vitro. *Gynecological Endocrinology*, *34*(2), 75–77.
Wang, S., Amidi, F., Beall, M., Gui, L., & Ross, M. G. (2006). Aquaporin 3 expression in human fetal membranes and its up-regulation by cyclic adenosine monophosphate in amnion epithelial cell culture. *Journal of the Society for Gynecologic Investigation*, *13*(3), 181–185.
Wang, S., Amidi, F., Yin, S., Beall, M., & Ross, M. G. (2007). Cyclic adenosine monophosphate regulation of aquaporin gene expression in human amnionepithelia. *Reproductive Sciences*, *14*(3), 234–240.
Wang, S., Chen, J., Au, K. T., & Ross, M. G. (2003). Expression of aquaporin 8 and its up-regulation by cyclic adenosine monophosphate in human WISH cells. *American Journal of Obstetrics and Gynecology*, *188*(4), 997–1001.
Wang, S., Chen, J., Beall, M., Zhou, W., & Ross, M. G. (2004). Expression of aquaporin 9 in human chorioamniotic membranes and placenta. *American Journal of Obstetrics and Gynecology*, *191*(6), 2160–2167.
Wang, S., Chen, J., Huang, B., & Ross, M. G. (2005). Cloning and cellular expression of aquaporin 9 in ovine fetal membranes. *American Journal of Obstetrics and Gynecology*, *193*(3 Pt. 1), 841–848.
Wang, S., Kallichanda, N., Song, W., Ramirez, B. A., & Ross, M. G. (2001). Expression of aquaporin-8 in human placenta and chorioamniotic membranes: Evidence of molecular mechanism for intramembranous amniotic fluid resorption. *American Journal of Obstetrics and Gynecology*, *185*(5), 1226–1231.
Watson, A. J., & Barcroft, L. C. (2001). Regulation of blastocyst formation. *Frontiers in Bioscience*, *6*, D708–D730.
Wooding, F. B., Wilsher, S., Benirschke, K., Jones, C. J., & Allen, W. R. (2015). Immunocytochemistry of the placentas of giraffe (*Giraffa cameleopardalis* giraffa) and okapi (*Okapi johnstoni*): Comparison with other ruminants. *Placenta*, *36*(1), 77–87.

Xiong, Y., Tan, Y. J., Xiong, Y. M., Huang, Y. T., Hu, X. L., Lu, Y. C., et al. (2013). Expression of aquaporins in human embryos and potential role of AQP3 and AQP7 in preimplantation mouse embryo development. *Cellular Physiology and Biochemistry, 31*(4–5), 649–658.

Yakata, K., Hiroaki, Y., Ishibashi, K., Sohara, E., Sasaki, S., Mitsuoka, K., et al. (2007). Aquaporin-11 containing a divergent NPA motif has normal water channel activity. *Biochimica et Biophysica Acta, 1768*(3), 688–693.

Yakata, K., Tani, K., & Fujiyoshi, Y. (2011). Water permeability and characterization of aquaporin-11. *Journal of Structural Biology, 174*(2), 315–320.

Yasui, M. (2009). pH regulated anion permeability of aquaporin-6. *Handbook of Experimental Pharmacology, 190*, 299–308.

Zeuthen, T. (1999). Transport of water and glycerol in aquaporin 3 is gated by H^+. *Journal of Biological Chemistry, 274*(31), 21631–21636.

Zhao, Y., Lin, L., & Lai, A. (2018). Expression and significance of aquaporin-2 and serum hormones in placenta of patients with preeclampsia. *Journal of Obstetrics and Gynaecology, 38*(1), 42–48.

Zhu, C., Jiang, Z., Bazer, F. W., Johnson, G. A., Burghardt, R. C., & Wu, G. (2015). Aquaporins in the female reproductive system of mammals. *Frontiers in Bioscience (Landmark Edition), 20*, 838–871.

Zhu, X., Jiang, S., Hu, Y., Zheng, X., Zou, S., Wang, Y., et al. (2010). The expression of aquaporin 8 and aquaporin 9 in fetal membranes and placenta in term pregnancies complicated by idiopathic polyhydramnios. *Early Human Development, 86*(10), 657–663.

Zhu, X. Q., Jiang, S. S., Zhu, X. J., Zou, S. W., Wang, Y. H., & Hu, Y. C. (2009). Expression of aquaporin 1 and aquaporin 3 in fetal membranes and placenta in human term pregnancies with oligohydramnios. *Placenta, 30*(8), 670–676.

FSC
www.fsc.org
MIX
Papier | Fördert
gute Waldnutzung
FSC® C083411

Kolibri 360 GmbH
Ferdinand-Jühlke-Straße 7
99095 Erfurt, Deutschland
produktsicherheit@kolibri360.de